# Person-Centred Care in Radiology

This edited book focuses on the application of patient care within the three specialisms: diagnostic radiography (including fluoroscopy, computed tomography, breast imaging, ultrasound, and magnetic resonance imaging), radiotherapy and oncology, and nuclear medicine and molecular imaging.

*Person-Centred Care in Radiology: International Perspectives on High-Quality Care* draws from recent publications and clinical expertise, supported with this trend of technological advances and how they are supposed to enhance patient care. The chapters seek to uncover the role and behavior of radiographers. This will be supported with chapters on a key aspect, which will impact both radiographers and patients, vis-à-vis advancing technology. These chapters include topics such as artificial intelligence, image acquisition, coincided with topics surrounding ethics. The edited volume includes contributions from the United States, Canada, the UK and Australasia to bring together for the first time those at the forefront of this growing field in medical imaging.

This book may be used to influence policymaking decisions and thus influence how healthcare delivery is offered in an ever-evolving imaging environment. In short, this text bridges the gap between what is advocated in the literature, with experience, as observed in practice. The targeted audience for this book is multifaceted. It will primarily be a book that facilitates undergraduate radiography students worldwide. It will offer a useful tool for academics delivering undergraduate (pre-registration) radiography programs.

This book will act as a 'primer' for undergraduate students, but importantly 'signpost' to other key texts within the field. Further, academics will find this text useful as it aims to enrich scholarly learning, teaching and assessment to healthcare programs nationally and internationally.

**Shayne Chau** is currently working as a Senior Lecturer at Charles Sturt University and an adjunct Senior Lecturer at University of Exeter, UK and University of Canberra, Australia.

**Emma Hyde** is Associate Professor in Diagnostic Imaging at the University of Derby, UK.

**Karen Knapp** is Associate Professor in Musculoskeletal Imaging and Head of the Department of Health and Care Professions in the Faculty of Health and Life Sciences at the University of Exeter, UK.

**Christopher Hayre** is Associate Professor of Medical Imaging at the University of Canberra, Australia.

**Medical Imaging in Practice**
Series Editor: Christopher Hayre, Assistant Professor in
Medical Imaging Institute of Applied Technology, UAE

**Research Methods for Student Radiographers: A Survival Guide**
*Christopher M. Hayre, Xiaoming Zheng*

**Person-Centred Care in Radiology**
International Perspectives on High-Quality Care
*Edited by Shayne Chau, Emma Hyde, Karen Knapp and Christopher Hayre*

For more information about this series, please visit: www.crcpress.com/Medical-Imaging-in-Practice/book-series/MIIP

# Person-Centred Care in Radiology

## International Perspectives on High-Quality Care

Edited by Shayne Chau, Emma Hyde, Karen Knapp and Christopher Hayre

CRC Press
Taylor & Francis Group
Boca Raton London New York

CRC Press is an imprint of the
Taylor & Francis Group, an **informa** business

Designed cover image: JohnnyGreig/istockphoto

First edition published 2024
by CRC Press
2385 NW Executive Center Drive, Suite 320, Boca Raton FL 33431

and by CRC Press
4 Park Square, Milton Park, Abingdon, Oxon, OX14 4RN

*CRC Press is an imprint of Taylor & Francis Group, LLC*

ISBN: 978-1-032-31529-4 (hbk)
ISBN: 978-1-032-30464-9 (pbk)
ISBN: 978-1-003-31014-3 (ebk)

DOI: 10.1201/9781003310143

Typeset in Sabon
by Newgen Publishing UK

The editors would like to thank the contributing authors. Your commitment to this book reflects the hard work and determination of generating the first person-centred care in radiology textbook encompassing international perspectives on high-quality care. It has been a pleasure for us to work with you and bring together this collection of highly informative chapters. We also thank all the patients who have shared their experiences both in this book, but in our wider careers. You will never know how much your stories have touched us and acted as a catalyst for us to strive for person centred care. Finally, the editors agree that this has been an exciting and prosperous project, which we hope readers will enjoy and utilize.

*Shayne Chau* would like to dedicate this book to his wife, Jo Davies, and son, Caspian Chau. This dedication is a small token of my appreciation for the immeasurable impact you have had on my life and my work.

*Associate Professor Emma Hyde* would like to dedicate this book to her husband, Martin, and her sons, Rocco and Milo. Thank you for your support.

*Associate Professor Karen Knapp* dedicates this book to her family, friends and the best colleagues she could hope to work with.

*Dr Christopher Hayre* would like to dedicate this book to his wife for her continued support and courage. He would also like to dedicate this book to his young daughters, Ayva and Ellena, and memory of Evelynn. Love to all.

# Contents

SECTION NINE
**Person-centred care in forensic imaging**

SECTION TEN
**Person-centred care in therapeutic radiography**

SECTION ELEVEN
**Person-centred care: The patient experience**

# About the editors

**Shayne Chau** is currently working as a Senior Lecturer at Charles Sturt University and an adjunct Senior Lecturer at University of Exeter, UK and University of Canberra, Australia. He also works clinically as diagnostic radiographer at Flinders Medical Centre. Shayne is a Fellow of the Australian Society of Medical Imaging and Radiation Therapy (ASMIRT), a Fellow of the Higher Education Academy (HEA) and an Associate Fellow of the Higher Education Research and Development Society of Australasia (HERDSA). Shayne has served on the Board of Directors and many committees from ASMIRT. Outside of his teaching, research and clinical time, Shayne is an accreditation assessor and committee member of the Medical Radiation Practice Board of Australia and an Editorial Board Member and Associate Editor of the *Journal of Medical Imaging and Radiation Sciences*.

**Emma Hyde** is Associate Professor in Diagnostic Imaging at the University of Derby, in the UK. She has published extensively on the topic of patient/person-centred care, in peer-reviewed scientific journals and trade publications. Emma has been the Clinical Director of the Personalised Care Institute since January 2022, and is an associate member of the International Community of Practice in Person-Centred Practice. Emma is President of the UK Imaging & Oncology Congress (UKIO) for 2024 and 2025. Emma has been a Fellow of the Higher Education Academy since 2008, and was awarded a National Teaching Fellowship by Advance HE in 2020. Emma is an editorial board member of the *Journal of Medical Imaging and Radiation Sciences*, and a peer reviewer for the *Radiography* journal.

**Karen Knapp** is Associate Professor in Musculoskeletal Imaging and Head of the Department of Health and Care Professions in the Faculty of Health and Life Sciences at the University of Exeter, UK. Karen has spent more than 20 years researching in the field of osteoporosis and new diagnostic techniques and keeps patient care at the centre of her work, engaging with patients from the beginning of projects. Karen has published 68 original research articles, 14 invited papers and editorials, in excess of 100 conference abstracts and has undertaken 35 conference presentations as an invited speaker. Karen has supervised 11 PhD students to successful completion, including one who explored compassionate care in radiography. Combining a passion for research and education with her clinical background, Karen utilises research-led teaching to inspire students. With a background in teaching undergraduate and post-graduate students from master's through to professional doctorates and supervising PhD students, Karen is keen to engage at all levels of higher education. Karen is chair of the Bone Densitometry Training and Accreditation panel for the Royal Osteoporosis Society and chaired the panel who developed the new Education and Career Framework for the Society and College of Radiographers.

**Christopher Hayre** is Associate Professor of Medical Imaging at the University of Canberra, Australia. He has published both qualitative and quantitative refereed papers and brought together several books in the field of medical imaging, health research, technology, and ethnography.

# Contributors

**Alberta Naa Afia Adjei,** Assistant Lecturer. Department of Medical Imaging, Kwame Nkrumah University of Science and Technology, Kumasi, Ghana.

**Benedict Apaw Agyei,** Department of Medical Imaging, Kwame Nkrumah University of Science and Technology, Senior Sonographer, Obstetrics and Gynecology Directorate, Komfo Anokye Teaching Hospital, Kumasi Ghana.

**Elio Arruzza** is a lecturer at University of South Australia with research interests in health science education, evidence-based practice and radiography.

**Jill Bleiker** is currently working as Honorary Senior Lecturer with the Department of Medical Imaging at the University of Exeter Medical School. Jill does research in Health Psychology, Positive Psychology and Teaching Methods in radiography.

**Harry Bliss,** London South Bank University, UK. Harry is Course Director of the Pre-Registration Diagnostic Radiography pathways and Senior Lecturer in the School of Allied and Community Health. Harry specialises in radiographic reporting with interests in pediatrics and trauma.

**Shayne Chau** is currently working as a Senior Lecturer at Charles Sturt University and an adjunct Senior Lecturer at University of Exeter, UK and University of Canberra, Australia. Shayne's research interests are in neuroimaging and oncology.

**Anja Christoffersen,** Disability Advocate; Founder & CEO, Champion Health Agency Pty Ltd; Founder & CEO, Women with Disabilities Entrepreneur Network.

**Geoff Currie,** PhD, is Professor in Nuclear Medicine at Charles Sturt University, Wagga Wagga, Australia and adjunct Professor in Radiology at Baylor College of Medicine in Houston, Texas, USA. He has broad research and teaching interests across the medical radiation sciences and, indeed, health generally with more than 200 peer reviewed journal papers,

185 conference presentations, 75 invited speaker presentations, and a reviewer for 25 international journals.

**Josie Currie,** UNSW Rural Medical School, University of New South Wales, Wagga Wagga, NSW, Australia. She has run several workshops and training programs in for university students and for clinical practitioners on emotional intelligence and cultural proficiency.

**Andrew Donkor,** PhD, Lecturer, Department of Medical Imaging, Kwame Nkrumah University of Science and Technology, Ghana; Professional Research Fellow (Improving Palliative, Aged and Chronic Care through Clinical Research and Translation), University of Technology Sydney, Australia.

**Chrissie Eade,** Royal Cornwall Hospitals NHS Trust, United Kingdom. Chrissie is a consultant plain film reporting radiographer who works in both the NHS and private healthcare sectors. She has a specialist interest in forensic radiography and paediatric imaging. She is a member of the European Society of Paediatric Radiology AI Taskforce and is a guest lecturer on the MSc and Advanced practice modules for Plymouth and Exeter, Universities.

**Joleen K. Eden,** Consultant Radiographer, Breast Imaging Service, East Lancashire Hospitals NHS Trust. Joleen is Consultant Radiographer with over 15 years of experience working in both symptomatic and NHS breast screening services. Devoted to advanced practice, Joleen is an educator and published researcher with an interest in clinical development and improved wellbeing for both patient and practitioner.

**Michael Fuller** has professional, teaching, and research interests in radiography clinical education, patient-centred care and the technical aspects of radiographic practice.

**Jill Griffin,** DCR (R), is Head of Clinical Engagement, Royal Osteoporosis Society, Bath, UK and Consultant Practitioner (visiting), Healthy Bones Service, University Hospitals Plymouth NHS Trust, Devon, UK. Her principal interests include quality clinical practice in osteoporosis, fragility fractures and bone densitometry.

**Amy Hancock** qualified as Therapeutic Radiographer in 2006. She is a Senior Lecturer for on the Medical Imaging (Diagnostic Radiography) programmes. Prior to this Amy was employed as the Principal Therapeutic Radiographer for Research and Development at Weston Park Cancer Centre and previously worked at Sheffield Hallam University as Senior Lecturer on the Radiotherapy and Oncology programmes.

**K. Elizabeth Hawk,** PhD, is Nuclear Medicine Physician and Neuroradiologist. Currently, Dr Hawk is faculty for the Stanford School of Medicine,

Department of Radiology, Division of Nuclear Medicine and Chair of Nuclear medicine at the University of San Diego.

**Christopher Hayre** is Associate Professor of Medical Imaging at the University of Canberra, Australia. He has published both qualitative and quantitative refereed papers and brought together several books in the field of medical imaging, health research, technology, and ethnography.

**Christine Heales,** Senior Lecturer in Medical Imaging, Department of Health and Care Professions, University of Exeter.

**Julie Hendry,** PhD, Associate Professor, St George's University of London, UK. Research interests: person-centred caring, student outcomes.

**Johnathan Hewis,** Charles Sturt University, Senior Lecturer in Medical Imaging; patient experience, distress and advanced practice

**Catherine A. Hill** is now an application specialist, having worked clinically as a mammographer for 25 of her 30-year radiography career. Catherine specialised in training and education with image quality and patient care as her focus.

**Peter Hogg,** University of Salford. Peter Hogg is Professor Emeritus at the University of Salford, UK. Until his retirement he led research teams in breast cancer diagnosis using mammography and also in radiation dose optimisation in medical imaging; he was also Research Dean of a large multi-professional healthcare school.

**Darren Hudson,** MRI Clinical Lead, InHealth Group, High Wycombe, UK.

**Emma Hyde,** PhD, is Associate Professor in Diagnostic Imaging at the University of Derby, in the UK. She has published extensively on the topic of patient/person-centred care, in peer-reviewed scientific journals and trade publications.

**Lynne Ingram** has professional and research interests in quality assurance, organisational learning, continuing professional development, and patient-centred care in medical imaging.

**Aruna Jago-Brown,** Canterbury Christ Church University, is Senior lecturer and Radiation Protection Supervisor in the Diagnostic Radiography course. Registered member of IPEM and BNMS. Passionate about Physics Applied to medicine, Nuclear Medicine and in particular Radiopharmacy. I have worked in the UK, Australia, USA and Brazil in NM and PET imaging departments and have many years of experience in these specific fields. Having worked with the care and management of oncology patients, high standards of patient centered care is essential.

**Judith Kelly,** Countess of Chester Hospital, UK. Judith has worked as a Consultant Breast Radiographer for the last 20 years in Chester and for many years undertook collaborative research in breast imaging projects with the University of Salford. She has also co-edited a Digital Mammography textbook (1st and 2nd Editions) and presented at International and National Conferences.

**Niabh Kirk,** Royal Belfast Hospital for Sick Children, UK. Niabh is the Imaging Site Lead for the Royal Belfast Hospital for Sick Children and the Belfast City Hospital. Niabh's background is in paediatric radiography with special interest in paediatric CT, dose optimisation and risk benefit dialogue.

**Karen Knapp,** PhD, is Associate Professor in Musculoskeletal Imaging and Head of the Department of Health and Care Professions in the Faculty of Health and Life Sciences at the University of Exeter, UK. Karen has spent more than 20 years researching in the field of osteoporosis and new diagnostic techniques and keeps patient care at the centre of her work, engaging with patients from the beginning of projects.

**Iain MacDonald,** PhD, Institute of Health, University of Cumbria, Carlisle, United Kingdom.

**Lyndal Macpherson,** CEO, Australasian Society for Ultrasound in Medicine (ASUM); Director ASUM Outreach; Graduate Australian Institute of Company Directors; Certified Association Executives; Accredited Medical Sonographer.

**Margot McBride,** PhD, is Honorary Research Fellow at the University of Dundee and member of the Impact Advisory Board at the University of St. Andrews. Her current research interests include the use of AI in diagnosing osteoporosis and she is Clinical Advisor for two diagnostic imaging product developers.

**Claire E. Mercer** is Head of Radiography, School of Health and Society, University of Salford. Claire joined the University of Salford in 2015 and is currently the Head of Radiography. A radiographer by background, specialising in mammography Claire has 20 years' experience in the NHS and a PhD by published works in mammography. Claire is passionate about educating and supporting the developing radiography workforce focused on a person-centred approach.

**Johanna E. Mercer,** Psychological Wellbeing Practitioner, Health in Mind. Johanna is a psychological wellbeing practitioner with a varied amount of experience working with adult and adolescent mental health over the years. Johanna is passionate about improving staff wellbeing and patient care and has worked on creating initiatives to address this in a variety of workplaces.

**Kathleen Naidoo,** PhD, is a diagnostic radiography educator at the Cape Peninsula University of Technology in South Africa. She is a qualitative research enthusiast who is passionate about radiography education and the concept of caring. Dr Naidoo considers herself an advocate for person-centred care.

**Charlotte Primeau**, WMG, University of Warwick, United Kingdom. Dr Primeau is a biological anthropologist, medical researcher and a diagnostic radiographer. Currently, she works as an assistant professor in forensic imaging with the University of Warwick, where she performs post-mortem forensic micro-CT imaging for UK and international police forces.

**Eric Rohren**, PhD, is Professor and Chairman of the Department of Radiology at Baylor College of Medicine in Houston, Texas. His research interests include novel radiopharmaceuticals for the imaging of cancer, emerging applications of PET/MR, and molecular brain imaging for evaluation of neurocognitive disorders.

**Emma Rose**, Great Ormond Street Hospital, UK. Emma is Consultant Radiographer specialising in Paediatric Interventional Radiology. She is an advocate for advanced practice radiographers and paediatric care.

**Kevin Strachan**, Circle Cardiovascular Imaging, Clinical MRI applications specialist and Cardiac MRI Reporting Radiographer; cardiac MRI and MRI safety.

**Ruth Strudwick**, Professor in Diagnostic Radiography, Head of Allied Health Professions, School of Allied Health Sciences, University of Suffolk.

**Samantha Thomas**, PhD University of Sydney; Researcher, Perinatal Imaging Research Group University of New South Wales; Accredited Medical Sonographer, SAN Ultrasound for Women, Ultrasoundcare, PRP Radiology.

**Shelley Thomson**, Patients For Life; Faculty, Australian Council on Healthcare Standards Improvement Academy; Facilitator, Design Thinkers Academy; Board member, Australian Podiatry Association; Co-founder & Director, Patient Experience Agency.

**Riaan van de Venter**, PhD, Lecturer and Research Associate, Department of Radiography, Faculty of Health Sciences, Nelson Mandela University, South Africa. Research interests: person centred-care, professional practice and development, trauma, workplace wellbeing and stress, role extension, artificial intelligence, and inclusive and affirming care for sexual and gender minority patients.

**Robert Whiteman**'s research interest is centred on the patient's journey through Interventional Radiology. He is dedicated to enhancing and developing non-medical staffing groups to ensure optimal service accessibility.

**Yaw Amo Wiafe**, PhD, Senior Lecturer, Department of Medical Imaging, Kwame Nkrumah University of Science and Technology, Kumasi, Ghana.

# Preface

In this preface I want to begin by congratulating my co-editors and contributing authors in producing this edited book volume. This book not only forms part of the Medical Imaging in Practice Book Series, with CRC Press, but remains a key text within our literature. Our book seeks to disseminate knowledge, understanding, practice and research in Radiology by thinking carefully about our patients' experiences, care, and pathway from an array of imaging approaches. This book not only provides a contemporary account, but also offers an international lens of person-centred care (PCC), providing a holistic account of applied person-centred approaches. Naturally, contributions demonstrate transnational collaboration with colleagues. Practitioners and researchers are brought together, at the forefront, whether utilising or critically evaluating person-centred care in Radiology. Our audience for this book is multifaceted. First, this book will be of value to medical imaging students, looking to learn and develop their own theoretical knowledge and understanding, albeit for undergraduate or postgraduate writing or study. Second, for practitioners, this book could be used departmentally to critically challenge or (re-)think about alternate person-centred approaches within their specialist imaging modality. Third, for researchers, this book offers both varied empirical and methodological value, whilst critically appraising the evidence base supported with an extensive reference list to either support or uncover novel areas of research concerning the topic of PCC.

In short, patients remain at the forefront of medical imaging practices, thus continuously reminded, by the work presented here, of our ongoing need to reflect and apply, whilst critically thinking about person-centred care in Radiology.

Christopher Hayre, PhD.

Section one

# Introduction to person-centred care

# 1 Person-centred care

*Emma Hyde*

Person-centred care is a concept that has been evolving since the 1980s and is defined by the Picker Institute (2023) as: 'An approach that puts people at the heart of health and social care services'.

Person-centred care challenges health care professionals to put the needs and preferences of individuals receiving care above the systems or processes operating within health or care organisations. By treating people as individuals and equal partners in their own care, people are encouraged to play an active role, and are listened to and respected.

Internationally, there is a growing emphasis on the provision of person-centred care. Publications such as the Australian Commission on Safety and Quality in Healthcare's 'Patient Centred Care: Improving quality and safety through partnerships with patients and consumers (2011)', and the Health Foundation's 'Person-Centred Care Made Simple (2014)' are supporting positive changes to health and care professionals' practice and providing frameworks to measure person-centred care with. Alongside this, campaigns such as 'Hello My Name is' and 'What Matters to You', have shown how simple changes to the way that health and care professionals communicate with individuals can have a massive impact on the quality of care provided.

The difficult balance between providing person-centred care and models of health and care delivery is apparent within many imaging and radiotherapy services. Rapid advances in imaging and radiotherapy technologies, such as artificial intelligence, are reshaping service delivery, and facilitating significant improvements in efficiency. Radiographers can often be focused on the technical aspects of imaging examinations or radiotherapy treatments. Combined with the tendency for radiographers to use a reductionist approach and focus on the pathology being imaged or treated, or the symptomatic body part, this can lead to a lack of attention to person-centred care. This is often attributed to a lack of time, but there is growing body of evidence that a reductionist approach may also be adopted due to a lack of capacity to provide the emotional support required. These tensions have been well documented by researchers such as Bleiker et al. (2018), Bleiker (2020), Bolderston et al. (2010), Bolderston (2016), Hayre et al. (2016), Hendry

DOI: 10.1201/9781003310143-2

(2019), Hyde & Hardy (2021a, b, c), Strudwick et al. (2018), Strudwick et al. (2011), Taylor & Hodgson (2020), and more.

As a global community of radiography professionals, we must ensure that as our services develop, and new technology is adopted, this is not at the expense of person-centred approaches to care. Imaging and oncology services must keep person-centred approaches at the core of day-to-day practice. In this book we deliberately use the term 'person-centred care' to encourage a holistic view of individuals we are interacting with and providing imaging or radiotherapy services for. We may on occasion use the term 'patient' or 'service user'; this will depend on the context of the chapter.

The uniqueness of this book is that it is based upon research conducted by the authors of each chapter. Through their research the authors have provided an evidence-base which is shaping radiographers' perspectives on person-centred care. As such, readers will gain knowledge that will help them to meet the expectations of both individuals receiving care, and their carers, in all areas of radiography.

## References

Australian Commission on Safety and Quality in Health Care. (2011) *Patient[1]Centred Care: Improving Quality and Safety through Partnerships with Patients and Consumers*, ACSQHC, Sydney.

Bleiker J. (2020) What radiographers talk about when they talk about compassion *Journal of Medical Imaging and Radiation Sciences* Dec;51 (4S): S44–S52. DOI: http://dx.doi.org/10.1016/j.jmir.2020.08.009

Bleiker J, Knapp KM, Morgan-Trimmer S & Hopkins SJ (2018) 'It's what's behind the mask': Psychological diversity in compassionate patient care. *Radiography* 24 (S1): S28–S32. DOI: https://doi.org/10.1016/j.radi.2018.06.004

Bolderston A (2016) Patient experience in medical imaging and radiation therapy. *Journal of Medical Imaging and Radiation Sciences* 47 (2016): 356–361 DOI: http://dx.doi.org/10.1016/j.jmir.2016.09.002

Bolderston A, Lewis D & Chai M (2010) The concept of caring: Perception of radiation therapists. *Radiography* 16 (2010): 198–208.

Granger K. (2013) *Hello My Name is Campaign*. Available at: www.hellomynameis.org.uk Accessed: 29/06/2023.

Hayre C.M. Blackman S & Eyden A. (2016) Do general radiographic examinations resemble a person-centred environment? *Radiography* 22 (4): e245–251.

Hendry J. (2019) Promoting compassionate care in radiography: What might be suitable pedagogy? A discussion paper. *Radiography* 25 (3): 269–273 DOI: http://doi.org/10.1016/j.radi.2019.01.005

Hyde E & Hardy M. (2021a) Delivering patient centred care (Part 1): Perceptions of service users and service deliverers. *Radiography* 27 (1): 8–13 DOI: https://doi.org/10.1016/j.radi.2020.04.015

Hyde E & Hardy M. (2021b) Delivering patient centred care (Part 2): a qualitative study of the perceptions of service users and deliverers. *Radiography* 27 (2): 322–331. DOI: https://doi.org/10.1016/j.radi.2020.09.008

Hyde E & Hardy M. (2021c) Delivering patient centred care (Part 3): Perceptions of student radiographers and radiography academics. *Radiography* 27 (3): 803–810 DOI: https://doi.org/10.1016/j.radi.2020.12.013

Picker Institute Europe. *Principles of Person-Centred Care*. Available at: www.picker. org/about-us/picker-principles-of-person-centred-care/ Accessed: 29/06/2023

Strudwick R, Mackay S and Hicks S. (2011) Is diagnostic radiography a caring profession? *Synergy*. June 2011, 4–7.

Strudwick R, Newton-Hughes A, Gibson S, Harris J, Gradwell M, Hyde E, Harvey-Lloyd J, O'Regan T & Hendry J. (2018) *Values-Based Practice (VBP) Training for Radiographers*. Available from: www.sor.org/learning/document-library?page=1

Taylor A & Hodgson D. (2020) The behavioural display of compassion in radiation therapy: Purpose, meaning and interpretation. *Journal of Medical Imaging and Radiation Sciences* 51 (2020): S59–S71. DOI: https://doi.org/10.1016/j.jmir.2020.08.003

The Health Foundation. (2014) *Person-Centred Care Made Simple*. Available at: www. health.org.uk/publications/person-centred-care-made-simple   Accessed:   29/06/2023.

# 2 Cultural competence in person-centred care

*Geoff Currie and Josie Currie*

## Introduction

The concepts of person-centred care and cultural competence are connected but ambiguous. One perspective purports that cultural competence is one of numerous factors that contribute to person-centred care (Saha et al, 2008). Conversely, person-centred care could be viewed as one of several factors driving cultural competence (Saha et al, 2008). At the same time, the role of emotional acuity in interpersonal interactions, conflict resolution and identity are key features of person-centred care and cultural competence (Currie and Currie, 2022). From a radiography, nuclear medicine and radiotherapy context, the connection between person-centred care and patient-centred care, personalised care/medicine and precision medicine need clarification. At their core, all of these principles aim to improve the quality of healthcare.

Cultural competence alone is not consistent with person-centred care. If person-centred care represents the unique individual, that individual identity can be diluted within broader cultural values. That is, culture is the common threads within a group of people that represent a collective identity, beliefs and values (Figure 2.1). It is emotional acuity that allows awareness and sensitivity to individual nuances in beliefs and values to deliver person-centred care (Figure 2.2). Optimising person-centred care requires understanding of the culture of the healthcare environment. This allows the meeting of individual patient needs with the socially and regulatory appropriate healthcare; and identification of the need to change institutional culture. Indeed, institutionalised bias is a key factor in continued inequality in healthcare and social asymmetry (Currie et al, 2022).

Patient-centred care and personalised medicine focus on the individual relationships with patients, their choices, beliefs and preferences to improve overall healthcare quality one individual at a time. Emotional acuity and cultural competence play an important role in driving productive relationships and safe environments to deliver person-centred care. Indeed, the differentiator of patient-centred and person-centred lies in emotional and cultural acuity that identifies the whole person rather than a patient. A patient may be synonymous with a disease, condition or procedure they are about to

DOI: 10.1201/9781003310143-3

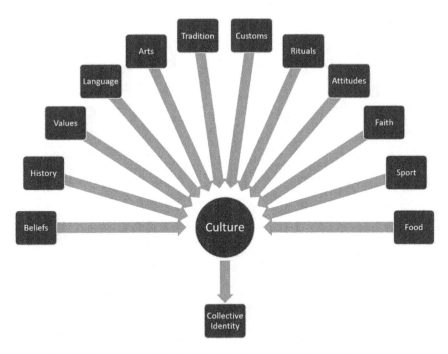

*Figure 2.1* Culture as the collective identity of a community.

undergo. If person-centred care is considered to be assessing, identifying and planning health needs of an individual and adapting that care based on those individuals needs and expectations, then cultural and emotional competence are tools by which health professionals create safe and productive environments that allow accuracy in the person-centred approach. That is, to deliver individualised care there is a need to deeply understand the individual. Indeed, person-centred care may create conflict, a disconnect between expected outcomes (e.g. cure disease) and personal preferences (e.g. type of therapy). Here, emotional acuity sharpens the blade of cultural competence to negotiate within a culturally and emotionally safe environment to optimise person-centred outcomes.

Cultural competence is also founded in social justice which then forms the foundation of addressing healthcare inequality associated with social asymmetry. Cultural competence is not limited to ethnicity. Cultural competence and emotional acuity are associated with skills and capabilities, fairness and justice and person-centred care that recognises and respects unique healthcare drivers, expectations and needs related to ethnicity, religion, beliefs, language, gender, sexual orientation, disability, age, socioeconomic factors, morbidity and geography.

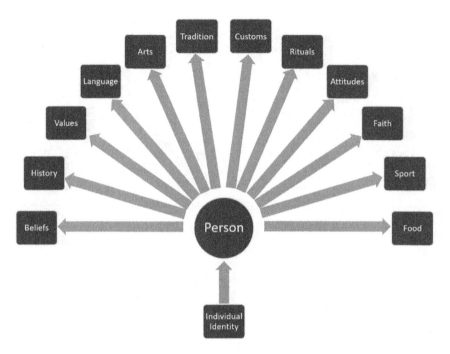

*Figure 2.2* The person is the individual identity within a collective cultural community.

## Emotional intelligence

Higher emotional intelligence among health professionals leads to better patient care (Karimi L et al, 2021). Emotional intelligence is the acuity associated with one's own emotions and with those of others (Jiménez-Picón et al, 2021). Emotional intelligence is used to shape actions and guide motivations to produce productive interactions (Jiménez-Picón et al, 2021). For healthcare professionals, emotional intelligence creates situational awareness and the ability to adjust interactions to produce emotionally safe environments and more positive interactions. While emotional intelligence can be developed organically (inherent skills), emotional intelligence can also be learned, enhanced or engineered for a specific purpose. Regardless of the individual, emotional intelligence requires a deep understanding of personality traits and motivators of oneself and others. Higher order emotional taxonomies are achieved by careful attention to the training and development of the health practitioners' (or students') insight into the dynamic of social interactions (Figure 2.3).

Health professionals, like those in radiography, nuclear medicine and radiotherapy, interact with patients on the basis of spontaneous and often unconscious impressions of motivation, circumstance and personality. Even

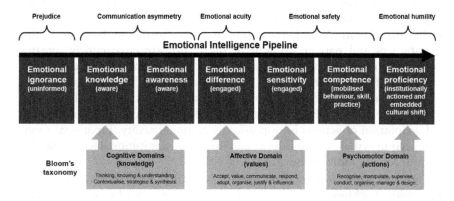

*Figure 2.3* The emotional intelligence pipeline including learning taxonomies (modified and reprinted with permission (Currie and Currie, 2022)).

in those with a high degree of cultural awareness, impressions rely on intuition that are subject to bias and prejudice, and susceptible to the confounding effects of the stressful environment. Person-centred care requires health professionals to possess an inventory of emotional intelligence tools, founded formally through undergraduate learning and reinforced through continuing education or continuing professional development. The value of honing both the science and the art of emotional intelligence is that emotional competence is important in a patient-facing, personalised medicine environment. While developing and refining emotional intelligence does not change personalities or the motivators of others, it does arm health professionals with the capability to adjust patient care and/or communication in real time to better fit the individual patient; personalised care or person-centred care. For example, in a study of women with breast cancer, investigators showed that understanding patient personality traits allowed more appropriate patient care and interventions (Cerezo et al, 2020).

The clinical setting is a complex and often stressful environment and needs to consider patient and practitioner in interactions. Emotional intelligence requires a framework within which health practitioners can use agility and flexibility to accommodate the unique mix of personality traits, clinical circumstances, stress and many other factors. While, these simplified tools or inventories provide part of that framework, there are varying levels of emotional intensity (magnitude with which emotions are experienced or felt), emotional flux (frequency with which different emotions are experienced or felt) and emotional density (concentration of different emotional stimuli in the environment) (Currie and Currie, 2022). Both patient and health professional could also be confronting acute or chronic emotional exhaustion (burnout). While simple in principle, emotional intelligence is complex and imperfect.

Emotionally safe environments require a foundation of emotional knowledge and awareness, scaffolded with values and attitude-driven emotional intelligence (Figure 2.1). The unequal power relationships between health practitioners and patients, typical of the healthcare environment, creates communication asymmetry. Success in building an emotionally competent health workforce requires mastery of the capacity for self-assessment, critical reflection, management of emotional dynamics, emotional knowledge, recognition of emotional differences, understanding the impact difference makes and adaptation of actions in response to emotional acuity. Emotional proficiency requires these capabilities at individual and institutional levels but also requires capacity for emotional humility.

There are numerous approaches to emotional intelligence building in the health workforce. Stand-alone initiatives using workshops are useful for building emotional awareness. More effective programmes to progress from emotional awareness to emotional competence need structured learning approaches. Traditionally, there has been a reliance on students to develop emotional intelligence organically as they progress through their studies. This approach has been widely reported as flawed and specifically, among radiography students, has shown no difference in emotional intelligence metrics across three years of study at multiple institutions de Galvao et al, 2017. At Charles Sturt University, undergraduate nuclear medicine and radiography students are provided a preliminary workshop (Currie and Currie, 2022) creating awareness and understanding associated with emotional intelligence (cognitive taxonomies). The knowledge is enriched with immersive exercises that are linked to reflective tasks to connect the learning with experience (affective taxonomies). Students are then released into the clinical environment where further application and reflection are used to hone emotional intelligence (psychomotor taxonomies).

### Cultural competence

From the health workforce context, cultural competence can be thought of as an individual's capability to respond to cultural diversity inside healthcare systems (Leininger and McFarland, 2006). More specifically, this includes understanding and respecting variations in patient beliefs, values, preferences, behaviours, thresholds for seeking care, expectations of healthcare, compliance, and attitudes to the procedures provided in radiology and nuclear medicine (Betancourt et al, 2003). Cultural competence is a key component in addressing inequities, including health, but requires more than basic cultural awareness (Bainbridge et al, 2015). Cultural competence involves attitudes and behaviours, reinforced through policy and practice (Bainbridge et al, 2015) (Figure 2.4).

Building a culturally competent health workforce is constrained by a lack of consistent definition and language around cultural competence, and lack of evidence of intervention impact or identifying appropriate performance indicators (Bainbridge et al, 2015, Currie and Yindyamarra, 2022). Similarly to emotional

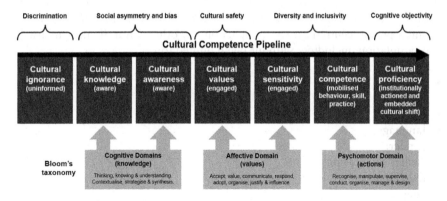

*Figure 2.4* The cultural competence pipeline including learning taxonomies (modified and reprinted with permission *11*).

intelligence, a culturally competent health workforce requires mastery of the capacity for self-assessment, critical reflection, management of cultural dynamics, cultural knowledge, recognition of cultural differences, understanding the impact difference makes and adaptation of actions in response to cultural acuity. Recognising cultural difference and understanding the value that those cultural differences bring to a community or team is an important part of cultural competence (Figure 2.4). Cultural proficiency requires these capabilities at individual and institutional levels but also requires capacity for cultural humility and effective and affirming responsiveness to cultural differences (Figure 2.4).

## Training in cultural competence

Like emotional intelligence, there are an array of initiatives targeting cultural competence workshops for healthcare workers. Given the diversity of cultural factors, a prototype approach has been adopted here. That is, an exploration of Indigenous cultural competence to identify translational themes and learnings to the many aspects of cultural diversity. While each culturally marginalised community will confront unique challenges and solutions, the underlying principles and practice of cultural competence are uniform and transferrable. One should also keep in mind that identifying cultures also dilutes individualism. Combining emotional intelligence and cultural competence is required to deliver unique person-centred healthcare. For example, a high degree of Indigenous cultural competence may overlook the unique health preferences of a patient who identifies as, among other things, Indigenous. Is there a typical set of traits that identify the health needs of every Indigenous person or even every Indigenous person from a specific region? The individual health needs and preferences of any single Indigenous person may more closely resemble those of another cultural sub-group (e.g. aged, disabled, religion) than any other single Indigenous person. It is crucial to recognise that this cultural generalisation can be culturally insensitive, undermine cultural safety, prohibit person-centred care and impact on any other culturally marginalised sub-population, including, without being limited to:

- ethnicity;
- religion;
- beliefs;
- language;
- gender;
- sexual orientation;
- disability;
- age;
- socioeconomic factors;
- morbidity; and
- geography.

Furthermore, superficial training can be seen as tokenistic. For example, research undertaken in Australia reported that the majority of industry-based Indigenous cultural training was at the "cultural awareness" level, well below the expected level of healthcare practitioners (Downing R et al, 2011). Conversely, university sector Indigenous cultural training tended to provide a foundation of knowledge and awareness that scaffolded to values, attitudes, sensitivity and empathy, and entry level cultural competence (Ewen et al, 2012). The research concluded that participants that developed confidence working with Indigenous patients should keep in mind that confidence does not equate to competence. Confidence in the absence of objective competence is likely to create cultural insensitivity and undermine cultural safety despite foundations of cultural knowledge. More effective training programmes might rely on cultural immersions because these activities develop richer insights into the connection between culture and health behaviours. Through these rich and deep culturally immersive experiences, a better understanding of barriers to health and communication can be gleaned, and more appropriate strategies to meet the health and cultural safety needs can be developed. Importantly, these cultural barriers contribute significantly to healthcare inequity and cultural competence is part of the solution.

Healthcare inequities result from a number of factors, including implicit and explicit cultural bias, and these effects are compounded by extrinsic and intrinsic barriers associated with accessing healthcare services among different cultural groups (Stephens et al, 2005, Anderson et al, 2016, Shapiro et al 2006). Cultural competence is a critical part of the curriculum for healthcare professions, including in radiology and nuclear medicine, yet health asymmetry continues to be problematic. This reflects cultural and process complexity associated with changing engrained culture using a bottom-up approach (i.e. student education) (Brach et al, 2000; Kumas-Tan et al, 2007; Butler et al, 2016). Another key factor is the lack of confidence among healthcare professionals in meeting the cultural needs of multi-cultural patients (Brach et al, 2000; Kumas-Tan et al, 2007; Butler et al, 2016). It is essential to address these barriers by developing in addition to undergraduate curriculum, cultural competence initiatives in professional, continuing education, continuing professional development and post-graduate activities.

Understanding cultural needs and barriers can create culturally safe clinical environments that facilitate person-centred healthcare. Social and health asymmetry is worsened by implicit, explicit and historical cultural bias. These biases reinforce the social injustice weaved through the culture and policies of "institutions" to drive systematic disadvantage to those in most need and, in turn, undermine person-centred care. Inequity and bias might manifest as a lack of diversity in the teams of healthcare professionals, further eroding person-centred care. Diversity in the healthcare team enables creative problem solving and implementation of solutions better suited to culturally marginalised groups; person-centred care.

Healthcare biases can be discriminatory and rely heavily on intuition shaped by personal experiences of the observer. Healthcare is neither neutral nor objective, but embedded in and driven by social, political and economic agendas (Currie et al, 2022). Healthcare policy can amplify social inequalities by reinforcing previously unknown bias. An essential element of professional development and undergraduate training in radiology and nuclear medicine is to develop the capability for critical reflection that reveals intrinsic bias to ensure that culturally marginalised sub-groups not only find radiology and nuclear medicine a culturally safe place but they encounter healthcare professionals who exhibit the attitudes and behaviours of cultural competence. In turn, learning from structured and hidden curricula, and through cultural mentoring, will re-engineer the cultural framework of the institution and produce cultural proficiency, debugging policy and practice from historical bias.

The journey through cultural awareness to enlightenment and cultural competence is tortuous and requires institutional and workforce commitment and perseverance. The task for health professionals is to challenge the ways of knowing with the goal of "deinstitutionalising" their individual and collective attitudes, beliefs and actions. There are challenges to confront as a result of historical injury or cultural incompatibilities. There are many examples that have caused culturally based injury that undermine confidence in and trust for the health sector (including fear of the health sector) that, in turn, decrease access to health services:

- age discrimination in the workforce and poor treatment of older patients;
- social racism institutionalised in policy, practice or data used to inform policy and practice;
- lack of services to accommodate language barriers among non-English speaking patients;
- lack of accessibility to services (or education about service availability) for refugee communities;

- failure to understand and/or accommodate religious practices within the health procedure or environment;
- ignoring individual beliefs that would impact patient compliance;
- making assumptions about minimum level of schooling or health literacy in health promotion;
- gender disparities in the workforce and failure to accommodate non-binary gender identification;
- social discrimination based on sexual orientation creating lack of work-force diversity and decreased quality of patient experiences;
- lack of accessibility (physical or virtual) to services based on a person's disability;
- lack of accessibility due to cost reinforced by a disparity in health resources available to lower socioeconomic groups;
- discrimination against patients perceived to have contributed to their own morbidity (e.g. obesity, smoking, alcoholism); and
- disparity of opportunity and access to health services for rural communities compounded by decreased resources to accommodate rural patient needs in metropolitan referral centres.

Short-term strategies or changes in policy do not instil trust and confidence among culturally marginalised communities. Success in strategy development requires engagement with, and input from, the key stakeholder groups.

Culturally competent healthcare, including in radiography, nuclear medicine and radiotherapy, addresses social asymmetry. When cultural ignorance or lack of cultural safety are perceived by any individual or group of individuals, it drives lower uptake of health services. Developing culturally safe and appropriate healthcare spaces will increase utilisation of health services. Inclusive of this is the obligation to make careers in health, including radiology and nuclear medicine, attractive and achievable for a culturally diverse population. The move towards gender equity is a simple example that displaces the male-centric medical decision-making model that can intimidate or discourage female participation as patient or health professional. More recently, cultural safety for members of the lesbian, gay, bisexual, transgender, intersex and queer (LBGTIQ+) community for patients has been driven by cultural safety for health workers with the aim of increasing the proportion of the workforce that identify as LBGTIQ+ and increasing the uptake of health services by other members of the LBGTIQ+ community (Bolderston et al, 2021). An important distinction should be made about identity and "identifying" which highlights the previous discussion with respect to culture (group trait identity) and emotional intelligence (individual identity). Providing a culturally safe environment is not only attractive to patients and potential health workers, but also supports existing members of the workforce to reveal their identity; something that may have been obscured in a culturally inappropriate environment. With cultural safety and true self-identity comes a workforce that is diverse, inclusive and productive;

a happy workforce is a productive workforce and both are driven by a workforce of individuals that are the best version of themselves. At Charles Sturt University, institutional cultural proficiency has seen the highest proportion of students in Australia that identify as Indigenous. This reflects a culturally safe environment that preferentially attracts Indigenous students but also an environment in which existing "non-identified" students are confident to identify as Indigenous (Currie et al, 2018).

An important consideration that links emotional intelligence to cultural competence is the idea of identity. Identity is used to be culturally appropriate yet can be cultural insensitive and at the same time emotionally unsafe. The very premise of person-centred care is the person, the individual. Their identity should also be individualised. Cultural competence can attribute group identity. This has important value, particularly in driving equity of health services generally, but can undermine emotional safety and overlook an individual's true identity for the purpose of person-centred care. An individual who identifies as Indigenous, for example, is really indicating that they identify with the beliefs, history, values, customs and traditions (Figure 2.1) of Indigenous peoples. Yet those beliefs, history, values, customs and traditions are different among different communities, and experienced differently for individuals. Moreover, one or more of these attributes may be replaced by attributes more typical of an alternative "identity." For example, beliefs may be influenced by identification with a Christian religion. The tendency in the health sector might be to identify an individual as Indigenous and fail to provide person-centred care, despite recognition of obvious co-identity (e.g. age, disability, gender) because other important co-identity (e.g. religion, sexuality, socioeconomic status) are either overlooked or assumed.

These principles are further highlighted in rural and remote Western Australia where Indigenous people have a 3 times higher myocardial infarction rate compared to non-Indigenous people and 50% lower five-year cancer survival (Currie and Yindyamarra, 2022). A study undertaken in Western Australia showed comparative hospital services between metropolitan and regional communities (900 beds with 5700 staff in the metropolitan hospital and 800 beds with 6000 staff in the regional hospital) and revealed disparities (Taylor et al, 2020). In metropolitan services only 0.9% of staff and 0.8% of the patient population were Indigenous. For the regional service 3.7% of staff and 8% of the patient population were Indigenous. While the regional service indicates under-representation of Indigenous people in the workforce, it also signals the lack of cultural preparedness the metropolitan service has for local Indigenous patients and those referred from lower tier regional sites for specialist services. Also of note, health policy, including funding/rebates, are driven by metropolitan teaching hospitals whose data not only overlooks the unique features of regional and rural communities but, expressly and grossly, under-represents Indigenous people. These issues are barriers to both health equity and person-centred care.

## Indigenous cultural competence

The 370 million *first nation* peoples in 70 countries around the world are substantially disadvantaged by colonialisation (Stephens et al, 2005; Anderson et al, 2016; Shapiro et al 2006, Currie et al 2018). A key factor in healthcare inequity relates to the cultural disconnect between the Indigenous people's paradigm (square peg) and the Westernised paradigm of health (round hole) resulting in inequitable access to health services, lack of Indigenous people in the healthcare workforce, under-utilisation of services (lack of trust), and decreased accessibility to healthcare (Bainbridge et al, 2015).

There are four key characteristics valued by Indigenous Australians that are important in the radiology and nuclear medicine department. The first relates to accessibility of services (Gomersall et al, 2017) but the nature of clinical practice makes co-location with other health services challenging, and the environments less than welcoming, particularly post-COVID (Currie, 2020a; Currie 2020b). The second issue highlights the value of person-centred care. Indigenous Australians and many culturally diverse sub-populations value healthcare that is appropriate and responsive to their holistic needs and beliefs; a whole person approach (Cerezo et al, 2020). Despite this, some cultural norms are not easy to accommodate within the social, economic and regulatory frameworks of Westernised healthcare and this will require a degree of emotional intelligence to negotiate person-centred outcomes within a culturally safe environment. Thirdly, again typical of a diverse cultural spectrum, Indigenous people value culturally safe places where ethnicity and beliefs are respected (Cerezo et al, 2020). These culturally safe environments require a foundation of cultural knowledge and awareness on which deep engagement through cultural values and attitudes is scaffolded (Figure 2.1). A fourth value would be more equal power relationships to health equity. This can be achieved in culturally safe environments by emotionally competent health professionals. Cultural safety needs to recognise the impact of historical, social and institutionalised disadvantage, discrimination, bias, prejudice and inequity in order to build trust with evolved or reengineered social, institutional and political structures (Kurtz et al, 2018). These principles are transferrable between Indigenous peoples and other cultural groups identifying based on ethnicity, religion, belief, age, gender, sexuality, disability, socioeconomic status or geography, and can be used to drive the equity pipeline to address lack of vertical and horizontal diversity in health (Figure 2.5).

## Taxonomies of cultural competence

Bloom's cognitive taxonomy can provide a valuable framework for learning cultural competence. Bloom's cognitive taxonomies are used to scaffold the learning from lower order capabilities like knowing and understanding through to higher order capabilities like evaluation and synthesis. This taxonomy of learning is the foundation of undergraduate education yet it simply affords

*Figure 2.5* The healthcare equity pipeline (modified and reprinted with permission *11*).

early capabilities in the cultural competence pipeline (Figure 2.4). Progression to cultural safety and sensitivity (acuity) along the pipeline requires focus on the affective domain of Bloom's taxonomy where feelings, attitudes and values are scaffolded from receiving and responding through to internalised values. The step to cultural competence and proficiency requires command of the capabilities of Bloom's taxonomies in the psychomotor domain. Here, the emphasis is on behaviours, skills and what individuals actually do, and has a scaffold from perception through to adaption and organisation (Figure 2.4). Similarly to emotional intelligence, Charles Sturt University radiography and nuclear medicine students are provided a preliminary workshop (Currie and Yindyamarra, 2022) creating awareness and understanding associated with cultural competence (cognitive taxonomies). The knowledge is enriched with immersive exercises that are linked to reflective tasks to connect the learning with experience (affective taxonomies). Students are then released into the clinical environment where further application and reflection are used to hone cultural competence (psychomotor taxonomies).

## Person-centred care

Person-centred care is a little analogous to other novel solutions-based strategies that are excellent in theory but seldom practised in reality. For example, problem-based learning is an excellent learning strategy in higher education but in reality, a modified version of problem-based learning, like scenario-based learning, is actually implemented. The philosophy of person-centred care is also not a new concept in healthcare. The idea of patient-centred care and shared decision making has been entrenched in Westernised medicine for more than 50 years (Britten et al, 2016). The three pillars of person-centred care put an emphasis on personhood (the individual identity, needs and expectations), partnership (respect, shared knowledge and trust), and a holistic framework (circumstances, principles, resources, enablers, regulations and needs) (Harding et al, 2015). As previously outlined, patient-centred care emphasises the focus on the patient but tends to identify the patient as a function of their symptoms, disease or procedure. Person-centred care is not only a holistic focus on the person but is inclusive of family and carers (Santana et al, 2018). There are a number of person-centred care frameworks (Santana et al, 2018; McCormack and McCance, 2006; McCormak and McCance, 2010) that share common components (Figure 2.6):

- Structure includes the regulatory and institutional environment in which healthcare is provided. This includes health workforce and healthcare system sub-domains.
- Process is associated with healthcare provision and represents the interface between person-centred care policy/strategy and the improved outcomes associated with person-centred care delivery.
- Outcome is associated with the patient and their satisfaction with outcomes and individual identity.

Within the framework, cultural competence and emotional intelligence reinforce the framework; the glue or mortar that links foundations and scaffold, and hold the model together. Cultural competence is an essential driver of structure-based changes to institutional culture, the environment and strategy development. The process of person-centred care implementation demands command of both cultural competence and emotional intelligence. Outcomes and patient perception of care is also enhanced by a culturally competent, emotionally intelligent workforce.

The importance of cultural competence in person-centred care has been evidenced in student education and the clinical environment. Developing undergraduate nursing curricula for person-centred care in older people emphasised the importance of cultural diversity and, indirectly, cultural competence (Markey et al, 2021). The capacity to measure cultural competence and the impact on person-centred healthcare has also been explored as a tool to monitor and enhance outcomes (Ahmed et al, 2018). Santana et al. (Santana et al, 2020) identified 26 quality indicators for person-centred care that reflected the three framework domains of structure, process and outcomes. Among the 26 quality indicators, cultural competence was a key feature in several (those with asterisk* also require emotional intelligence):

- person-centred policy
- education programmes*
- culturally competent care*
- care in partnership*
- accommodating support persons*
- compassionate care*
- equitable care*
- relationships of trust*
- communication*
- patient involvement*

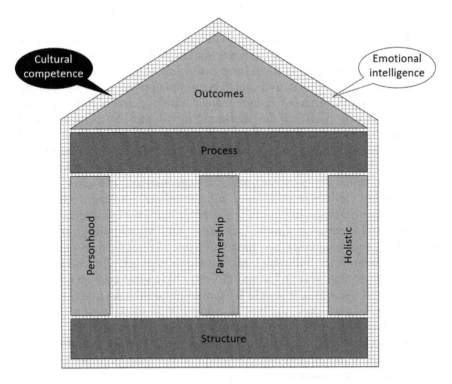

*Figure 2.6* The person-centred care framework including the three pillars and the reinforced support of cultural competence and emotional intelligence.

Ekman et al. (Ekman et al, 2012) examined the impact of person-centred care in patients with chronic heart failure and found that hospital stays were shortened without increasing patient risk of re-admission. In radiology, there is an array of research conducted in patient-centred care but a paucity in person-centred care. This is perhaps highlighted by research examining radiographer perceptions of patient-centred care and issues related to care communication (Hyde and Hardy, 2021a; Hyde and Hardy, 2021b). While altruistic agreement with the philosophy of person-centred care is widespread among radiographers, there remains concern regarding time commitment for implementation. Perhaps key to implementing person-centred care in radiology and nuclear medicine will be the concurrent development of curricula and continuing education activities that not only support the pillars and framework of person-centred care (Figure 2.6) but also foster cultural competence and emotional intelligence.

Communication individualised for each patient was also identified as a key tool in striving for person-centred care for autistic patients undergoing MRI (Stogiannos et al, 2022). Central to success will be both cultural competence and emotional intelligence. In this circumstance, cultural competence affords cultural sensitivity and creation of safer clinical environments for autistic patients (collective identity). Emotional intelligence, however, allows specific nuances of individual patients to be identified and tools and strategies adjusted in real time to better meet patient needs and expectations, and to optimise outcomes.

Radiography provides an interesting case study to highlight the barriers and challenges to person-centred care. The production line perception of the medical imaging department is inconsistent with a person-centred model, yet time/efficiency need not be sacrificed in adopting a person-centred approach. Society has evolved to be one that values instant gratification and patient perceived quality markers may favour short waiting times and faster procedures (Hayre et al, 2016). The craft health practitioners must wield provides a balance of quality care, accurate outcomes and efficiency.

A culturally competent workforce with emotional intelligence not only supports implementation of person-centred care but is well placed to provide leadership and mentoring among colleagues, and to assimilate care-based outcomes as patient quality benchmarks. While the emphasis on precision medicine, theranostics, radiomics and radiation dosimetry in higher order medical imaging (e.g. SPECT/CT and PET/CT) have some consistencies with person-centred outcomes, it should be kept in mind that the journey to precision medicine outcomes can be one riddled with social asymmetry and/ or that reduces the patient to symptoms or disease. The power of precision medicine might be fully realised when the journey is truly person-centred and navigated in partnership with cultural and emotional competent health practitioners (communicator) (Figure 2.7).

### Self-centred care

Health professionals diminish their value to patients if they overlook their own health and wellbeing. Health professionals are more prone to burnout than other professions because the environment is complex, stressful and high stakes. A pre-COVID investigation reported high burnout scores for emotional exhaustion and depersonalisation among radiographers and radiologists (Singh et al, 2017). This is incompatible with person-centred care. More recent investigations post-COVID report 47% of radiologists self-identifying burnout (Baggett et al, 2022) while another reported symptoms of burnout in as many as 72% of radiologists (Canon et al, 2022). It is entirely appropriate that health workers apply person-centred care to themselves; including both physical and mental health and wellbeing. Maintaining person-centred care in radiology and nuclear medicine requires introspection that harnesses both cultural competence and emotional intelligence.

### Summary

Culturally appropriate person-centred healthcare requires understanding and respect for health beliefs, values, preferences, behaviours, symptom

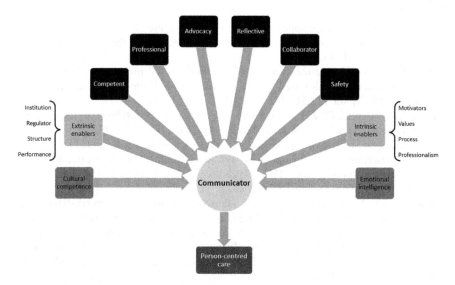

*Figure 2.7* The person-centred practitioner that encapsulates core professional capabilities driven by intrinsic and extrinsic enablers and informed by cultural and emotional competence.

recognition, thresholds for seeking care, expectations of healthcare, compliance and attitudes to procedures. Cultural competence is also a key strategy for tackling healthcare inequities but cultural acuity needs a degree of emotional acuity. Emotional intelligence improves patient care and contributes to person-centred healthcare. Development of skills and capabilities in cultural competence and emotional intelligence should be deliberate and formalised rather than rely on organic or incidental/accidental development. Cultural competence and emotional intelligence is an essential scaffold for establishing person-centred care.

## Acknowledgement

Parts of this chapter have been reproduced or extracted, with permission, from:

- Currie, J., & Currie, G. (2022). Emotional intelligence and productive relationships with patients and colleagues, *Journal of Nuclear Medicine Technology*, vol. 50; pp. 357–365.
- Currie, G. (2022). Yindyamarra Winhanganha: a Conduit to Indigenous Cultural Proficiency. *Journal of Nuclear Medicine Technology*, vol. 50; pp. 66–72.

# References

Saha, S, Beach, MC, & Cooper, L. (2008) Patient centeredness, cultural competence and healthcare quality. *Journal of the National Medical Association.* 100(11):1275–1285.

Currie, J, Currie, GM. (2022) Emotional intelligence and productive relationships with patients and colleagues. *Journal of Nuclear Medicine Technology.* May 24. doi: 10.2967/jnmt.122.264052

Currie, G, Rohren, E. (2022) Social asymmetry and artificial intelligence: the nuclear medicine landscape. *Seminars in Nuclear Medicine.* 52: 498–503.

Karimi L, Leggat SG, Bartram T, Afshari L, Sarkeshik S, Verulava T. (2021) Emotional intelligence: predictor of employees' wellbeing, quality of patient care, and psychological empowerment. *BMC Psychology.* 9:93, https://doi.org/10.1186/s40359-021-00593-8 Accessed 15 Feb 2022.

Jiménez-Picón N, Romero-Martín M, Ponce-Blandón J.A, Ramirez-Baena L, Palomo-Lara JC, Gómez-Salgado J. (2021) The relationship between mindfulness and emotional intelligence as a protective factor for healthcare professionals: Systematic review. *International Journal of Environmental Research and Public Health.* 18:5491. https://doi.org/10.3390/ijerph18105491 Accessed 15 Feb 2022.

Cerezo MV, Blanca MJ, Ferragut M. (2020) Personality profiles and psychological adjustment in breast cancer patients. *International Journal of Environmental Research and Public Health.* 17:9452. doi:10.3390/ijerph17249452 Accessed 15 Feb 2022.

de Galvao A, Medeiros B, Lewis S, McNulty J, White P, Lane S, Mackay S. (2017) Emotional intelligence development in radiography curricula: results of an international longitudinal study. *JMIRS.* 48:282–287.

Leininger MM, McFarland MR. (2006) *Culture care diversity and universality: A worldwide nursing theory.* Sudbury: Jones & Bartlett Learning.

Betancourt J, Green A, Carrillo J. (2003) Defining cultural competence: A practical framework for addressing racial/ethnic disparities in health and health care. *Public Health Reports.* /July–August 118:293–302.

Bainbridge R, McCalman J, Clifford A, Tsey K. (2015) Australian Government. Cultural competency in the delivery of health services for Indigenous people. Issues paper no. 13 produced for the Closing the Gap Clearinghouse.

Currie. G. Yindyamarra Winhanganha: (2022) A Conduit to indigenous cultural proficiency. *Journal of Nuclear Medicine Technology.* 50:66–72.

Downing R, Kowal E, Paradies Y. (2011) Indigenous cultural training for health workers in Australia. *International Journal for Quality in Health Care.* 23:247–257.

Ewen SC, Paul DJ, Bloom GL. (2012) Do indigenous health curricula in health science education reduce disparities in health care outcomes? *Medical Journal of Australia.* 197:50–52.

Stephens C, et al. (2005) Indigenous peoples' health – why are they behind everyone, everywhere? *Lancet.* 366:9479.

Anderson I, et al. (2016) Indigenous and tribal peoples' health (The Lancet–Lowitja Institute Global Collaboration): a population study. *Lancet.* DOI:https://doi.org/10.1016/S0140-6736(16)00345-7

Shapiro J, et al. (2006) "That never would have occurred to me": A qualitative study of medical students' views of a cultural competence curriculum. *BMC Medical Education.* 6(1):1.

Brach C, Fraserirector I. (2000) Can cultural competency reduce racial and ethnic health disparities? A review and conceptual model. *Medical Care Research and Review.* 57:181–217.

Kumas-Tan Z, et al. (2007) Measures of cultural competence: examining hidden assumptions. *Academic Medicine.* 82(6):548–57.

Butler M, et al. (2016) *Improving Cultural Competence to Reduce Health Disparities.* Rockville (MD): Agency for Healthcare Research and Quality.

Bolderston A, Middleton J, Palmaria C, Cauti S, Fawcett S. (2021) Improving lesbian, gay, bisexual, transgender, queer and two-spirit content in a radiation therapy undergraduate curriculum. *Journal of Medical Imaging and Radiation Sciences.* 52(2):160–163.

Currie, G, Wheat, J & Wess, T. (2018) Building foundations for Indigenous cultural competence: an institutions' journey toward "close the gap." *Journal of Medical Imaging and Radiation Sciences.* 49 (1):6–10.

Taylor EV, Lyford M, Parsons L, Mason T, Sabesan S, Thompson SC. (2020) "We're very much part of the team here": A culture of respect for Indigenous health workforce transforms Indigenous health care. *PLoS ONE.* 15(9):e0239207. https://doi.org/10.1371/journal.pone.0239207

Gomersall, J. S., Gibson, O., & Dwyer, J., et al. (2017) What Indigenous Australian clients value about primary health care: a systematic review of qualitative evidence. *Australian and New Zealand Journal of Public Health.* 41(4):417–423.

Currie, G. (2020a) A lens on the post-COVID19 "new normal" for imaging departments, *Journal of Medical Imaging and Radiation Sciences.* 51;361–363.

Currie, G. (2020b) Post-COVID19 "new normal" for nuclear medicine practice: an Australasian perspective (invited commentary). *Journal of Nuclear Medicine Technology.* 48(3):234–240.

Kurtz D, Janke R, Vinek J, Wells T, Hutchinson P, Froste A. (2018) Health Sciences cultural safety education in Australia, Canada, New Zealand, and the United States: a literature review. *International Journal of Medical Education.* 9:271–285. ISSN: 2042-6372 doi: 10.5116/ijme.5bc7.21e

Britten N, Moore L, Lydahl D, Naldemirci O, Elam M, Wolf A. (2016) Elaboration of the Gothenburg model of person-centred care. *Health Expectations.* 20: 407–418

Harding E, Wait S, Scrutton J. (2015) *The State of Play in Person-centred Care: A Pragmatic Review of How Person-Centred Care is Defined, Applied and Measured.* London: Health Foundation.

Santana MJ, Manalili K, Jolley RJ, Zelinsky S, Lu M. (2018) How to practice person-centred care: A conceptual framework. *Health Expectations.* 21:429–440.

McCormack B, McCance TV. (2006) Development of a framework for person-centred nursing. *Journal of Advanced Nursing.* 56:472–479.

McCormack, B. and McCance, T. (2010) *Person-centred Nursing: Theory and Practice.* Oxford: Wiley Blackwell.

Markey K, Brien BO, O'Donnell CO, Martin C, Murphy J. (2021) Enhancing undergraduate nursing curricula to cultivate person-centred care for culturally and linguistically diverse older people. *Nurse Education in Practice.* 50:102936. https://doi.org/10.1016/j.nepr.2020.102936

Ahmed S, Siad FM, Manalili K, *et al*. (2018) How to measure cultural competence when evaluating patient-centred care: a scoping review. *BMJ Open*. 8:e021525. doi:10.1136/ bmjopen-2018-021525.

Santana M-J, Manalili K, Zelinsky S, *et al*. (2020) Improving the quality of person-centred healthcare from the patient perspective: development of person-centred quality indicators. *BMJ Open*. 10:e037323. doi:10.1136/ bmjopen-2020-037323.

Ekman I, Wolf A, Olsson L-E, et al. (2012) Effects of person-centred care in patients with chronic heart failure: the PCC-HF study. *European Heart Journal*. 33(9):1112–1119

Hyde E, and Hardy M. (2021a) Patient centred care in diagnostic radiography (part 2): a qualitative study of the perceptions of service users and service deliverers. *Radiography*. 27(2):322–331.

Hyde E, and Hardy M. (2021b) Patient centred care in diagnostic radiography (Part 1): Perceptions of service users and service deliverers. *Radiography*. 27(1):8–13.

Stogiannos N, Carlier S, Harvey-Lloyd JM, Brammer A, Nugent B, Cleaver K, McNulty J.P, Dos Reis CS, and Malamateniou C. (2022) A systematic review of person-centred adjustments to facilitate magnetic resonance imaging for autistic patients without the use of sedation or anaesthesia. *Autism*. 26(4):782–797.

Hayre CM, Blackman S, and Eyden A. (2016) Do general radiographic examinations resemble a person-centred environment?. *Radiography*. 22(4):e245–e251.

Singh N, Knight K, Wright C, Baird M, Akroyd D, Adams RD, Schneider ME. (2017) Occupational burnout among radiographers, sonographers and radiologists in Australia and New Zealand: Findings from a national survey. *Journal of Medical Imaging and Radiation Oncology*. Jun;61(3):304–310.

Baggett SM, Martin KL. (2022) Medscape radiologist lifestyle, happiness & burnout report 2022. Medscape. Available at: www.medscape.com/slideshow/2022-lifest yle-radiologist-6014784#1. Accessed March 24, 2022.

Canon CLK, Chick JFB, DeQuesada I, Gunderman RB, Hoven N, Prosper AE. (2022) Physician burnout in radiology: perspectives from the field. *American Journal of Roentgenology*. 218(2):370–374.

# 3 Affirming and inclusive diagnostic imaging and radiotherapy services for sexual and gender minority patients

*Riaan van de Venter*

## Introduction

The World Health Organization (WHO) (WHO, 1948) describes health as being in a state of equilibrium between individuals' physical, psychological, and social wellbeing. One can thus argue that this definition of health asserts that individuals' health and subsequent healthcare needs are multifaceted. The intersectionality theoretical framework becomes a useful tool to advance health equity and person-centred care because it requires radiographers (i.e., diagnostic and therapeutic), and other healthcare professionals (HCPs), to consider a variety of social dynamics, identities, and factors that may shape patients' perspectives about health and their lived realities in healthcare settings (Bowleg, 2012). Being mindful of this theoretical premise is important because sexual and gender minority (SGM) patients collectively is a heterogenous group, with overlapping, yet unique healthcare needs which requires personalised healthcare provisions.

However, the status quo in contemporary healthcare settings is heteronormative where SGM patients are expected to conform to gender and sexual orientation binaries, because the predominant ideology is still that individuals' biological sex, sexual orientation and gender identity should be linearly congruent. This cis-heteronormative approach, and accompanying microaggressions, have serious negative consequences for SGM patients relative to their heterosexual, cisgender counterparts (Pieri & Brilhante, 2022; Müller, 2016). Discrimination, prejudice, verbal and physical abuse, harassment about their sexuality and gender identity, lack of privacy, being disrespected, as well as refused access to healthcare services and insurances, are some of the barriers encountered by SGM patients. The lack of inclusive language usage on information brochures and intake forms is another hindrance to affirming healthcare provision for SGM patients. The limited availability of appropriate healthcare services and competent healthcare providers also impede SGM patients' health. These impediments can lead to SGM patients' fearing to disclose sensitive information, be distrusting of HCPs and noncompliant to HCPs' advice, reluctant to seek care when they feel ill, (re-)traumatisation, social exclusion, a decline in quality of life, and

DOI: 10.1201/9781003310143-4

late diagnosis of diseases such as cancer. These experiences can give rise to mental health-related conditions such as anxiety, depression, suicidal ideation, and suicide attempts, as well as other conditions like chronic stress, substance abuse and cardiovascular-related disorders (Igual, 2023; National LGBT Cancer Network, 2023; Pride in Health + Wellbeing, 2023; Abboud et al., 2022; Pieri & Brilhante, 2022; Silva & Costa, 2020; Müller, 2016; Dean et al., 2016; Utamsingh et al., 2016; Mayer et al., 2008).

It is against this backdrop that this chapter makes a case for the creation of inclusive and affirming caring praxes in diagnostic imaging and radiotherapy services, to promote SGM patients' health and wellness. The linkage between human rights and professional codes of conduct are explored in the context of person-centred care (PCC). Strategies to promote PCC in diagnostic imaging and radiotherapy settings for SGM patients are also discussed. The strategies discussed broadly relate to inclusive language, creating an affirming physical environment, including and honouring the patient voice, and radiographer education.

## Patients' rights and professional practice

All individuals have the right to an environment conducive to healing and wellbeing, and to be treated with respect and dignity. Radiographers are professionally responsible and accountable to maintain patients' dignity during diagnostic imaging and radiotherapy examinations. Patients' health status and records should also remain confidential. This is particularly important so that SGM patients are not carelessly "outed" to other people which could promote further prejudice and discriminatory behaviour towards them, including abuse. Additionally, the patient–radiographer relationship should be built on trust, and promote the best interests of the patient. These rights apply to every individual, regardless of ethnicity, disability, language spoken, religious conviction, socioeconomic status, political views, origin, biological sex, sexual orientation, and gender identity. These patient rights are enshrined in regulatory documents outlining the professional conduct and standards of proficiency for radiographers (Wood et al., 2022; International Society of Radiographers and Radiological Technologists, 2021; Health Professions Council of South Africa, 2016; Society of Radiographers, 2013; Health & Care Professions Council, 2007; Universal Declaration of Human Rights, 1948). Hence, every radiographer, regardless of discipline or place of practice, has an ethico-legal duty to uphold patients' rights and adhere to patient's rights charters that may exist in their professional contexts, for example the European Charter of Patients' Rights and South African Patients' Rights Charter (National Department of Health, 2007; Active Citizenship Network, 2002).

One can argue that adopting a person-centred care (PCC) approach to diagnostic imaging and radiotherapy practice is key to upholding patients' rights (Fridberg et al., 2021). PCC requires radiographers to be cognisant of patients' unique healthcare needs and treatment goals, belief and value systems, and their contexts; whilst considering the impact that treatment plans and examinations can have on their family and caregivers too. This requires radiographers to adopt a supportive and empowering disposition when engaging with patients in their care, so as to treat their patients as whole persons beyond their clinical manifestations (Santana et al., 2018; Coulter & Oldham, 2016).

The following section in this chapter provides some strategies that radiographers can consider using in their own clinical practice to provide PCC-oriented services to SGM patients.

## Strategies to promote PCC for SGM patients

A recent literature review found that strategies that could be implemented to create inclusive and affirming healthcare environments could be classified in three categories, namely: inclusive language, education and affirming physical healthcare environments (van de Venter & Hodgson, 2020). Strategies in these areas are also underscored by a joint statement by the Association for Cancer Physicians, Royal College of Physicians, and Royal College Radiologists (2021). Below are specific approaches that could be used to address the three broad categories of strategies identified above to provide PCC for SGM patients in the diagnostic imaging and radiotherapy contexts.

### *Using inclusive language during patient interactions and in official documentation*

Understanding the basic terminology associated with SGM healthcare is fundamental to facilitate inclusive language usage. In Table 3.1 a non-exhaustive list of common terminology and an explanation of each term is outlined (Rioux et al., 2022; Kruse et al., 2022; Davison et al., 2021; Society of Radiographers, 2021; Rosati, Pistella, Nappa, & Baiocco, 2020; Clements, 2018; Smith, 2018; Sowinski & Gunderman, 2018; Unicef, 2017; American Psychological Association, 2015). This can be a primer for cultivating inclusive language use in clinical practice among radiographers.

From Table 3.1, one can appreciate that many terms are interrelated and some nuancedly different. Incorrect uses of terms can thus easily happen. Therefore, it is important that radiographers educate themselves about the terminology associated with SGM patients and their healthcare so that the correct terms are used in the appropriate context. This includes learning derogatory and offensive terms so that these can be avoided, some of which are included in Table 3.1. When mistakes inadvertently happen, simply apologise for the error, correct yourself and move on. Do not belabour the issue. It may make the situation more awkward and uncomfortable.

*Table 3.1* A glossary of common terminology associated with SGM healthcare

| Term | Explanation |
| --- | --- |
| Biological sex | A biological status reference assigned at birth. The typical categories are male, female, and intersex. Predominantly based on external and internal reproductive organs/genitalia, sex chromosomes, and hormone levels. |
| Sexual orientation | This term is used to describe a person's sexual, emotional, and romantic attraction to another person. Sexual and romantic orientations are independent from biological sex and gender, i.e., sexual and romantic orientation is not gender specific. The derogatory counterpart of sexual orientation is sexual preference. |
| Gender/sex binary | The gender/sex binary relates to the belief system that sex and gender is binary with two mutually exclusive categories representing men and women with no grey area in between. Where in fact sex and gender should be viewed as existing on a fluid, dynamic continuum where the constructs are not necessarily mutually exclusive. |
| Gender | A socio-cultural construct which differentiates the different attributes (e.g., roles, responsibilities, aptitudes, feelings, ways of being and behaviours) of masculinity and femininity with reference to societal norms and expectations in a given context. |
| Gender diversity | This concept refers to the extent that an individual's gender identity, expression or role is different from the contextual societal norms in a given context for individuals of a particular biological sex. |
| Gender identity | Aspects of a person's gender that correlates to their sense of self as male, female, both or neither, relative to their felt and inner sense of gender, which may be different from their sex assigned at birth. Gender identity may be different from a person's gender expression as they may not be comfortable to present as their felt gender. |
| Gender expression | This concept refers to how an individual expresses their gender. It also refers to their gendered image and may be communicated through a person's pronouns, names, or gendered language as well as their behaviours and roles, for example. |
| Gender dysphoria | This term is used to label the distress or discomfort experienced by an individual due to the incongruence between their gender identity and sex assigned at birth. This is not necessarily experienced by all transgender and gender diverse individuals. |
| Sex assigned at birth | The practice of assigning a sex to a new-born based on their external anatomy. Two abbreviations may be encountered in practice related to this, namely: AMAB (assigned male at birth) and AFAB (assigned female at birth). This is usually seen in health records when providing healthcare to transgender patients. |

*Table 3.1* (Continued)

| Term | Explanation |
|---|---|
| Gender modality | This concept is used to refer to an individual's experience relative to their sex assigned at birth. It can be congruent or incongruent. |
| Cisgender (or cis) | Refers to persons whose gender identity is congruent with their sex assigned at birth. |
| Transgender (or trans) | Refers to persons whose gender identity is incongruent with their sex assigned at birth. There are sub-classifications of transgender individuals based on their gender identity and sex assigned at birth.<br>• Transgender male = male gender identity and female sex assigned at birth. Also sometimes referred to as FTM (female to male).<br>• Transgender female = female gender identity and male sex assigned at birth. MTF (male to female) is also used often.<br>Avoid using derogatory terms like tranny and transvestite. |
| Transitioning | This term is used to describe the multifaceted process that transgender individuals undergo to align their felt gender with their gender expression. This process involves bodily changes, name, gender, and pronoun changes as well as how they present themselves and behave. |
| Gender nonconforming or genderqueer or nonbinary | This is an umbrella term used to describe people whose gender identity, role and expression differs from the societal gender norms associated with their sex assigned at birth. |
| Gay | A term used to identify men that are romantically and/or sexually attracted to other men. Avoid terms like faggot, limp-wristed, sod, and poof as these are offensive. |
| Lesbian | A term used for females that are romantically and/or sexually attracted to other females. Offensive terms to avoid include dyke and lesbo. |
| Bisexual | This term is used to refer to a person who is attracted (romantically and/or sexually) to individuals of similar and opposite gender to their own. |
| Intersex | Intersex people are born with atypical biological sex characteristics relative to traditional assumptions of male and female anatomy. These individuals' chromosomes and hormonal levels may also be atypical. Using words like hermaphrodite, transsexual, ambiguous etc. are derogatory and must be avoided. |
| Queer | This term is used to identify individuals that neither conform to dominant societal norms nor identify as heterosexual or cisgender. In some settings this may be deemed derogatory so one's context should be used as a guide on the use of this term. |

*(Continued)*

*Table 3.1* (Continued)

| Term | Explanation |
| --- | --- |
| Asexual | This concept is used to identify a person who is romantically attracted to one or more genders or sexes. Asexual individuals are not sexually attracted to anyone. |
| Pansexual | A term used to describe a person who identifies as any gender, including nonbinary and transgender and is sexually and/or romantically attracted to any gender. |
| Two-spirited | A term used almost exclusively to identify individuals of the indigenous first nations people in Northern America who have two spirits comprising their gender identity: masculine and feminine. |
| Pronouns | Pronouns are words used in English to indicate someone talking or being talked about. Pronouns are very gendered. There is a move towards using gender neutral pronoun forms. Below is a non-exhaustive list of commonly used pronouns.<br>• Masculine pronouns: he/his/him<br>• Feminine pronouns: she/hers/her<br>• Gender-neutral pronouns: they/their/them |
| Honorifics | These are salutations used to address SGM patients and their significant others, such as Mr, Mrs, Ms, Ma'am, and Sir. The neutral form of this in many cases is Mx especially when addressing gender non-binary individuals. It is best to be guided by patients on how they would like to be addressed. |
| Legal and chosen names | A legal name is the one assigned to a person at birth and is reflected on one's identity document. On the other hand, a chosen name is a name individuals use in engagements with others. |
| Deadname | This term is used to denote a transgender individual's birth name. It is considered inappropriate and offensive to use this name to refer to a trans person. |
| Misgendering | This concept is used to describe a situation where one person purposely, or accidentally, refers to another individual using language that is incongruent with their affirmed gender. |
| Ally | This term is used to describe people identifying as heterosexual and/or cisgender who are supportive of SGM individuals. |
| Gender affirming interventions | This term is used to collectively refer to the medical and/or surgical interventions that transgender individuals undergo to affirm their gender identity. Specific gender affirming interventions have specific names depending on the procedure or process. |

*Table 3.1* (Continued)

| Term | Explanation |
| --- | --- |
| Coming out and being outed | These terms are used to refer to the process that SGM individuals undergo to disclose their sexual orientation and gender identity to other people. It originated from the idea of coming out of the closet. It is not a one-time event; it occurs multiple times across the lifespan. Considering the healthcare context, when SGM patients see a new HCP for the first time they may need to come out to them. However, when someone else makes an SGM individual's sexuality and/or gender identity public, usually without the consent of the person, it is then considering being outed or outing the SGM person. |

Sources: Rioux et al., 2022; Kruse et al., 2022; Davison et al., 2021; Society of Radiographers 2021; Rosati, Pistella, Nappa & Baiocco, 2020; Clements, 2018; Smith, 2018; Sowinski & Gunderman, 2018; Unicef, 2017; American Psychological Association, 2015.

### Direct patient communication and interaction

When communicating directly with SGM patients, for example before a radiographic examination, always introduce yourself by indicating your name and preferred pronouns. Thereafter, you should ask the patients how they would like to be addressed and which pronouns you should use (Society of Radiographers, 2021). In my experience it makes patients more comfortable and trusting in a radiographer's ability to competently care for them. During communicative interactions with SGM patients always reassure them that their health records and associated documentation will be handled with the required confidentiality. This is important so that SGM patients feel reassured that their information is safe, that they will not be "outed", and their fears related to sensitive information disclosure, victimisation, abuse and discrimination may also be allayed. Radiographers should also take cues from their patients in terms of use of language. This requires radiographers to actively listen to how patients refer to themselves, and their medical conditions, so that radiographers can use these terms too. In the radiotherapy context, linking patients to appropriate support groups and giving specific written information related to the cancer that a patient has, or the treatment plan proposed, will assist them to make informed decisions and help them make sense of the verbal information that has been shared. This may also be useful to their partners, family and/or caregivers (Kano et al., 2022).

Gathering the appropriate information about SGM patients' sex and gender can inform their clinical management. For example, it can allow for appropriate screening to detect complications of cancer in a timely manner, or ensure that the correct screening is done before, during and after chemotherapy and radiation therapy. Radiographers should ask about a patient's gender and sex if such information is not available in their health records. It is important to remember to explain why this information is necessary, so that patients understand the relevance of their answers (Quinn et al., 2015). In a similar vein, SGM patients should be advised on the most appropriate ways to maintain a healthy and safe sexual lifestyle during cancer treatment, so that they do not suffer from additional complications due to inappropriate self-care during and after cancer treatment. Guidelines in therapeutic radiographers' contexts should guide these conversations with SGM patients in congruence with their biological sex, sexual orientation, and gender identity. For example, guidelines exist in the UK about the abstinence periods before, during and after cancer treatment for patients with prostate cancer (Ralph, 2021). Other authors have also published screening recommendations for transgender patients to consider (Sowinski & Gunderman, 2018).

*Gathering and capturing patient information*

Gender and sex binary terminology is still commonplace in patients' electronic health records (EHRs), policies and other administrative documents like hospital admission forms and diagnostic imaging and radiotherapy requests. There is a need to allow for more representativeness and inclusivity of the continuum of sexes and genders. Firstly, this will allow for accurate data gathering about the SGM patient populace using healthcare services, which can inform service planning, policy, research, and quality improvement endeavours. Secondly, having this information can mitigate the probability to misgender a patient, because a patient's preferred names, pronouns and even gender expression and position of gonads/reproductive organs, could be available before a radiographer calls a patient for their examination or radiotherapy. This information would help radiographers to utilise appropriate language to create an affirming healthcare environment, as well as build rapport with and gain trust from a patient (Davison et al., 2021; Pedersen & Sanders, 2018). In this regard the sex, identity, gender, expression (SIGE) form is an example of good practice (Pedersen & Sanders, 2018). The inclusive pregnancy status (IPS) form is another example which uses different language and there are two versions: one for diagnostic imaging and nuclear medicine, and another for radiotherapy (Society of Radiographers, 2021). Table 3.2 provides a summary of the SIGE and IPS forms. Thirdly, only ask questions relevant to the examination or treatment, so as to avoid discomfort and distress for the patient. It is very important to always explain the rationale for asking particular information and indicating how it will be utilised (Society of Radiographers, 2021). Considering the experiences, identified in the introduction, of SGM patients in healthcare settings, one can appreciate why this is imperative.

*Table 3.2* Summary of the SIGE and IPS forms

| SIGE form | IPS form |
| --- | --- |
| • This form is used to gather information about patients' sex, identity, gender, and expression to inform radiation protection practices.<br>• The information gathered relates to patients' names, pronouns, position of reproductive organs, last menstrual period and pregnancy status.<br>• Allows for asking sensitive information in a respectful manner. | This form is used to gather information about patients' pregnancy status in a non-threatening and affirming manner. The purpose of this form is to inform radiation protection measures applicable to pregnant or possibly pregnant patients.<br><br>The IPS for diagnostic imaging and nuclear medicine<br><br>• This form gathers information about patients' sex assigned at birth, sex characteristics, previous surgery impacting fertility, last menstrual period, and possibility of pregnancy.<br>• It is used for patients between 12 and 55 years<br><br>The IPS for radiotherapy<br><br>• Collects patient identification information, sex assigned at birth, sex characteristics, pregnancy status, and requires patients to acknowledge their awareness that they should not become pregnant during the course of radiotherapy treatment as well as that taking testosterone and undergoing chemotherapy do not preclude them from becoming pregnant.<br>• It is used for patients between 12 and 55 years. |

Sources: Society of Radiographers, 2021; Pedersen & Sanders, 2018.

Lastly, when designing EHR systems and intake forms, and even research tools, never use "other" as an option for identifying sex and/or gender. Allow for self-identification instead through an open-ended question. Using other is offensive, stigmatising and even ostracising because it invalidates SGM persons' existence (Davison et al., 2021). The question could be phrased in the line of: "Specify your sex" or "Indicate your gender identity", for example. In a similar vein there should also be an option for SGM patients to indicate whether they do not wish to disclose information about their sex and/or gender. This is to allow them privacy and comfort in knowing that they will not be outed (i.e., their sex/gender disclosed to others against their will or without their consent) (Society of Radiographers, 2021). If this type of information is crucial to the examination or treatment, the radiographer should discuss why the information is required and how it will be stored with the patient in a private setting, and then allow the patient to make the decision. Some patients may not want their details to be stored in any way but may be willing to share details verbally. Remember any details shared in confidence should be kept private, to avoid potential patient harm.

### Creating physical healthcare environments that are affirming for SGM patients

Literature advocates for the creation of physical healthcare environments that demonstrate inclusivity, belonging and affirmation. The environment should be welcoming and represent safety, acceptance, and representativeness (Davison et al., 2021; van de Venter & Hodgson, 2020). A study found that introducing inclusivity indicators in the clinical setting had a positive impact on SGM patients' experience and were barely noticed by non-SGM-identifying individuals (Zhu et al., 2022). In diagnostic imaging and radiotherapy departments the use of pride flags, posters that use imagery of sexual and gender diverse people, information leaflets that use inclusive language and representative imagery of SGM patients and having nongendered bathrooms can all cultivate affirmation for SGM patients. Moreover, having gowns that are neutrally coloured with no gendered connotation is also an example of SGM inclusive and affirming care.

The departmental culture around the language used can also be considered a strategy to create an inclusive environment (Davison et al., 2021; van de Venter & Hodgson, 2020;Bolderston, 2018). For example, use a phrase like "good morning, everyone!" instead of "good morning, guys!" when addressing a mixed group of individuals. The neutral language is more inclusive. Another example of this can be seen in the way we speak about others. In cases where one is not sure of another person's preferred pronouns the neutral form they can be used in both singular and plural formats. Speaking up for SGM patients when their rights are infringed is another strategy that diagnostic and therapeutic radiographers can implement to create affirming healthcare spaces for these patients. This will show allyship from the radiographers, regardless of their sex and/or gender. In the UK, the NHS Rainbow Badge project is another example of showing allyship for SGM patients. This demonstrates to patients who their support structure is in the healthcare space and who SGM patients can confide in (LGBT Foundation, n.d.).

From the above examples one can appreciate that small gestures of kindness, and considered changes to the physical healthcare environment, can make a big difference for SGM patients and promoting PCC.

### Patient partnerships

Patient involvement is used in health professions education to assist in developing students' professionalism, communication, and other clinical skills (Butani, Sweeney & Plant, 2020; Suikkala, Koskinen, Leino-Kilpi, 2018; Naylor, Harcus & Elkington, 2015; Lown, Sasson & Hinrichs, 2008). Typically, patients teach using stories of their health and care journeys to stimulate students to reflect and learn from those experiences. This sharing allows students to gain a greater understanding of patients' experiences, decreases anxiety about working with patients, increases students' respect for patients and assists them in appreciating that patients are a central part of their own healthcare (Butani et al., 2020; Suikkala et al., 2018; Lown et al., 2008). Public and patient involvement in health policy development is becoming more common, and there is evidence that this contributes to greater democratic decision making and accountability among all stakeholders (Souliotis, 2016). It can therefore be argued that including patient partners can enable the development of more holistic PCC practices, and more inclusive policies for SGM patients. Hence, diagnostic imaging and radiotherapy departments should consider including patient partners, including patients who are SGM, on their advisory boards to contribute to the planning and management of services. The patient partner roles should be clearly delineated and negotiated so that there is effective and appropriate management of expectations. Patient partners contributions should be recompensed for their time and any expenses incurred.

### Radiographer education and training about SGM healthcare provision

Education and training should be structured so that it promotes culturally competent healthcare services for SGM patients. Hitherto currently there are very limited opportunities for education in this area of medical imaging and radiation sciences, and other health sciences curricula (Bolderston et al., 2021; Silva & Costa, 2020; van de Venter & Hodgson, 2020; Gordon & Mitchell, 2019; Obedin-Maliver et al., 2011). Educational interventions should be part of formal qualifications at pre-registration and/or post-registration level and continuing professional development (CPD) activities. Education opportunities should focus on developing culturally competent radiographers to provide PCC to SGM patients (Antonio et al., 2022). Radiographers need to have a thorough knowledge of the terminology and concepts associated with SGM health, as well as the unique healthcare needs of this specialised population. Moreover, the link between PCC for SGM patients and radiographers' duty towards patients, patients' rights and scope of practice, should be highlighted. Radiographers should also be empowered to create inclusive and affirming healthcare spaces for SGM patients (van de Venter, 2022). One example of an online, continuing professional development (CPD) educational opportunity focused on SGM patients and their healthcare needs is offered by the Canadian Association for Medical Radiation Technologists (CAMRT). It is accessible to radiographers, in all disciplines, worldwide (Canadian Association for Medical Radiation Technologists, 2023).

## Conclusion

In this chapter, the barriers to accessing inclusive and affirming healthcare that SGM patients face were used as a background to outline strategies that can be used to overcome these challenges. These strategies were considered from a PCC perspective. Diagnostic and therapeutic radiographers have an ethico-legal duty to provide appropriate and effective healthcare to SGM patients that is aligned with the patients' unique needs and context. Radiographers can only fulfil this requirement if they are properly educated and enabled to develop inclusive and affirming practices for SGM patients.

## References

Abboud, S., Veldhuis, C., Ballout, S., Nadeem, F., Nyhan, K., and Hughes, T., 2022. Sexual and gender minority health in the Middle East and North Africa region: a scoping review. *International Journal of Nursing Studies Advances* [online], 4:100085. Available from: https://doi.org/10.1016/j.ijnsa.2022.100085 [Accessed 12 January 2023].

Active Citizenship Network., 2002. European Charter of Patients' Rights. Available from: http://health-rights.org/index.php/cop/item/european-charter-of-patients-rights [Accessed 30 January 2023].

American Psychological Association., 2015. Key terms and concepts in understanding gender diversity and sexual orientation among students. Available from: www.apa.org/pi/lgbt/programs/safe-supportive/lgbt/key-terms.pdf [Accessed 30 January 2023].

Antonio, M., Lau, F., Davison, K., Devor, A., Queen, R., and Courtney, K., 2022. Toward an inclusive digital health system for sexual and gender minorities in Canada. *Journal of the American Medical Informatics Association* [online], 29 (2):379–384. Available from: https://doi.org/10.1093/jamia/ocab183 [Accessed 30 January 2023].

Association for Cancer Physicians, Royal College of Physicians and Royal College of Radiologists., 2021. Joint statement: Improving cancer care for sexual and gender minorities. Available from: www.rcr.ac.uk/posts/improving-cancer-care-sexual-and-gender-minorities [Accessed 10 February 2023].

Bolderston, A., 2018. Diversity, identity and healthcare. *Radiography* [online], 24 (Suppl 1):S7–S8. Available from: https://doi.org/10.1016/j.radi.2018.04.006 [Accessed 20 January 2023].

Bolderston, A., Middleton, J., Palmaria, C., Cauti, S., and Fawcett, S., 2021. Improving lesbian, gay, bisexual, transgender, queer and two-spirit content in a radiation therapy undergraduate curriculum. *Journal of Medical Imaging and Radiation Sciences* [online], 52 (2):160–163. Available from: https://doi.org/10.1016/j.jmir.2021.01.001 [Accessed 10 January 2023].

Bowleg, L., 2012. The problem with the phrase women and minorities: intersectionality – an important theoretical framework for public health. *American Journal of Public Health* [online], 102 (7):1267–1273. Available from: https://doi.org/10.2105%2FA JPH.2012.300750 [Accessed 10 January 2023].

Butani, L., Sweeney, C., and Plant, J., 2020. Effect of patient-led educational session on pre-clerkship students' learning of professional values and on their professional development. *Medical Education Online* [online], 25 (1):1801174. Available from: https://doi.org/10.1080/10872981.2020.1801174 [Accessed 14 February 2023].

Canadian Association of Medical Radiation Technologists (CAMRT)., 2023. Identity Matters: LGBTQ2S+ Education for MRTs. Available from: www.camrt.ca/ident ity-matters-lbgtq2s-education-for-mrts [Accessed 10 April 2023].

Clements, K.C., 2018. What Does It Mean to Misgender Someone? Available form: www.healthline.com/health/transgender/misgendering [Accessed 13 February 2023].

Coulter, A., and Oldham, J., 2016. Person-centred care: what is it and how do we get there? *Future Hospital Journal* [online], 3 (2):114–116. Available from: https://doi. org/10.7861%2Ffuturehosp.3-2-114 [Accessed 23 January 2023].

Davison, K., Queen. R., Lau. F., and Antonio, M., 2021. Culturally competent gender, sex, and sexual orientation information practices and electronic health records: rapid review. *JMIR Medical Informatics* [online], 9 (2):e25467. Available from: https://doi.org/10.2196%2F25467 [Accessed 30 January 2023].

Dean, M.A., Victor, E., and Grimes, L.G., 2016. Inhospitable healthcare spaces: why diversity training on LGBTQIA issues is not enough. *Bioethical Inquiry* [online], 13 (4):557–570. Available from: https://doi.org/10.1007/s11673-016-9738-9 [Accessed 12 January 2023].

Fridberg, H., Wallin, L., and Tistad, M., 2021. The innovation characteristics of person-centred care as perceived by healthcare professionals: an interview study employing a deductive-inductive content analysis guided by the consolidated framework for implementation research. *BMC Health Services Research* [online], 21(904). Available from: https://doi.org/10.1186/s12913-021-06942-y [Accessed 25 January 2023].

Gordon, C., and Mitchell, V., 2019. Risks and rewards in sexual and gender minority teaching and learning in a South African health sciences medical curriculum. *Education as Change* [online], 23 (1):3757. Available from: http://dx.doi.org/ 10.25159/1947-9417/3757 [Accessed 10 January 2023].

Health & Care Professions Council., 2007. Standards of proficiency: radiographers. Available from: www.hcpc-uk.org/standards/standards-of-proficiency/radiograph ers/ [Accessed 9 February 2023].

Health Professions Council of South Africa., 2016. General ethical guidelines for the health care professions – booklet 1. Available from: www.hpcsa.co.za/Uploads/ Professional_Practice/Ethics_Booklet.pdf [Accessed 9 February 2023].

Igual, R., 2023. "God is against you guys": LGBTQ+ stigma in Gauteng clinics. Available from: www.mambaonline.com/2023/01/24/god-is-against-you-guys-lgbtq-stigma-in-gauteng-clinics/ [Accessed 25 January 2023].

International Society of Radiographers and Radiological Technologists., 2021. ISRRT position statement: the radiographer/radiological technologist's role in patient care and patient safety. Available from: www.isrrt.org/patient-care-and-safety [Accessed 10 January 2023].

Kano, M., Jaffe, S.A., Rieder, S., Kosich, M., Guest, D.D., Burgess, E., Hurwitz, A., Pankratz, V.S., Rutledge, T.L., Dayao, Z., and Myaskovsky, L., 2022. Improving sexual and gender minority cancer care: patient and caregiver perspectives from a multi-methods pilot study. *Frontiers in Oncology* [online], 12:833195. Available from: https://doi.org/10.3389/fonc.2022.833195 [Accessed 10 February 2023].

Kruse, M.I., Bigham, B.L., Voloshin, D., Wan, M., Clarizio, A., and Upadhye, S., 2022. Care of sexual and gender minorities in the emergency department: a scoping review. *Annals of Emergency Medicine* [online], 79 (2):196–212. Available from: https://doi.org/10.1016/j.annemergmed.2021.09.422 [Accessed 30 January 2023].

LGBT Foundation. NHS Rainbow Badge. Available from: https://lgbt.foundation/howwecanhelp/nhs-rainbow-badge [Accessed 14 February 2023].

Lown, B.A., Sasson, J.P., and Hinrichs, P., 2008. Patients as partners in radiology education: an innovative approach to teaching and assessing patient-centered communication. *Academic Radiology* [online], 15 (4):425–432. Available from: https://doi.org/10.1016/j.acra.2007.12.001 [Accessed 14 February 2023].

Mayer, K.H., Bradford, J.B., Makadon, H.J., Stall, R., Goldhammer, H., and Landers, S., 2008. Sexual and gender minority health: what we know and what needs to be done. *American Journal of Public Health* [online], 98 (6):989–995. Available from: https://doi.org/10.2105%2FAJPH.2007.127811 [Accessed 12 January 2023].

Müller, A., 2016. Health for All? Sexual Orientation, Gender Identity, and the Implementation of the Right to Access to Health Care in South Africa. *Health and Human Rights Journal* [online], 18 (2):195–208. Available from: www.ncbi.nlm.nih.gov/pmc/articles/PMC5395001/ [Accessed 10 January 2023].

National Department of Health., 2007. Patients' Rights Charter. Available from: www.justice.gov.za/vc/docs/policy/patient%20rights%20charter.pdf [Accessed 15 January 2023].

National LGBT Cancer Network., 2023. Barriers to health care. Available from: https://cancer-network.org/cancer-information/cancer-and-the-lgbt-community/barriers-to-health-care/ [Accessed 12 January 2023].

Naylor, S., Harcus, J., and Elkington, M., 2015. An exploration of service user involvement in the assessment of students. *Radiography* [online], 21 (3):269–272. Available from: https://doi.org/10.1016/j.radi.2015.01.004 [Accessed 12 April 2023].

Obedin-Maliver, J., Goldsmith, E.S., Stewart, L., White, W., Tran, E., Brenman, S., Wells, M., Fetterman, D.M., Garcia, G., and Lunn, M.R., 2011. Lesbian, gay, bisexual, and transgender-related content in undergraduate medical education. *Journal of the American Medical Association* [online], 306 (9):971–977. Available from: https://doi.org/10.1001/jama.2011.1255 [Accessed 25 January 2023].

Pedersen, S, and Sanders, V., 2018. A new and inclusive intake form for diagnostic Imaging departments. *Journal of Medical Imaging and Radiation Sciences* [online], 49 (4):371–375. Available from: https://doi.org/10.1016/j.jmir.2018.10.001 [Accessed 20 January 2023].

Pieri, M., and Brilhante, J., 2022. "The light at the end of the tunnel": experiences of LGBTQ+ adults in Portuguese healthcare. *Healthcare* [online], 10 (1):146. Available from: https://doi.org/10.3390/healthcare10010146 [Accessed 10 January 2023].

Pride in Health + Wellbeing., 2023. Infographic of LGBT barriers to care and health disparities. Available from: www.prideinhealth.com.au/blog/infographic-of-lgbt-barriers-to-care-and-health-disparities/ [Accessed 12 January 2023].

Quinn, G.P., Sanchez, J.A., Sutton, S.K., Vadaparampil, S.T., Nguyen, G.T., Green, B.L., Kanetsky, P.A., and Schabath, M.B., 2015. Cancer and Lesbian, Gay, Bisexual, Transgender/Transsexual, and Queer/Questioning Populations (LGBTQ). *CA: A Cancer Journal for Clinicians* [online], 65 (5):384–400. Available from: https://doi.org/10.3322/caac.21288 [Accessed 10 February 2023].

Ralph, S., 2021. Developing UK guidance on how long men should abstain from receiving anal sex before, during and after interventions for prostate cancer. *Clinical Oncology* [online], 33 (12):807–810. Available from: https://doi.org/10.1016/j.clon.2021.07.010 [Accessed 10 February 2023].

Rioux, C., Paré, A., London-Nadeau, K., Juster, R-P., Weedon, S., Levasseur-Puhach, S., Freeman, M., Roos, L.E., and Tomfohr-Madsen, L.M., 2022. Sex and gender terminology: a glossary for gender-inclusive terminology. *Journal of Epidemiology & Community Health* [online], 76:764–768. Available from: http://dx.doi.org/10.1136/jech-2022-219171 [Accessed 30 January 2023].

Rosati, F., Pistella, J., Nappa, M.R., and Baiocco, R., 2020. The coming-put process in family, social and religious contexts among young, middle, and older Italian LGBQ+ adults. *Frontiers in Psychology* [online], 11:617217. Available from: https://doi.org/10.3389/fpsyg.2020.617217 [Accessed 14 April 2023].

Santana, M.J., Manalili, K., Jolley, R.J., Zelinsky, S., Quan, H., and Lu, M., 2018. How to practice person-centred care: a conceptual framework. *Health Expectations* [online], 21 (2):429–440. Available from: https://doi.org/10.1111/hex.12640 [Accessed 23 January 2023].

Silva, J.F., and Costa, G.M.C., 2020. Health care of sexual and gender minorities: an integrative literature review. *Revista Brasileira de Enfermagem* [online], 73(Suppl 6):e20190192. Available from: https://doi.org/10.1590/0034-7167-2019-0192 [Accessed 12 January 2023].

Smith, L., 2018. Glossary of transgender terms. Available from: www.hopkinsmedicine.org/news/articles/glossary-of-terms-1 [Accessed 13 February 2023].

Society of Radiographers., 2013. Code of professional conduct. Available from: www.sor.org/learning-advice/professional-body-guidance-and-publications/documents-and-publications/policy-guidance-document-library/code-of-professional-conduct [Accessed 9 February 2023].

Society of Radiographers., 2021. Inclusive pregnancy status guidelines for ionising radiation: Diagnostic and therapeutic exposures. Available from: www.sor.org/learning-advice/professional-body-guidance-and-publications/documents-and-publications/policy-guidance-document-library/inclusive-pregnancy-status-guidelines-for-ionising [Accessed 20 January 2023].

Souliotis, K., 2016. Public and patient involvement in health policy: A continuously growing field. *Health Expectations* [online], 19 (6):1171–1172. Available from: https://doi.org/10.1111/hex.12523 [Accessed 14 February 2023].

Sowinski, J.S., and Gunderman, R.B., 2018. Transgender patients: what radiologists need to know. *American Journal of Roentgenology* [online], 210 (5):1106–1110. Available from: https://doi.org/10.2214/ajr.17.18904 [Accessed 30 January 2023].

Suikkala, A., Koskinen, S., and Leino-Kilpi, H., 2018. Patients' involvement in nursing students' clinical education: a scoping review. *International Journal of Nursing Studies* [online], 84:40–51. Available from: https://doi.org/10.1016/j.ijnurstu.2018.04.010 [Accessed 14 February 2023].

Unicef., 2017. Gender equality: glossary of terms and concepts. Available from: www.unicef.org/rosa/reports/gender-equality [Accessed 30 January 2023].

Universal Declaration of Human Rights.,1948. Available from: www.un.org/en/about-us/universal-declaration-of-human-rights [Accessed 27 January 2023].

Utamsingh, P.D., Richman, L.S., Martin, J.L., Lattaner, M.R., Chaikind, J.R., 2016. Heteronormativity and practitioner–patient interaction. *Health Communication* [online], 31 (5):566–574. Available from: https://doi.org/10.1080/10410 236.2014.979975 [Accessed 12 January 2023].

van de Venter, R., 2022. Fostering inclusive and affirming care practices for sexual and gender minority patients among radiography students: transformational leadership in the classroom. *Journal of Medical Imaging and Radiation Sciences* [online], 53 (4):S53–S56. Available from: https://doi.org/10.1016/j.jmir.2022.09.013 [Accessed 10 January 2023].

van de Venter, R., and Hodgson, H., 2020. Strategies for inclusive medical imaging environments for sexual and gender minority patients and radiographers: an integrative literature review. *Journal of Medical Imaging and Radiation Sciences* [online], 51 (4):S99–S106. Available from: https://doi.org/10.1016/j.jmir.2020.05.009 [Accessed 10 January 2023].

Wood, C., Ringrose, K., Gutierrez, C., Stepanovich, A., and Colson, C., 2022. The role of data protection in safeguarding sexual orientation and gender identity information. Available from: https://fpf.org/wp-content/uploads/2022/06/FPF-SOGI-Report-R2-singles-1.pdf [Accessed 12 April 2023].

World Health Organization (WHO)., 1948. WHO Constitution. Available from: www. who.int/about/governance/constitution [Accessed 10 January 2023].

Zhu, X., Gray, D., and Brennan, C.S., 2022. Impact of implementing welcome spacing indicators on sexual and gender minority patients' health care experience. *AIDS Patient Care and STDs* [online], 36 (S2):S127–S142. Available from: https://doi.org/10.1089/apc.2022.0098 [Accessed 23 January 2023].

# 4 Achieving person-centred diagnostic imaging care in low- and middle-income countries

## Barriers and strategies

*Andrew Donkor, Alberta Naa Afia Adjei and Yaw Amo Wiafe*

## Introduction

The World Bank Group (2023) categorises low- and middle-income countries (LMICs) into: low-income countries (those with a gross national income (GNI) per capita of $1,085 or less); low-middle income countries (those with a GNI per capita between $1,086 and $4,255); and upper-middle-income countries (those with a GNI per capita between $4,256 and $13,205). A total of 136 countries are combined to represent LMICs, with: 28 low-income countries; 54 low-middle income countries; and 54 upper-middle-income countries (World Bank Group, 2023).

Person-centred diagnostic imaging care is seen as an opportunity to foster safety and quality of patient care in LMICs. However, the concepts in person-centred diagnostic imaging care are not well articulated in the literature from LMICs. Elements of person-centred diagnostic imaging care that are considered essential from the perspective of LMICs include: treating patients as individuals; recognising that all individuals accessing diagnostic imaging services have a unique clinical history, physical, social and economic needs; recognising that individuals accessing diagnostic imaging services need a comfortable environment; communicating effectively with each individual; and respecting the dignity, privacy and confidentiality of individuals accessing diagnostic imaging services.

This chapter discussed the various barriers and enablers to delivering person-centred diagnostic imaging care in LMICs (see Figure 4.1). Four important and useful lessons from LMICs which have helped to support delivery of person-centred diagnostic imaging care are outlined and discussed, namely:

1. Strengthening partnerships to increase access to diagnostic imaging services
2. Providing leadership to improve responsiveness

DOI: 10.1201/9781003310143-5

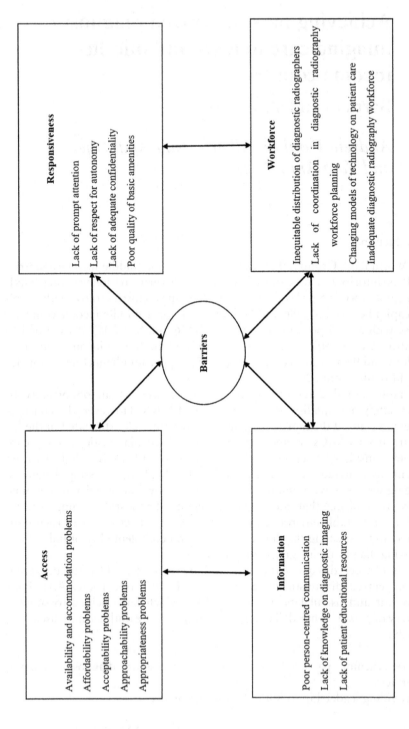

*Figure 4.1* A conceptual framework of barriers to person-centred diagnostic imaging care in LMI.

3. Promoting diagnostic imaging information and education through improvement in patients' health literacy
4. Developing and allocating diagnostic radiographers to promote quality and safety

### Strengthening partnerships to increase access to diagnostic imaging services

Access to diagnostic imaging services is essential for timely diagnosis and treatment. Despite the important role of diagnostic imaging services, gaining access is a major challenge in most LMICs. The following subsections explore the meaning of access, barriers to access and strategies to improve access to diagnostic imaging services in LMICs.

#### Meaning of access

Generally, access to healthcare service has been described as a complex concept (Gulliford et al., 2002; Hoe, 2007). Access to diagnostic imaging services is often considered the ability to access the service easily and affordably. Easy access to diagnostic imaging services contributes to the improvement of health. There are several dimensions of access, which include: availability and accommodation; affordability; acceptability; appropriateness; and approachability (Levesque, Harris, & Russell, 2013).

To summarise, access is the ability of the client to seek and obtain safe and high-quality diagnostic imaging services. Diagnostic imaging services should be organised and delivered according to the preferences, needs and values of the individual patient. However, most people in LMICs usually do not have access to diagnostic imaging services because access remains unaffordable. Patients in rural areas often encounter transport difficulties in their attempt to access diagnostic imaging services in urban areas. Barriers to accessing diagnostic imaging services in LMICs are discussed in the subsection below.

#### Access barriers to diagnostic imaging services

Studies have looked at a range of barriers to accessing diagnostic imaging services in LMICs (Gavahi et al., 2022; Piersson & Gorleku, 2017; Thambura & Swindon, 2019). Using the Levesque et al. (2013) patient-centred framework of access to health, the barriers to accessing diagnostic imaging services were categorised into: availability and accommodation problems; affordability problems; accessibility problems; acceptability problems; approachability problems; and appropriateness problems.

*Availability and accommodation problems*

Availability and accommodation problems are often rooted in poor diagnostic imaging infrastructure, human resources issues and the patients' ability to physically attend for imaging in a timely manner. Most LMICs are faced with insufficient diagnostic imaging infrastructure. Challenges related to diagnostic imaging infrastructure vary in nature and intensity across geographical regions even within the same LMIC. In many LMICs, there is unequal procurement and distribution of diagnostic imaging supplies and equipment. Diagnostic imaging supplies and equipment are often concentrated in the urban areas due to inadequate investment into health facilities and hard-to-reach communities. However, rural areas in LMICs often have a higher proportion of poor people and older persons who may have a greater need for diagnostic imaging services. People living in rural areas are forced to travel long distances to access diagnostic imaging services.

Some evidence has been generated about the unequal procurement and distribution of diagnostic imaging equipment such as magnetic resonance imaging (MRI) in Ghana (Gavahi et al., 2022; Piersson & Gorleku, 2017). For example, in Ghana out of the 12 MRI scanners in the country eight are found in the capital city. It means that around 13 million or 40% of the Ghanaian population, mainly in rural and remote areas, have no or limited access to MRI services because there are fewer providers (Gavahi et al., 2022). A Zimbabwean-based study also identified shortfall and inequitable distribution of existing diagnostic imaging equipment. It found that 57% (215/ 380 units) of diagnostic imaging equipment were located in two major cities (Harare and Bulawayo) in Zimbabwe (Maboreke, Banhwa, & Pitcher, 2019). A similar study in Uganda also recently audited registered diagnostic imaging equipment and found an unequal distribution, with most equipment located in urban areas. The results suggest Uganda lags behind the World Health Organization recommended ratio of equipment versus the population, which is 20 equipment per million population (Kiguli-Malwadde et al., 2020).

The diagnostic imaging workforce in most LMICs faces the following challenges in delivering person-centred care: understaffing; undertraining; underfunding; and underappreciating. Chronic underfunding has left many diagnostic imaging facilities understaffed. Rural populations in LMICs suffer most directly from staff shortages as the diagnostic imaging personnel that are available tend to cluster in urban areas. Compounding existing shortages and inequitable distribution of diagnostic imaging workforce is the migration of radiologists and radiographers from LMICs to high-income countries. A study from Hungary identified the reasons behind the migration of radiographers to high-income countries (Sipos, Csaba, & Petone Csima, 2017). The authors established a significant correlation between migration and a desire for higher wages, as well as an opportunity for career advancement as a radiographer. Similar results from South Africa show that 25% of diagnostic radiographers who had migrated attributed their decision to migrate to high workload, poor remuneration and lack of professional development (Thambura & Swindon, 2019).

*Affordability problems*

There is limited evidence on the cost of diagnostic imaging services in LMICs. Affordability reflects the economic capacity for patients to spend financial resources and time to access appropriate diagnostic imaging services. Affordability problems emerge from direct costs of diagnostic imaging services, indirect expenses such as travel costs, and opportunity costs related to loss of income (Donkor et al., 2022).

The lack of health insurance for the poor in LMICs is generally high. Studies from Ghana have demonstrated that patients without health insurance face catastrophic levels of spending for diagnostic imaging services that are largely operated by private providers (Kyei et al., 2016; Piersson & Gorleku, 2017). For MRI services in Ghana, Piersson and Gorleku (2017) have found evidence indicating that patients accessing public diagnostic imaging services paid on average $156 out-of-pocket payments compared to those accessing private diagnostic imaging services, who spend $176 on out-of-pocket payments.

Patient's ability to pay for diagnostic imaging services remains a major policy concern in many LMICs. Paying out-of-pocket for diagnostic imaging services is a significant part of the total health expenditure. Many patients in LMICs are forced to utilise their income, savings, borrow money or sell their properties to make direct payments for diagnostic imaging services. The high out-of-pocket payments often push households into poverty or indebtedness, which restricts the capacity of patients to pay for the needed diagnostic imaging services. Research from Iran, for example, showed that the proportion of patients facing catastrophic out-of-pocket diagnostic imaging payment was 32% (Marzban et al., 2015). They found out-of-pocket payments to vary with the type of diagnostic imaging services, with computed tomography (CT) scan services accounting for 41% (Marzban et al., 2015). The authors concluded that the out-of-pocket payment burden was disproportionately distributed among various socio-economic and demographic groups. Male patients experienced nearly 79% out-of-pocket payment burden compared to a female patient, who experienced 21.2% out-of-pocket payment burden, which may be a consequence of using more healthcare services (Marzban et al., 2015).

*Acceptability problems*

Research suggests that one of the major access barriers to better person-centred care is the lack of acceptability of diagnostic imaging services in LMICs. The term acceptability is a multi-factorial concept that reflects the extent to which people delivering or receiving healthcare services consider it to be appropriate based on cultural, societal factors, emotional responses, experienced cognitive and emotional responses to the service (Levesque et al., 2013; Sekhon et al., 2017). Acceptability means the information and explanations provided should take into account patients' cultural and social value (Obrist et al., 2007). For example, individuals accessing diagnostic imaging services should feel welcome, cared for and have trust in the competence of their diagnostic radiographers. Measures such as patient experience and satisfaction are commonly used to assess acceptability of diagnostic imaging services in the literature from LMICs (Efeoghene et al., 2021; Gadeka & Esena, 2020; Wahed et al., 2017). Patient experience is a process measure, which reflects the interpersonal aspects of quality of diagnostic imaging care received. The literature from LMICs indicate that a positive patient experience is underpinned by concepts of effective communication, dignity and respect for patients (Efeoghene et al., 2021; Gadeka & Esena, 2020). However, patient satisfaction is an outcome measure of patient's experiences with diagnostic imaging care. It reflects whether or not the diagnostic imaging care provided has met the patient's needs and expectations (Gadeka & Esena, 2020).

Box 4.1 summarises common acceptability barriers to diagnostic imaging services in LMICs using the theoretical framework of acceptability (Sekhon et al., 2017). Individual patients' attitudes towards diagnostic imaging services have been the subject of discussions for decades. There is reasonable evidence on the negative attitudes and feeling towards diagnostic imaging services in Ethiopia, showing 23% of patients felt diagnostic imaging staff were unfriendly towards them and 20% indicated that their opinion were ignored (Mulisa et al., 2017) . A Nigerian study, for example, reported that patients attach considerable importance to diagnostic imaging providers attitude towards time management (Efeoghene et al., 2021). The authors indicated that 97.9% of patients attributed their dissatisfaction to the late arrival of diagnostic imaging staff. Clearly, dissatisfied patients are reluctant to refer people to the diagnostic imaging service (Gadeka & Esena, 2020).

The amount of effort that patients are required to make to access diagnostic imaging services is a key factor in guiding acceptability. Over 95% of patients in Nigeria reported that they had difficulty locating and obtaining the diagnostic imaging services that they needed to access (Efeoghene et al., 2021). Similar problems and complaints have been expressed by patients in Egypt in trying to obtain an appointment for diagnostic imaging services (Wahed et al., 2017).

Evidence on the extent to which diagnostic imaging services have a good fit with patients' value system show that this may be lacking in LMICs.

Intervention coherence refers to the extent to which patients understand diagnostic imaging services and how they work. For example, Kotian et al., (2013) explored South Indian patients' perceptions of radiation awareness based on qualitative data. Kotian and colleagues found barriers affecting intervention coherence to include: poor awareness about the benefits of medical radiation; lack of knowledge about the potential risks related to cumulative radiation exposure; and misconceptions about basic aspects of radiation protection.

---

**Box 4.1 Summary of common acceptability barriers to diagnostic imaging services in LMICs**

- Affective attitude: how an individual feels about the service.
  - Diagnostic imaging staff not listening/accepting opinions of patients
  - Unfriendly diagnostic imaging staff
  - Unsupportive diagnostic imaging staff during and after the imaging procedure
  - Late arrival of diagnostic imaging staff
- Burden: the perceived amount of effort that is required to participate in the service.
  - Difficulty locating and obtaining the needed diagnostic imaging service
  - Difficulty obtaining appointment date
- Intervention coherence: the extent to which the patient understands the service and how it works.
  - Poor awareness about the benefits of medical radiation
  - Lack of knowledge about the potential risks related to cumulative radiation exposure
  - Misconceptions about basic aspects of radiation protection
- Perceived effectiveness: the extent to which the intervention is perceived as likely to achieve its purpose.
  - Lack of information on the benefit of diagnostic imaging

---

*Approachability problems*

Approachability can be defined as the ability to identify the existence of services, reach them, and have an impact on the health of the user (Levesque et al., 2013). Diagnostic imaging facilities can brand themselves to make visible the available services and their benefits. However, there is limited evidence on approachability barriers to diagnostic imaging services in LMICs. The lack of open and transparent process for operating diagnostic imaging services decreases public trust, which contributes to making services less approachable. Poor access to information is a major impediment to the use of diagnostic imaging services. In Ghana and Nigeria, about 60% of patients who visit the diagnostic imaging centre are first timers in the department (Efeoghene et al., 2021; Kyei et al., 2016).

*Appropriateness problems*

Most LMICs have difficulty ensuring access to appropriate, safe diagnostic imaging services for a large part of their population. Poor maintenance culture and lack of quality assurance systems are major technical barriers that affect the appropriateness of diagnostic imaging equipment (Daki et al., 2020; Inkoom et al., 2011; Ngoye et al., 2015; Ofori et al., 2013). In Tanzania, for example, a survey of 84 diagnostic radiographers practising in 54 hospitals identified poor implementation of quality assurance strategies such as: tube output and timer (94%); densitometry and sensitometry (87.7%); and collimation test (53.5%) (Ngoye et al., 2015), coupled with observations in the United Kingdom (Hayre et al., 2019). The authors concluded that quality assurance programmes are useful, but the poor implementation is a major cause of poor image quality and/or higher dose to patients. A related recent cross-sectional study revealed that the most common causes of diagnostic imaging equipment breakdown in Northwest and Southwest Regions of Cameroon were: electrical faults (75%); lack of professionals skills (10%) and age of equipment (10%) (Daki et al., 2020). A further cause of inappropriate diagnostic imaging services can be linked to the shortage of diagnostic radiographers, which affects the provision of appropriate and timely diagnostic imaging care (Nwobi et al., 2014; Wuni et al., 2020).

*Strategies*

When governments play an active role through partnership with private and charitable sectors, it often leads to improved access to diagnostic imaging services in LMICs. In practice, to develop a strong partnership involves creating a stable political environment, maintaining political will, developing the relevant legal framework, and providing equipment resources to deliver diagnostic imaging services. Many LMICs have realised that the public sector cannot shoulder the burden of directly providing the needed diagnostic imaging services alone. Studies have shown that contributions by the private health sector have improved access to diagnostic imaging services. For example, in Ghana, many patients choose private over public diagnostic imaging providers because they are more satisfied with the services provided (Kyei et al., 2016). The private diagnostic imaging providers consistently care for people from a wide range of incomes, including rural and poor populations. In Nigeria, 93% of people who seek diagnostic imaging services do so from self-financing and private providers (Ochonma et al., 2015).

Relatively little is known about the details of the ways that government, private and non-charitable sectors can work together more effectively to deliver affordable, appropriate and acceptable diagnostic imaging services. A recent systematic review outlined key success factors for organisational partnership, which included: motivation and purpose; resources and capabilities; governance and leadership; relationships and cultures; and external factors (Alderwick, Hutchings, Briggs, & Mays, 2021).

The partnership should focus on person-centred diagnostic imaging services, equity, affordable services and population-based national plans. The first strategic step is for LMICs to gather information for advocacy and investment to facilitate access to appropriate diagnostic imaging services. All persons should receive diagnostic imaging services they require, which should be of adequate quality and without experiencing financial hardship. Innovative financing, financial risk protection and improved efficiency strategies should be implemented to ensure access to diagnostic imaging services in LMICs.

## Providing leadership to improve responsiveness

Responsiveness is a key outcome on which health systems, including diagnostic imaging services, are judged. The following subsections explore the meaning of responsiveness, barriers to responsiveness and strategies to improve diagnostic imaging services' responsiveness.

### Meaning of responsiveness

Responsiveness involves the experience of patients' interactions with their health systems, which confirms or disconfirms their initial expectations (Mirzoev & Kane, 2017). Patient satisfaction is fundamental to diagnostic imaging services' responsiveness because it gives information on the provider's success at meeting the patient's values and expectations. A responsive diagnostic imaging service contributes to health enhancement by being more conducive for individuals to seek timely care and to better assimilate health information. Multiple domains characterise responsiveness, namely: prompt attention; respect for autonomy; respect for confidentiality; choice of care provider; and quality of basic amenities (Mirzoev & Kane, 2017).

### Barriers to diagnostic imaging services responsiveness

Four main barriers to diagnostic imaging services responsiveness in LMICs were identified in the literature, which were grouped into: lack of prompt attention; lack of respect for autonomy; lack of adequate confidentiality; and poor quality of basic amenities.

### Lack of prompt attention

Diagnostic imaging services are perhaps the most challenging components of the health system with respect to patient delay. Patients can arrive at the diagnostic imaging department from multiple different referral pathways, including ambulance and walk-in. Long patient waiting times adversely affect the willingness of the patient to return to the diagnostic imaging facility. Several studies have explored the multitude of factors responsible for long waiting times at the diagnostic imaging facility (Hamed & Salem, 2014; Omar et al., 2022; Wahed et al., 2017). These factors include: the long distance between given points within the healthcare facility, particularly at the payment point; insufficient diagnostic radiographers; problems accessing radiologists to timely read and interpret images; and inadequate diagnostic imaging equipment capacity (Efeoghene et al., 2021; Gadeka & Esena, 2020; Hamed & Salem, 2014; Nwobi et al., 2014; Omar et al., 2022; Ugwu et al., 2009; Wahed et al., 2017). In Saudi Arabia, the large number of non-emergency patients and high cost of private diagnostic imaging services are key contributing factors to long patient waiting times in the public diagnostic imaging centres, with 47% of patients waiting for 1–3 hours (Omar et al., 2022). Overcrowding at diagnostic imaging centres has also been linked to poorly functioning referral mechanism and the lack of appointment system to rationalise patient flows (Nwobi et al., 2014; Onwuzu et al., 2014; Wahed et al., 2017). Lack of prompt attention can also be linked to poor management supervision that ensures that diagnostic imaging staff are punctual and work starts on time to reduce long waiting times (Efeoghene et al., 2021).

*Lack of respect for autonomy*

Respecting patients' autonomy means acknowledging that patients who have decision-making capacity have the right to make decisions regarding their care, even when their decisions contradict the recommendation from the diagnostic imaging professional (Sedig, 2016).

The radiology department always seems busy with crowded patients and diagnostic radiographers trying to attend to patients before their shifts end. The main contributing factors that affect respect for autonomy are the lack of patient participation in the diagnostic imaging examination process and patient preferences in the decision-making process (Hamed & Salem, 2014; Mulisa et al., 2017; Ochonma et al., 2015). For example, a study conducted in Nigeria indicated that half of the patients that visited the diagnostic imaging department were not engaged in the planning of their imaging procedure (Ochonma et al., 2015). For instance, patients could have been engaged in a pre-visit planning such as: a meaningful conversation about their allergies to contrast for CT examination with contrast; information on the importance of creatinine blood test for patients undergoing either a CT or MRI examination that will require intravascular contrast; and education on the use of powder, deodorant or any skin products on the chest or breast on the day of the appointment for patients scheduled for mammogram.

*Lack of adequate confidentiality*

Confidentiality is respecting the privacy of patients (Byock & Palac, 2011). Major barriers to ensuring adequate patient confidentiality in LMICs include lack of: privacy in changing and examination rooms; capacity to secure medical records and patient information; and privileged communications (Chand et al., 2012; Chiegwu et al., 2017; Efeoghene et al., 2021; Gadeka & Esena, 2020; Mulisa et al., 2017; Nwobi et al., 2014; Wahed et al., 2017). A survey conducted by Wahed et al. (2017) in Egypt revealed that 65% of the patients who accessed diagnostic imaging services were dissatisfied with privacy. Mulisa et al. (2017) also observed that 99.70% of patients in Hawassa University Teaching and Referral Hospital (South Ethiopia) were dissatisfied with their privacy during their examination. A prospective cross-sectional study concluded that an insufficient number of reception staff affected the timely retrieval of radiographs and the efficiency of patient registration (Chiegwu et al., 2017; Efeoghene et al., 2021; Nwobi et al., 2014).

*Poor quality of basic amenities*

Basic amenities such as toilets, water and furniture are necessities in a hospital for patient comfort especially in the diagnostic imaging department. Available studies from LMICs have revealed a lack of basic amenities such as: inadequate furniture in the waiting area; poor ventilation; limited space in the waiting room; and unclean toilet rooms, examination rooms and diagnostic imaging equipment (Chand et al., 2012; Efeoghene et al., 2021; Hamed & Salem, 2014). Hamed and Salem (2014) of Egypt revealed in their study that 17.6% of patients were uncomfortable with the seats in the diagnostic imaging waiting area. Concerns about the cleanliness of the department and restrooms were expressed by some patients. Patients also voiced concerns about the overcrowded spaces and inadequate ventilation (Chand et al., 2012; Efeoghene et al., 2021; Hamed & Salem, 2014).

**Strategies**

The quality of diagnostic imaging leadership directly and indirectly affects the responsiveness and quality of person-centred care. Leadership plays a critical role in mobilising people towards a common vision. Research from LMICs have not focused on interventions to strengthen leadership skills for diagnostic imaging professionals. The design and implementation of a strategy to provide safe, effective and responsive diagnostic imaging services will need shared and participatory leadership of diverse groups, including radiologists, radiographers and non-clinical staff. The South Western Sydney Local Health District (SWSLHD) has developed a six-point leadership framework to positively transform how patients, family members and communities experience health services, which provide reasonable strategies

(SWSLHD, 2017). The six-point leadership framework consists of: setting direction; developing self; developing and enabling others; fostering and building relationships; communicating with influence; and innovating and leading change (SWSLHD, 2017).

Interventions that take a leadership approach should be designed to improve the personal qualities of diagnostic imaging staff, eliminate diagnostic imaging process wastes and standardise imaging pathways. Such interventions may help efficiently manage and improve diagnostic imaging services in LMICs.

## Promoting diagnostic imaging information through improvement in patients' health literacy

Many patients in LMICs report difficulties in obtaining relevant information about their diagnostic imaging procedures. The following subsections explore the meaning of health literacy, barriers and strategies to improve patient information and education on diagnostic imaging.

### Meaning of health literacy

Health literacy has been defined as the capacity of an individual to access, process and comprehend health information and services required to make effective health decisions (Sørensen et al., 2012). Globally, the interest in patients' health literacy has strongly increased in recent years. Diagnostic imaging techniques are frequently complex and without the proper knowledge, patients may find it challenging to manage the services. Patients' health-seeking attitudes, behaviours and beliefs can be supported by meeting their information needs and fostering their active involvement in their diagnostic imaging procedure.

### Barriers to patient information and education

The barriers to patient information and education were categorised into poor patient-centred communication; lack of patient educational resources; and lack of knowledge on diagnostic imaging and its related benefits.

### Poor person-centred communication

Benefits of effective patient-centred communication include: increased patient satisfaction; enhanced compliance; provision of emotional support; support for informed decision-making; and reduced anxiety (Mcintosh, 2022). However, several studies in LMICs have highlighted communication issues related to patient care in diagnostic imaging, including: language barrier; poor attitude of staff; limited time; excessive workload; and cultural and ethnicity issues (Antwi et al., 2014; Chingarande et al., 2013; Hamed & Salem,

2014; Kyei et al., 2016; Mokavelaga & Pape, 2021; Nghipukuula et al., 2021; van Vuuren et al., 2021; Wahed et al., 2017).

In Namibia, 18% of patients with limited or no English language proficiency that accessed diagnostic imaging services were dissatisfied with the technical language the radiographers used despite the availability of an interpreter (Nghipukuula et al., 2021). A study from Egypt observed that 70.7% of the patients did not receive accurate communication to enable them to prepare before their diagnostic imaging examination, which often result in patients facing long waiting times (Wahed et al., 2017). Sometime poor communication may result in cancellation or postponement of appointment because the patient may not be well prepared for the diagnostic imaging procedure as in the case of a diabetic patient taking medication that contains Metformin and scheduled for a CT examination with intravascular contrast.

## Lack of patient educational resources

Patient education is an organised, structured process or programme with the goal of imparting information to facilitate learning (Falvo, 2010). Education in any setting is complex and it consists of organised resources to enhance patient teaching. The lack of availability of patient educational resources, such as videos, posters and brochures about diagnostic imaging procedures, serves to act as a barrier to patient teaching (Wahed et al., 2017). Patient educational resources are often overlooked by diagnostic imaging policymakers. Lack of funding and limited research are major contributory factors to the lack of educational resources that address the diagnostic imaging information needs of patients (Chingarande et al., 2013; van Vuuren et al., 2021). Patient educational resources enhance patients' understanding and increase cooperation with the diagnostic imaging staff. For example, it was observed that patients in Nigeria and Ethiopia were afraid of x-ray exposures because of lack of access to diagnostic imaging information in (Efeoghene et al., 2021; Mulisa et al., 2017).

## Lack of public knowledge about relevance of diagnostic imaging

Diagnostic imaging misinformation in the media, lack of awareness on diagnostic imaging benefits and lack of understanding on radiation protection measures pose major knowledge barriers to the use of diagnostic imaging services in most LMICs (Anim-Sampong, Opoku, Addo, & Botwe et al., 2015; Maharjan, Parajuli, Sah, & Poudel et al., 2020; Tanha et al., 2019). In an online cross-sectional survey conducted by Tanha et al. (2019), the data from 1,200 Afghans indicated that more than 70% of the public have inadequate knowledge on radiation safety and had just heard about radiation from one source. Diagnostic imaging staff must be prepared to explain to the public in general and patients why the diagnostic imaging procedures are important, what radiation dose they will receive, and how the diagnostic images are used to benefit patient diagnosis, treatment and follow-up care.

### Strategies

The diagnostic imaging staff and patients that visit diagnostic imaging departments in LMICs have busy schedules, which limits their opportunities for health teaching. Critical opportunities for patient information and education include: during the explanation of the diagnostic imaging procedures; when responding to the patient concerns; as part of the instructions needed to prepare for the diagnostic imaging procedure; and during instruction for follow-up care (Ehrlich & Coakes, 2020). Training interventions should be designed to improve diagnostic imaging professionals' communication skills to help involve patients in their diagnostic imaging care.

There are several communication models that diagnostic imaging professionals can use to help structure conversations with patients. Examples of these communication models include AIDET (Acknowledge, Introduce, Duration, Explanation and Thank you); RESPECT (Rapport, Empathy, Support, Partnership, Explanations, Cultural competence and Trust); PREPARED (Prepare for the discussion, Relate to the person, Elicit patient and caregiver preferences, Provide information, Acknowledge emotions and concerns, Realistic hope, Encourage questions and Document) (Clayton et al., 2007; Mutha et al., 2002; Studer Group, 2010). All these communication models are of value, although they were not designed specifically for diagnostic imaging professionals.

Explanation of the diagnostic imaging procedure is critical for all patients. Patients cooperate with the radiographer when they know what to expect during the diagnostic imaging procedure. Diagnostic radiographers should be trained to keep explanations simple, with good listening skills to respond to patient concerns. Diagnostic radiographers should reassure patients that careful consideration has been given to the benefit of the procedure, and that this outweighs potential risks. A hurried appearance by diagnostic radiographers to complete a procedure can discourage patients from asking questions.

To engage more patients in the educational activities and appeal to visual learners, custom-designed posters with images should be displayed in the waiting area. Video and printed information about diagnostic imaging services is useful to educate patients and their family members. Video messages can be played in the patient reception area. Diagnostic imaging brochures can be placed in the reception area or given to patients before the diagnostic imaging procedure. The video and brochures should explain the value of diagnostic imaging as a diagnostic tool as well as highlighting the benefits of diagnostic imaging services. The combination of traditional and social media such as YouTube and mobile applications can be an effective method of educating the public and patients about diagnostic imaging. Frequently asked questions about the need for and safety of diagnostic imaging services, x-ray exposure and other concerns can be addressed by diagnostic radiographers.

## Developing and allocating diagnostic radiographers to promote quality and safety

The diagnostic imaging workforce should be planned, funded, educated, deployed and regulated in ensuring appropriate clinical standards, quality and safety objectives. The diagnostic imaging workforce challenges in LMICs are complex and multifaceted. The subsections explore the various barriers and strategies for strengthening diagnostic imaging workforce in LMICs.

### Barriers to diagnostic radiography workforce

Demand for diagnostic imaging services is evolving rapidly and continues to challenge the diagnostic radiographers' ability to deliver safe and high-quality services in LMICs. Insufficient diagnostic radiographers, undertraining, underfunding, underappreciating and high migration of diagnostic radiographers have been discussed earlier. Three additional barriers are discussed in the following sections: inequitable distribution of diagnostic radiographers; lack of coordination in diagnostic radiography workforce planning; and changing models of technology on patient care.

### Inequitable distribution of diagnostic radiographers

No person should be disadvantaged when accessing the services of diagnostic radiographers and radiologists. There is a continued shortage of diagnostic radiographers and radiologists in rural areas of LMICs. There is lack of financial or non-financial incentives to encourage diagnostic radiographers to locate and practice in rural areas (Ashong, 2021; Mark et al., 2014). Lack of clinical support, lower economic conditions and standards of living prevailing in rural areas are critical barriers to attracting and retaining diagnostic radiographers in rural areas. Other contributing factors to the inequitable rural–urban distribution are the lack of training positions and professional development in rural areas; exposure to rural diagnostic imaging practice; and targeted recruitment policy of diagnostic radiographers who grew up in rural areas. For example, in 2018, less than 20% of the 300 registered diagnostic radiographers in Ghana were working in rural areas (Ashong, 2021).

### Lack of coordination in diagnostic radiography workforce planning

Diagnostic imaging stakeholders – governments, academic institutions, regulators and service providers – have not aligned their workforce development and allocation objectives to meet the demand for diagnostic radiographers and radiologists in most LMICs. The major contributing factor to the poor planning of diagnostic radiography workforce is the lack of accurate and reliable data. Governments in most LMICs do not have reliable data to conduct supply-and-demand forecasting for diagnostic radiography workforce financial clearance. Industrial action by healthcare workers demanding posting and better conditions of services is problematic, given its consequences on the healthcare system. For example, lack of leadership and management (92%), demand for higher salaries (82%) and infrastructural issues (63.3%) are seen as important underlying causes of strikes among Nigerian healthcare professionals, including diagnostic imaging staff (Oleribe et al., 2016).

### Changing models of technology on patient care

Changes in technology and tasks of diagnostic imaging professionals, particularly radiographers, are resulting in paradigm shifts on the purpose of person-centred diagnostic imaging care. The introduction of digital Picture Archive Communication System (PACS) has led to greater demands on diagnostic radiographers during the examination but limited opportunities to acquire high technological competence. Artificial intelligence (AI) is an inevitable part of our current and future diagnostic imaging system, with great impact on the diagnostic imaging workforce dynamics and readiness. AI offers analytical, connectivity and automation of some aspects of diagnostic imaging care to help improve efficiency. The pivotal role of AI in diagnostic imaging has resulted in an increased need for trained diagnostic radiographers and radiologists to deliver new levels of care coordination and management that are essential in person-centred care. Negative attitude towards AI has been highlighted as a barrier for AI acceptance in diagnostic imaging. A cross-sectional study involving 1,020 diagnostic radiographers across 28 African countries found 61.3% of the respondents indicating that AI tools could replace diagnostic imaging professionals and negatively affect the profession in Africa (Botwe et al., 2021).

Task shifting for diagnostic radiographers helps maximise clinical efficiency and job satisfaction (Riaan van de Venter & Friedrich-Nel, 2021). Having diagnostic radiographers undertake advanced practices, specialist and consultant-level roles presents many benefits to patients, such as reducing patient waiting times and higher accuracy of reporting. However, there are several barriers to task shifting, including: lack of clarity regarding the nature of extended role within the diagnostic imaging profession; lack of clear leadership; inadequate education and training to support role development and extension; lack of regulatory and legal framework; lack of appropriate renumeration; and clinical restrictions from radiologists (Bwanga, 2020; Bwanga et al., 2019; Bwanga et al., 2020; R. van de Venter & ten Ham-Baloyi, 2019; Wuni et al., 2020). Evidence from Zambia suggests that there is resistance from radiologists for fear of losing part of their role of reporting to diagnostic radiographers (Bwanga et al., 2019).

### Strategies

Appropriate management of developing and allocating diagnostic radiographers should be a major priority in LMICs. Training and education strategies should be well-coordinated and focused on career progression, particularly leading to advanced and specialist practitioners. The following strategies should be considered:

- Pre-diagnostic imaging school initiatives should be designed and implemented to select students who are more likely to practice and advance their career in diagnostic radiography, particularly in rural areas.
- Undergraduate and postgraduate education and training in diagnostic imaging should be designed and implemented. Formal relationships should be built between urban and rural diagnostic imaging facilities to rotate training students and build local capacity. Diagnostic radiography workforce planners must align their objectives to prevent fragmentation.
- Incentive and non-incentive strategies should be designed to encourage diagnostic imaging professionals to locate and remain in rural areas. Designing and implementing incentive and non-incentive strategies will require an understanding of the various factors to influence motivation and retention of diagnostic radiography workforce in LMICs.
- Diagnostic radiography students should be exposed to rural diagnostic imaging practice during their training. Students should be mentored by successful diagnostic radiographers and encouraged to pursue practice in rural diagnostic imaging.
- Fellowship training in diagnostic radiography can improve the confidence of diagnostic radiographers and their preparedness to assume leadership level of responsibility as specialists and consultants.

### Conclusion

This chapter has discussed the barriers to delivering person-centred diagnostic imaging care in LMICs, and offered practical strategies to try to address this. Fundamental barriers to delivering person-centred diagnostic imaging care are related to access, workforce, responsiveness and information. It can be seen that the barriers are complex. However, by strengthening partnerships, providing leadership, promoting health literacy and investing in the training of diagnostic radiographers, person-centred diagnostic imaging care can be achieved in LMICs.

Further research on person-centred diagnostic imaging care employing more rigorous study design is needed to co-design, implement and evaluate person-centred interventions in LMICs. There is also the need to understand how well integrated person-centred care has become in the education, training and core competences of diagnostic imaging professionals.

## Acknowledgement

This chapter would not be complete without acknowledging the significant proofreading contribution of Esther Oparebea Ofori (BA Economics)

## References

Alderwick, H., Hutchings, A., Briggs, A., & Mays, N. (2021). The impacts of collaboration between local health care and non-health care organizations and factors shaping how they work: a systematic review of reviews. *BMC Public Health, 21*(1), 753–768. doi:10.1186/s12889-021-10630-1

Anim-Sampong, S., Opoku, S. Y., Addo, P., & Botwe, B. O. (2015). Nurses knowledge of ionizing radiation and radiation protection during mobile radiodiagnostic examinations. *Educational Research, 6,* 39–49.

Antwi, W. K., Kyei, K. A., & Quarcoopome, L. N. (2014). Effectiveness of multicultural communication between radiographers and patients and its impact on outcome of examinations. *World Journal of Medical Research, 3*(4), 1–2.

Ashong, G. (2021). *Rural radiography practice: exploration of experiences of radiographers in Ghana.* Cardiff University.

Botwe, B. O., Akudjedu, T. N., Antwi, W. K., Rockson, P., Mkoloma, S. S., Balogun, E. O., ...Arkoh, S. (2021). The integration of artificial intelligence in medical imaging practice: Perspectives of African radiographers. *Radiography, 27*(3), 861–866. doi:https://doi.org/10.1016/j.radi.2021.01.008

Bwanga, O. (2020). Barriers to Continuing Professional Development (CPD) in radiography: A review of literature from Africa. *Health Professions Education, 6*(4), 472–480. doi:https://doi.org/10.1016/j.hpe.2020.09.002

Bwanga, O., Mulenga, J., & Chanda, E. (2019). Need for image reporting by radiographers in Zambia. *Medical Journal of Zambia, 46*(3), 215–220.

Bwanga, O., Mwansa, E., Sichone, J., & Kafwimbi, S. (2020). Establishment of postgraduate education and training in the specialised areas of diagnostic imaging in Zambia. *African Journal of Health, Nursing and Midwifery, 3*(4), 55–64.

Byock, I. R., & Palac, D. (2011). Confidentiality. In G. Hanks, N. I. Cherny, N. A. Christakis, & S. Kaasa (Eds.), *Oxford textbook of palliative medicine* (pp. 281–289): Oxford University Press.

Chand, R., Pant, D., & Joshi, D. (2012). An evaluation of patients care in radio diagnosis Department of Tribhuvan University Teaching Hospital, Kathmandu, Nepal. *Nepal Medical College Journal, 14*(2), 133–135.

Chiegwu, H., Eze, J., & Okoli, M. (2017). Assessment of patients' waiting time in the radiology department in Owerri,Imo state, Nigeria. *European Journal of Biomedical, 4*(5), 597–602.

Chingarande, G. R., Estina, M., Mukwasi, C., Majonga, E., & Karare, A. (2013). A comparative analysis of the effectiveness of communication between radiographers and patients at two hospitals. *International Journal of Advanced Research in Management and Social Science, 2*(6), 21–29.

Clayton, J. M., Hancock, K. M., Butow, P. N., Tattersall, M. H., & Currow, D. C. (2007). Clinical practice guidelines for communicating prognosis and end-of-life issues with adults in the advanced stages of a life-limiting illness, and their caregivers. *Medical Journal of Australia, 186*(12), S77–S108.

Daki, A. M., Mugop, N. S., Berinyuy, F. C., & Mbinkong, S. R. (2020). An assessment of measures medical imaging personnel use to avoid the breakdown of diagnostic radiologic equipment in some health facilities in Northwest and Southwest regions of Cameroon. *International Journal of Innovation, Science, Research and Technology, 5*(8), 36–42.

Donkor, A., Atuwo-Ampoh, V. D., Yakanu, F., Torgbenu, E., Ameyaw, E. K., Kitson-Mills, D., ...Khader, O. (2022). Financial toxicity of cancer care in low-and middle-income countries: A systematic review and meta-analysis. *Supportive Care in Cancer*, 1–32.

Efeoghene, M., Eze, J., Idigo, F., Ohagwu, C., Okpalaeke, M., & Ezechukwu, U. (2021). Assessment of patients' satisfaction level with radio-diagnostic services in a Nigerian tertiary hospital. *Assessment, 40*(25), 8–15.

Ehrlich, R., & Coakes, D. (2020). *Patient care in radiography: with an introduction to medical imaging* (10 ed.). Missouri: Elsevier.

Falvo, D. (2010). *Effective patient education: A guide to increased adherence.* Ontario: Jones & Bartlett Publishers.

Gadeka, D. D., & Esena, R. K. (2020). Quality of care of medical imaging services at a teaching hospital in Ghana: clients' perspective. *Journal of Medical Imaging and Radiation Sciences, 51*(1), 154–164. Retrieved from www.jmirs.org/article/S1939-8654(19)30763-5/fulltext

Gavahi, S. S., Hosseini, S. M. H., & Moheimani, A. (2022). An application of quality function deployment and SERVQUAL approaches to enhance the service quality in radiology centres. *Benchmarking: An International Journal, 30*(5), 1649–1671.

Gulliford, M., Figueroa-Munoz, J., Morgan, M., Hughes, D., Gibson, B., Beech, R., & Hudson, M. (2002). What does' access to health care'mean? *Journal of Health Services Research & Policy, 7*(3), 186–188.

Hamed, M. A. G., & Salem, G. M. (2014). Factors affecting patients' satisfaction in nuclear medicine department in Egypt. *Egyptian Journal of Radiology and Nuclear Medicine, 45*(1), 219–224.

Hayre, C.M. Blackman, S. Carlton, K. Eyden, A. (2019). The use of cropping and digital side markers (DSM) in digital radiography. *Journal of Medical Imaging and Radiation Sciences*, 50(2), 234–242.

Hoe, J. (2007). Quality service in radiology. *Biomedical Imaging and Intervention Journal, 3*(3), 1–6.

Inkoom, S., Schandorf, C., Emi-Reynolds, G., & Fletcher, J. J. (2011). Quality assurance and quality control of equipment in diagnostic radiology practice-the Ghanaian experience. *Wide Spectra of Quality Control*, 291–308.

Kiguli-Malwadde, E., Byanyima, R., Kawooya, M. G., Mubuuke, A. G., Basiimwa, R. C., & Pitcher, R. (2020). An audit of registered radiology equipment resources in Uganda. *Pan African Medical Journal, 37*(295), 1–12.

Kotian, R. P., Sukumar, S., Kotian, S. R., & David, L. R. (2013). Perception of radiation awareness among patients in South Indian population: a qualitative study. *Medical Science, 2*(11), 373–376.

Kyei, K., Antwi, W., & Brobbey, P. (2016). Patients' satisfaction with diagnostic radiology services in two major public and private hospitals in Ghana. *International Journal of Radiology and Radiation Therapy, 1*(1), 1–5.

Levesque, J.-F., Harris, M. F., & Russell, G. (2013). Patient-centred access to health care: conceptualising access at the interface of health systems and populations. *International Journal for Equity in Health, 12*(1), 18–26. doi:10.1186/1475-9276-12-18

Maboreke, T., Banhwa, J., & Pitcher, R. D. (2019). An audit of licensed Zimbabwean radiology equipment resources as a measure of healthcare access and equity. *Pan African Medical Journal, 34*, 60–69.

Maharjan, S., Parajuli, K., Sah, S., & Poudel, U. (2020). Knowledge of radiation protection among radiology professionals and students: A medical college-based study. *European Journal of Radiology Open, 7*, 100287–100291. doi:https://doi.org/10.1016/j.ejro.2020.100287

Mark, O. C., Ugwuanyi, C. D., & Thomas, A. (2014). Radiographers' willingness to work in rural and underserved areas in Nigeria: a survey of final year radiography students. *Journal of the Association of Radiographers of Nigeria, 28*(1), 6–10.

Marzban, S., Rajaee, R., Gholami, S., Keykale, M. S., & Najafi, M. (2015). Study of Out-of-Pocket Expenditures for Outpatient Imaging Services in Imam-Khomeini Hospital in 2014. *Electronic Physician, 7*(4), 1183–1189. Retrieved from www.ncbi.nlm.nih.gov/pmc/articles/PMC4578538/pdf/epj-07-1183.pdf

Mcintosh, J. (2022). Communication and patient care in rmadiography. *South African Radiographer, 60*(1), 25–31.

Mirzoev, T., & Kane, S. (2017). What is health systems responsiveness? Review of existing knowledge and proposed conceptual framework. *BMJ Global Health, 2*(4), e000486–e000496. doi:10.1136/bmjgh-2017-000486

Mokavelaga, A., & Pape, R. (2021). Assessment on the effectiveness of communication between radiographers and patients during general radiographic examinations at Port Moresby General Hospital, PNG. *Pacific Journal of Medical Sciences, 22*(1), 13–24.

Mulisa, T., Tessema, F., & Merga, H. (2017). Patients' satisfaction towards radiological service and associated factors in Hawassa University Teaching and referral hospital, Southern Ethiopia. *BMC Health Services Research, 17*(1), 1–11.

Mutha, S., Allen, C., & Welch, M. (2002). *Toward culturally competent care: a toolbox for teaching communication strategies.* San Francisco, CA: Center for the Health Professions, University of California.

Nghipukuula, J. S., Daniels, E. R., & Karera, A. (2021). Effectiveness of communication between student radiographers and patients before, during and after radiographic procedures. *South African Radiographer, 59*(2), 7–14.

Ngoye, W. M., Motto, J. A., & Muhogora, W. E. (2015). Quality control measures in Tanzania: Is it done? *Journal of Medical Imaging and Radiation Sciences, 46*(3, Supplement), S23–S30. doi:https://doi.org/10.1016/j.jmir.2015.06.004

Nwobi, I., Luntsi, G., Ahmadu, M., Nkubli, F., Kawu, H., Dauda, F., ...Tahir, M. (2014). The Assessment of patients'' perception and satisfaction of radiology waiting time in university of Maiduguri teaching hospital. *Kanem Journal of Medical Sciences, 8*(1), 19–26.

Obrist, B., Iteba, N., Lengeler, C., Makemba, A., Mshana, C., Nathan, R., ...Mayumana, I. (2007). Access to health care in contexts of livelihood insecurity: a framework for analysis and action. *PLoS medicine, 4*(10), e308–e312.

Ochonma, O. G., Eze, C. U., Eze, S. B., & Okaro, A. O. (2015). Patients' reaction to the ethical conduct of radiographers and staff services as predictors of radiological experience satisfaction: a cross-sectional study. *BMC Medical Ethics, 16*(1), 1–9.

Ofori, E. K., Antwi, W. K., & Scutt, D. (2013). Current status of quality assurance in diagnostic imaging departments in Ghana: peer reviewed original article. *South African Radiographer, 51*(2), 19–25.

Oleribe, O. O., Ezieme, I. P., Oladipo, O., Akinola, E. P., Udofia, D., & Taylor-Robinson, S. D. (2016). Industrial action by healthcare workers in Nigeria in 2013–2015: an inquiry into causes, consequences and control – a cross-sectional descriptive study. *Human Resources for Health, 14*(1), 46–55. doi:10.1186/s12960-016-0142-7

Omar, W. A. A., Al-Shahrani, R. M., Almushafi, M. A., & Boraie, H. M. (2022). Factors affecting patients' waiting time at the Radiology Department. *World Family, 20*, 62–68.

Onwuzu, S. W., Ugwuja, M. C., & Adejoh, T. (2014). Assessment of patient's waiting time in the radiology department of a teaching hospital. *ARPN Journal of Science and Technology, 4*(3), 183–186.

Piersson, A., & Gorleku, P. (2017). Assessment of availability, accessibility, and afford-ability of magnetic resonance imaging services in Ghana. *Radiography, 23*(4), e75–e79. Retrieved from www.radiographyonline.com/article/S1078-8174(17)30075-5/fulltext

Sedig, L. (2016). What's the role of autonomy in patient-and family-centered care when patients and family members don't agree? *AMA Journal of Ethics, 18*(1), 12–17. Retrieved from https://journalofethics.ama-assn.org/sites/journalofethics.ama-assn.org/files/2018-05/ecas2-1601.pdf.

Sekhon, M., Cartwright, M., & Francis, J. J. (2017). Acceptability of healthcare interventions: an overview of reviews and development of a theoretical framework. *BMC Health Services Research, 17*(1), 88–100. doi:10.1186/s12913-017-2031-8

Sipos, D., Csaba, V., & Petone Csima, M. (2017). The attrition and migration behaviour among Hungarian radiographers. *Global Journal of Health Science, 10*(1), 1–10.

Sørensen, K., Van den Broucke, S., Fullam, J., Doyle, G., Pelikan, J., Slonska, Z., & Brand, H. (2012). Health literacy and public health: a systematic review and inte-gration of definitions and models. *BMC Public Health, 12*(1), 1–13.

Studer Group. (2010). *The nurse leader handbook* (1 ed.). Parkway: Fire Starter Publishing.

SWSLHD. (2017). *Our leadership strategy: A shared approach to leadership 2017-2021.* Sydney: ZEST Health Strategies.

Tanha, M. R., Khalid, F. R., & Hoeschen, C. (2019). Assessment of radiation pro-tection and awareness level among radiation workers and members of the public in Afghanistan – a pilot study. *Journal of Radiological Protection, 39*(3), N1–N7.

Thambura, J. M., & Swindon, L. (2019). Occupational risk factors and their impact on migration of radiographers from KwaZulu-Natal, South Africa. *Journal of the Ergonomics Society of South Africa, 31*(1), 1–15.

Ugwu, A. C., Shem, S. L., & Erondu, F. (2009). Patients' perception of care during special radiological examinations. *African Journal of Primary Health Care and Family Medicine, 1*(1), 1–3.

van de Venter, R., & Friedrich-Nel, H. (2021). An opinion on role extension, and advanced practice, in the South African radiography context. Where are we heading and what should we aspire to? *South African Radiographer, 59*(1), 45–48. doi:10.10520/ejc-saradio-v59-n1-a12

van de Venter, R., & ten Ham-Baloyi, W. (2019). Image interpretation by radiographers in South Africa: A systematic review. *Radiography, 25*(2), 178–185. doi:https://doi.org/10.1016/j.radi.2018.12.012

van Vuuren, C. J., van Dyk, B., & Mokoena, P. L. (2021). Overcoming communication barriers in a multicultural radiography setting. *Health SA Gesondheid, 26*( 1), 1–8.

Wahed, W., Mabrook, S., & Abdel Wahed, W. (2017). Assessment of patient satisfaction at Radiological Department of Fayoum University Hospitals. *International Journal of Medicine in Developing Countries, 1*(3), 126–131.

World Bank Group. (2023). *World Bank country and lending groups.* Washington, D.C: The World Bank Group Retrieved from https://datahelpdesk.worldbank.org/knowledgebase/articles/906519-world-bank-country-and-lending-groups

Wuni, A.-R., Courtier, N., & Kelly, D. (2020). Opportunities for radiographer reporting in Ghana and the potential for improved patient care. *Radiography, 26*(2), e120–e125.

# Section two

# Person-centred care in general radiography

# 5  Patient care in general radiography

*Ruth Strudwick*

## Introduction

This chapter will take a detailed look at patient care in general radiography, which is defined as projection imaging or plain radiography. The focus of this chapter will be on the imaging of outpatients. Outpatients are patients who attend the imaging department from their own home or place of residence and are not therefore patients from within the hospital. In most imaging or radiology departments there will be X-ray rooms which are used for out-patient imaging and patients will come into the department via a reception area where they 'book in', and a waiting area where they will wait to be called for their X-ray examination.

In this area of diagnostic radiography, the full range of patients attend for imaging examinations, with the full range of ages, from birth to older people and the full range of abilities and disabilities. In this area of practice, the diagnostic radiographer needs to adapt their technique and communication skills to be able to interact with many different patients. It is important to acknowledge that as diagnostic radiographers the examinations performed in this area of the imaging department may be routine to us, but to a patient, nothing is routine, and so we need to ensure that we do not treat the inter-action as routine, but that we tailor our interaction and communication to each individual patient. In radiography, the highly technical nature of the role of the radiographer can lead to a tension between image production, and the time and personal resources available for patient care. However, the quality of the interaction between radiographer and patient can make a diffe-rence to the individual patient's experience.

Outpatients are generally referred for X-ray examinations by either their General Practitioner (GP) or referrers from outpatient clinics who can either be medical referrers or non-medical referrers such as nurses and physiotherapists.

The imaging examinations performed include projection imaging of the appendicular and axial skeleton, the chest and the abdomen. The clinical indications for imaging can range from short-term acute issues; minor trauma or possible infections to long-term chronic problems such as osteoarthritis.

DOI: 10.1201/9781003310143-7

Patients who attend outpatient imaging departments are usually reasonably mobile and will either be walking unaided, walking with some assistance such as walking sticks or frames, or they might attend in a wheelchair.

In some outpatient imaging departments, there is an 'open access' system, where patients can attend at any time during the working day and wait their turn for their X-ray examination. In other departments, there are booked appointment slots for patients. Each of these systems have their advantages and disadvantages in terms of patient-centred care, waiting times and speed of access to the X-ray examination. With an open access system, patients can attend as soon as the referral has been made and do not need to wait for an appointment time; they can attend at a time of day that suits them better, however, they may have to wait when they arrive in the department. With an appointments system, the patient may need to wait for a few days for an appointment time, and the time given may not suit them, however when they arrive in the department, they should be seen at their appointment time.

This chapter will follow the patient pathway from arrival in the out-patient imaging department to the after care provided to the patient, giving an outline of person-centred care throughout the patient's journey through the imaging department. A person-centred approach means that all patients receive the highest quality care that they need and that their individual needs are met always (Picker Institute Europe, 2019).

## Patient pathway – arrival in department

When a patient arrives in the imaging department reception, they should be made to feel welcome and there should be good signage so that they know where the reception desk is. Reception desks should cater for all patients and ensure that anyone with a disability can still access the reception staff; for example, there should be a higher and lower counter for walking patients and wheelchair users, there should be a hearing loop for anyone using a hearing aid and there should be written material available in braille for those with sight impairment and in other languages for non-English speakers. There should be sufficient privacy for the patient to be able to discuss the examin-ation that they have attended for with the receptionist, so that their details are not broadcast to the entire waiting area.

Patient details should be thoroughly checked to ensure that they are accurate on the imaging request. This needs to include the patient iden-tity, the area of the body to be imaged and the referrer's details. Once these details have been checked by the receptionist, the patient will be asked to wait. Waiting areas should be welcoming spaces with a choice of seating options, and spaces where wheelchair users can be accommodated without having to move furniture. Ideally there should be a space for children to play and a quiet area for anyone who might find noise disturbing, for example neurodiverse patients. There should be clear signage so that patients know where to find the toilets, baby changing facilities, changing areas and the way out of the department. Patients should be well-informed and feel welcome.

Patients may need to remove clothing and change into a gown for some projection imaging examinations; for example, chest, abdomen, pelvis, hips, spine, knees. It is important to maintain the dignity of the patient if they need to undress or change. The requirement to remove clothing and why this is necessary should be clearly explained to the patient, and they should be given a choice about where this can happen, for example a patient may feel more comfortable remaining in their own clothing whilst they wait and removing clothes or changing into a gown in the X-ray room, rather than changing in a cubicle. We need to ensure that patients are well-informed and can make a choice based on what is important to them. Removal of clothing should take place in a private space, for example a cubicle. If a patient needs to wear a gown, it is important that they know how to put the gown on correctly in order maintain their dignity. I am sure that we can all recall examples of patients putting gowns on in an interesting way that did not maintain their dignity. Patients may require help to change their clothing, and consent should be gained from the patient to assist with this, they may have specific requirements about who helps them, for example they may prefer to be assisted by someone from the same gender, or they may have someone with them who can assist them. Patients need to know what to do with their clothing once they have changed, i.e. do they bring it with them, or leave it in the cubicle? Are there baskets they can use to put their clothes in. There should be a separate seating area for patients wearing gowns, or they can stay in the cubicle. Ideally patients in gowns should not be left in communal areas where they will feel vulnerable; they also need to be warm. Maslow's hierarchy of needs (Maslow, 1954) is crucial here: we need to consider the patient's basic needs before we can expect them to be able to focus on our instructions and answer specific questions. Patients need to be comfortable; if a patient is cold or uncomfortable, they are less likely to concentrate or cooperate.

### Radiographer and X-ray request form

Before the radiographer calls the patient into the X-ray room from the waiting area, they can review the X-ray request. It is the radiographer's role as the operator to justify the imaging request (IR(ME)R, 2017). This involves ensuring that the benefits of the X-ray examination outweigh the risks to the patient and that the X-ray examination requested has the potential to answer the clinical question posed by the referrer. Justification of requests constitutes good person-centred care, as the radiographer is acting as an advocate for the patient, ensuring that they will not undergo an unnecessary X-ray examination and radiation dose (IR(ME)R, 2017).

When reviewing the X-ray request the radiographer, will also start to 'build a picture' of the patient, for example reviewing their name which will give some idea of their ethnicity and cultural background, their date of birth so that they know the age of the patient, any known disabilities such as mobility issues, sensory disabilities or long-term conditions which might have an impact on the examination and may require some adaptation of technique, and finally the clinical history of the patient and the reason for the examination. In 'building a picture' before meeting the patient, the radiographer can begin to consider how to go about the X-ray examination; they can consider if they might need any additional help or equipment and if they might need to adapt their radiographic technique. For example, if the patient is a wheelchair user, or has a hearing impairment – there will be differences in the examination from someone who can walk into the room and hear well. If the patient requires some help mobilising, the radiographer may decide that they need a colleague with them.

Radiographers, in common with other health professionals, are sometimes guilty of labelling or categorising patients based on information such as: age; gender; the examination or treatment for which they are attending; the nature of the injury or pathology for which they are being investigated; and the circumstances of the acquisition of the injury or illness (Murphy, 2009; Reeves and Decker, 2012). Goffman (1963) argues that these are related to unconscious expectations and norms, and although in some instances can provide useful 'shortcuts' for busy radiographers under pressure of time, may also lead to formation of negative or prejudicial attitudes.

The ethics of labelling and categorising patients is a sensitive issue in current healthcare, particularly when standards of care are under scrutiny (Francis, 2013).

It is part of any culture or group to have 'types' of people and to categorise people into groups (Atkinson and Housley, 2003). When people meet for the first time, they categorise one another and a decision is made about the type of person they are, then it appears to be easier to predict how they will behave and understand their actions. Davis (1959), in his paper 'the cabdriver and his fare', says that cabdrivers develop a typology of cab users based on their appearance, demeanour and conversation. In healthcare this also applies (Holyoake, 1999).

Categorising the patient into a typology can assist the radiographer in their decision making and planning for the radiographic examination (Reeves, 2009). Categorisation is about workload, typifying patients helps radiographers to decide how the examination or treatment would go, how to address the patient and more crucially gives them some idea of how long the examination or treatment might take so that they could plan. In categorising the patient, based on previous experiences radiographers can make judgements about what to expect.

Radiographers also refer to patients by the examination they have attended for, e.g. the next one is a foot, or it is a chest. This reductionist language is common within radiography (Reeves and Decker, 2012; Strudwick, 2016).

The radiographer will scrutinise an X-ray examination request, which normally begin with the examination being requested, a body part.

It is also important that the radiographer looks at the patient's previous medical history and previous radiographic images. This is particularly useful if the patient has had previous surgery such as a joint replacement so that the radiographer knows where this is located and can make the necessary adaptations to the imaging examination, for example centring the X-ray beam in a different place or extending the collimation.

### Patient pathway – radiographer calls patient from waiting room

Once the radiographer has carried out all the necessary pre-examination checks, they can prepare the X-ray room for the patient, ensuring that the correct patient and examination is entered onto the computer system in the X-ray room, and that the equipment is ready; for example, placing the X-ray tube in the correct orientation, having the imaging receptor and any other ancillary equipment ready to be used. The X-ray room should be cleaned before the patient is called into the room and the radiographer should wash their hands. Infection control constitutes good patient-centred care. When the room is ready the radiographer can call the patient from the waiting area.

When calling the patient from the waiting area the radiographer can tie together their initial 'picture' of the patient which was formulated from the X-ray request and their initial observation of the patient. The radiographer can make a more detailed judgement at this point about the patient's mobility, hearing ability, communication skills and their general wellbeing. This can all be added to their initial thoughts about the patient and can be used in planning the examination and ensuring that the radiographer operates in a person-centred way.

The radiographer should move from the doorway to the X-ray room towards the patient and assist them as required, for example the patient may require some assistance in standing and walking, they may be in a wheelchair and therefore need help in moving the wheelchair, or they may need assistance with other walking aids such as sticks or crutches. Many times, I have seen the radiographer stand in the doorway of the X-ray room, waiting for the patient to come to them – this is not welcoming or person-centred and does not create a good first impression. X-ray examinations are short interactions between the radiographer and the patient, and it is important for the radiographer to initiate this relationship, create a good first impression, develop a rapport with the patient and gain their trust. Research suggests that the first brief moment you interact with the patient is central to their experience in your care (Taylor et al., 2017; Bleiker, 2020). How you introduce yourself and create a relationship with the patient is crucial; you may find out information in those few minutes which will help you empathise with them and make an appraisal of the situation, and how to interact with the patient, for example is the use of humour appropriate? The radiographer can also offer their assistance with carrying clothing or the patient's belongings. It is important to ask if the patient needs help, rather than making assumptions, as the patient may not require assistance and may wish to maintain their own independence.

### Accompanying people

At this point it is important to acknowledge that the patient may have someone accompanying them to the imaging department. These could be family members or friends, or professional escorts such as carers.

Rather than making assumptions, it is always best to ask the patient who the person is who is with them. I would ask something like 'Who is this with you?'. This avoids making any errors of judgement; assuming someone is a partner or relative could be offensive to both the patient and the accompanying person, so it is best to establish the relationship early on. You can also observe the communication and interaction between the patient and the accompanying person as this will give a clue about their relationship.

Once the relationship is established then the radiographer can ask the patient if they wish this person to accompany them into the X-ray room. This should always be the patient's choice. The accompanying person may have an important role to play in assisting with communication, for example if the patient has a hearing impairment or perhaps if they have dementia a familiar face may be of help. Always take your cues from the patient.

If someone does accompany the patient into the X-ray room, the radiographer should ensure that they still address the patient directly and not the accompanying person.

We should also consider the use of interpreters at this point. If a patient does not speak English, they may bring a member of the family with them to translate. This is not always the best idea as it does not maintain the confidentiality of the patient and a family member may not interpret exactly what has been said. In general, it is best to use a professional translator – this could be another member of staff or a professional service such as language line (Language Line UK, 2022).

### In the X-ray room

Once within the privacy of the X-ray room there will be a closer interaction between the radiographer and the patient. This should begin with the radiographer saying, 'Hello my name is…' (Granger, 2013), and then introducing their role; for example, 'Hello my name is Ruth and I am a radiographer, I am going be taking your images today'. Anyone else present in the room should also be introduced, for example colleagues and students. The radiographer should then ask the patient what they would like to be called, for example would they like to be called by their first name, or do they have a nickname? Or would they prefer a more formal name such as Mr Smith or Mrs Smith? Then the radiographer needs to check that they have the correct patient in the room with them; this will be done by checking the patient's identity – most department protocols will state that the radiographer needs to check the patient's full name, date of birth and the first line of their address. This is important as the radiographer needs to check that the correct patient has entered the X-ray room with them, and not someone with a similar name, or someone who thinks it must be their turn next. We do not want to carry out an X-ray examination on the wrong patient.

Once the radiographer is sure that they have the correct patient in the room with them, they need to check the patient's clinical history and the examination requested. This should be done sensitively and tactfully, for example the patient may not be aware that the referrer is looking for cancer, but they will be aware of their own symptoms such as a persistent cough. The side of the body, left or right, should also be confirmed with the patient. Referrers can make errors on requests and can request the incorrect limb to be imaged. It is the role of the radiographer to check this information with the patient and if they are unsure, they should check with the referrer before proceeding.

If relevant to the examination and the patient, pregnancy status should be checked at this stage and local rules complied with. This needs to be carried out in a sensitive and tactful way, particularly with younger patients.

Next, a clear explanation of the examination should be given to the patient in language that they understand. The radiographer will need to adapt their explanation of the examination to the individual patient, and use appropriate language; for example, an explanation for a child will be different from an explanation given to an adult. The patient should be involved in making decisions about how to proceed with the examination based on their capabilities. This is good values-based practice (Fulford, Peile and Carroll, 2012). The radiographer will seek to develop a rapport with the patient and begin to understand what is important to the patient, so that their values are given due consideration and the examination is undertaken in the best way for them. The patient should always give consent to proceed – this can be verbal or implied.

It is part of a radiographer's legal and ethical responsibility to ensure that consent is given by the patient before starting any treatment or investigation. This reflects the right that patients have over what happens to their own body and underpins person-centred care. For consent to be valid, it must be given voluntarily by an appropriately informed person who has the capacity to consent to the intended examination or treatment. Generally, this will be the service user or if under 18 years of age someone with parental responsibility. In other cases, it may well be someone authorised to give consent on behalf of someone else under a Lasting Power of Attorney (LPA), as a consultee or someone who is a court-appointed deputy.

The patient should always give consent unless they lack capacity to do so, but a carer may be involved in communicating this consent. However, the patient may wish to go against the wishes of the carer, and their voice must always be heard and prioritised.

Values-Based Practice (VBP) is the consideration of the individual patient's values in making decisions about their care. By patient's values we mean the unique preferences, concerns and expectations each patient brings to a practice encounter which must be integrated into any decisions about the care of the patient (Fulford, Peile and Carroll, 2012). VBP considers and highlights what matters, and therefore is important to the patient. As radiographers, we should not be making assumptions about what the patient wants or indeed, reflect our own values upon the patients we image as radiographers.

We can do this by asking the patient to tell us what is important to them and providing them with enough information so that they can make informed choices. This is a critical aspect of true person-centred care and VBP. Values can, and do vary, sometimes widely between individuals and between patient and practitioner. They are not fixed and may change over time and as life experiences accumulate. The crucial thing to remember is that as a radiographer you should not make assumptions about your patient's values; instead take the time if needed to ascertain what matters to them at that moment in time regarding the task in hand, in this case the acquisition of a diagnostic image.

Communication is an important skill. The radiographer needs to be aware of their verbal communication; voice, tone expression and use of language. They need to speak so that the patient can hear them and understand them. Their voice should be clear, use appropriate intonation and expression and the language used should be appropriate for and understandable by the patient. The radiographer will need to make adaptations to the language used depending on the individual patient and their needs, for example paediatric patients, older people, people with dementia, people with disabilities, this demonstrates person-centred care when the language used is adapted to the individual. Non-verbal communication is also important; the radiographer should face the patient, have an open posture and maintain eye contact with the patient when speaking to them. This demonstrates attentive communication. It is very easy to multitask and to try to move equipment whilst addressing the patient; this should be avoided as it will be distracting for the patient and does not convey attentive communication.

We always need to remember that this may be a routine interaction and a routine X-ray examination for the radiographer. They may have undertaken thousands of chest X-ray examinations, but it will not be routine for the patient and indeed it may be their first ever X-ray examination, they therefore may not know what to expect. We should never assume that the patient knows what the examination entails and what will happen next.

In order to carry out the X-ray examination successfully, the radiographer will need to touch the patient. Touch is used to position the patient for the examination and can also be used to provide reassurance. The radiographer should always ask permission from the patient before touching them and this touch should be professional, firm and reassuring for the patient. They should not feel vulnerable in this situation. If the patient becomes distressed, touch may also be used to comfort and reassure the patient, for example holding their hand, or touching their arm.

Patients need to be involved in their own positioning wherever possible, rather than being forced into position. The examination needs to be a partnership between the patient and the radiographer, with the radiographer guiding the patient and explaining what is required for the best position. Patients need to be involved, 'no decision about me without me'; patients should not have things 'done to' them (Coulter and Collins, 2011). The radiographer can also demonstrate the position they would like the patient in to aid in cooperation.

During the X-ray examination there should be a continuous dialogue. This can involve the patient's clinical history, a conversation about what is important to the patient (values-based practice), a discussion about what is happening now and next, and a general chat to gain the patient's trust and to develop a rapport. During the short time that the radiographer spends with the patient, it is important that they gain the patient's trust and engage with the patient (Murphy, 2009).

At the start of the examination, when reviewing the imaging request and meeting the patient for the first time, the radiographer will have developed a 'picture' of the patient and will have already made some decisions about how to undertake the examination. These decisions will include if they will be using the routine technique or making some adaptations, for example if the patient is a wheelchair user, they might decide to image the patient in the wheelchair. These decisions will be patient-focused and depend on the patient's capabilities. The radiographer needs to make sure that they observe the patient's reaction and make adaptions accordingly, e.g. looking to see if they are in any pain. This may require a change in technique if things are not working well. The radiographer needs to ensure that the patient feels special, valued, important and not just being rushed through the department on a 'conveyer belt'; they need to be given individual attention. As previously mentioned, the radiographer will have undertaken thousands of X-ray examinations but this may be the first one for the patient.

It is important to take time to explain and answer questions; time taken at the start of the X-ray examination to do this is time saved in the long run as it will reduce any misunderstandings and unexpected incidents. We should always make time to listen to the patient. It is important to note that for some outpatients you may be the only person that they have spoken to that day, or the only person who has given the patient their undivided attention in a one-to-one interaction.

There have been occasions where the patient has used this one-to-one time with a professional to disclose a safeguarding concern, so this is something that radiographers need to be prepared to deal with. The patient may have the opportunity to talk to the radiographer outside of the earshot of a perpetrator. The radiographer also needs to be aware of anything suspicious, such as unexplained bruising or information not tying up – for example, the mechanism of injury stated does not tally with the injury seen.

The examination needs to be completed whilst all the time giving attention to the patient. The radiographer should face the patient when they speak to them. During the examination it is important to check that the patient is okay, observe their reactions and keep them safe throughout by concentrating on the task at hand.

The next important task is to check the resultant images – are they the best quality you can obtain for this patient? Do they answer the clinical question from the referrer? Are any further projections needed, e.g. to demonstrate a fracture, abnormality or foreign body? We need to strive to do the best we can; there is always a balance between producing the best images and patient tolerance. However, we should not be so kind to the patient that the images are not good quality and vice versa. Our job is to produce good quality diagnostic images that answer the clinical question from the referrer. The 10-point critique should be routinely used to check the quality of the resultant image (McQuillen-Martensen, 2019).

Radiographers should take pride in the work; it may seem 'easy' or 'routine' to you, but it isn't to the patient. The radiographer needs to produce the best possible images so that a diagnosis can be made; they need to be proud of the images they have produced, knowing that they have done their best for the patient. In recent years the implementation of direct digital radiography has done away with the showing of images to colleagues in the viewing area, as the radiographer tends to view their own images in the X-ray room. This may have contributed to a lack of pride in good images and lower standards, e.g. collimation, anatomical marker placement, and this is something to be aware of (Hayre et al., 2019).

If there is an abnormality present on the image, the radiographer will need to highlight this using their departmental protocols. For example, if the patient has a new fracture, an infection that needs urgent treatment or another abnormal appearance that they have concerns about, the patient should not be sent away from the department until the images have been reviewed. It may be that the patient requires some attention or may need to be admitted to hospital, and it is good patient care to ensure that the patient is looked after whilst they are still in the hospital, rather than being sent home. Also, if the radiographer has concerns about the wellbeing of the patient, they should not send them home without raising this concern. As the patients we are discussing here are outpatients, their follow-up care will normally be several days later from either their GP or via an outpatient hospital clinic, so we need to be reassured that the patient will be okay waiting for the results of the X-ray examination and does not need more urgent treatment.

## After the examination is completed

The patient needs to know where to get changed, where to go next, how to leave the department, where and when to obtain results. It is the role of the radiographer to convey all this information to the patient, giving clear instructions. As radiographers we need to be aware that we know our way around our department and hospital, but to a newcomer, it may not be easy to navigate. Patients may be anxious, and worried about the results of the X-ray examination so it may be best to pass the information on in stages and ensure that they understand.

The radiographer needs to ensure that all the relevant paperwork is completed and that the resultant images are correctly labelled and annotated. The images then need to be sent to the picture archiving and communications system (PACS) for reporting and viewing by the referrer.

The radiographer should also report any concerns they might have about the patient.

## References

Atkinson P & Housley W (2003) *Interactionism*. Sage, London.

Bleiker J (2020) An Inquiry into Compassion in Diagnostic Radiography. Unpublished thesis http://hdl.handle.net/10871/121267.

Coulter A & Collins A (2011) *Making Shared Decision-making a reality. No decision about me, without me.* The King's Fund. London.

Davis F (1959) The cabdriver and his fare: facets of a fleeting relationship. *American Journal of Sociology*, Vol. 65, No.2, pp158–165.

Francis R (2013) *Report of the Mid Staffordshire NHS Foundation Trust Public Inquiry.* London: The Stationery Office; 2013.

Fulford KWM, Peile E and Carroll H (2012) *Essentials of Values-based Practice: Clinical Stories Linking Science with people.* Cambridge University Press, Cambridge.

Goffman E (1963) *Stigma*. Engelwood Cliffs NJ: Prentice Hall Inc.

Granger K (2013) *Hello My Name is Campaign*. Available at: www.hellomynameis. org.uk/ Accessed 22.01.2021.

Hayre CM, Blackman S, Eyden A and Carlton K (2019) The use of digital side markers (DSMs) and cropping in digital radiography. *Journal of Medical Imaging and Radiation Sciences*, Vol. 50, No.2, pp234–242.

Holyoake D (1999) Favourite patients: exploring labelling in inpatient culture. *Nursing Standard*. Vol. 13, No. 16, pp44–47.

Ionising Radiation (Medical Exposure) Regulations (2017). HMSO Crown Copyright.

Language Line UK (2022) Language Line UK. Available at: www.languageline.com Accessed 31.12.2022.

Maslow AH (1954). *Motivation and personality*. New York: Harper and Row.

McQuillen-Martensen K (2019). *Radiographic image analysis.* 5th edn. St. Louis: Elsevier.

Murphy F (2009) Act, scene, agency: the drama of medical imaging. *Radiography* 15:34e9.

Picker Institute Europe (2019) *Principles of person-centred care*. Available at: www.pic ker.org/about-us/picker-principles-of-person-centred-care/ Accessed: 11.10.2019.

Reeves PJ (2009) *Models of care for Diagnostic Radiography and their use in the education of undergraduate and postgraduate students*. Bangor: University of Wales, 1999.

Reeves PJ & Decker S (2012) Diagnostic radiography: a study in distancing. *Radiography,* Vol. 18, p78e83.

Strudwick R (2016) Labelling patients. *Radiography*, Vol 22. Issue 1, February 2016, p50–55.

Taylor A, Hodgson D, Gee M, Collins K (2017) Compassion in Healthcare: a concept analysis. *Journal of Radiotherapy in Practice*, Vol. 16, No.4, pp350–360.

# 6 General radiography

## The heart of radiology, where person and machine come together

*Kathleen Naidoo*

## Introduction

In the words of Gladys Knight "Sometimes the best things are right in front of you; it just takes some time to see them". These words hold true for all healthcare personnel from around the world, because it first took a pandemic for the world to realise who the real superheroes are. The figurative capes that healthcare personnel wear empower them to successfully undertake their duties and enables them to connect to a person's heart and wellbeing. When a radiographer walks into a radiology department, they should pause for a moment, take a deep breath in and then don themselves with a number of powerful capes to assist them through the day. The cape of a friendly smile to welcome and reassure all patients, the cape of a listening ear, to listen to patient's uncertainties and address their concerns, the cape of strength to provide a supportive arm for patients who are unable to move by themselves, the cape of compassion to provide care in a dignified and respectful manner and the cape of knowledge to provide high-quality x-rays. Now they are ready to start the day.

The concept of person-centred care is a shift from the traditional mode of how we view patients and requires us to treat the patient as a person first and they should not be characterised by their health or disease. So, in order to create a truly person-centred experience for patients, radiographers are required to transform their way of thinking from focusing solely on high-quality image production to providing the patient with a holistic service. By virtue of being human, all radiographers are considered to have the ability to provide care to others (Boykin & Schoenhofer 2013:1). However, the way in which we demonstrate this might differ based on our individual understandings of caring. As a result of this it is important for all radiographers to understand person-centred caring within a radiography context.

Throughout this chapter, a number of questions will be posed to the reader to consider how they would overcome some of the barriers general radiographers face. As the chapter unfolds, the concept of person-centred care will be highlighted and practical suggestions will be provided for

DOI: 10.1201/9781003310143-8

radiographers to reflect on in order to promote the development of a truly person-centred caring radiography environment.

## Overview of the general radiography environment

Within diagnostic radiography there are a number of different modalities, from general radiography to more advanced modalities such as mammography, interventional studies, magnetic resonance imaging (MRI), computed tomography (CT), theatre imaging and trauma radiography. Therefore, it is considered one of the first-line investigations for many patients that visit a hospital. The number of patients that attend a radiology department varies and could range anything from 100 to more than 200 a day. With such a large number of patients visiting the general department, radiographers are expected to work swiftly and effectively through their workload. Additionally, diagnostic radiographers are expected to work different schedules in order to provide a 24-hour service to patients.

In South Africa, if a service (medical or non-medical) is disrupted for whatever reason and it causes harm or damage to the health and safety of a person, it is viewed as an essential service (South African Government 2018). According to the Labour Relations Act of 1995, emergency medical health services and medical and paramedic services are considered essential services (Department of Labour South Africa 1997). Diagnostic radiography is a medical health service that plays a paramount role in emergency medicine as it offers radiographic imaging expertise for both trauma and medical emergency patients, hence it is considered an essential service. Based on this, at any given time the general radiography atmosphere is often filled with feelings of anxiety, apprehension, fear and distress. These feelings could be associated with a lack of understanding of what to expect from the radiology procedure or due to the fear of a possible negative diagnosis (Carlsson & Carlsson 2013:3226; Björkman, Enskär & Nilsson 2016:71). In order to alleviate these feelings and provide a valuable, efficient service to the patient, it is imperative that a radiographer be considerate and caring towards the needs of their patients (Bolderston, Lewis & Chai 2010:198; Munn et al 2014:246).

The question then arises, if a radiographer is expected to provide imaging services to such a large number of patients, "how do they ensure that they afford each patient high quality care and optimal imaging services within a limited timeframe so as not to compromise on their workload?"

## Radiography education

Firstly, let's take a look at where it all begins, the undergraduate education of radiographers. In South Africa, the South African Qualifications Authority (SAQA) is a legal board appointed by the Minister of Higher Education and

Training to oversee matters related to the National Qualifications Framework Act. One of the roles of the SAQA is to develop and implement policies and criteria for the registration of qualifications (SAQA 2014). The SAQA has developed exit-level outcomes for Diagnostic Radiography, which must be followed by all higher education institutions offering the programme. One of these exit-level outcomes states:

> Perform safe and effective patient care in accordance with the patient's needs and departmental protocol to provide a quality service and to maintain the welfare of the patient.

This implies that during the undergraduate education and training of radiography students, the principles of patient care are learned and radiographers are expected to deliver safe and effective patient care to their patients. However, it is important to note the fact that there is a difference between patient care and caring (Paulson (2004:359). Patient care is regarded as a professional responsibility, whereas the concept of caring focuses on the qualities that are more inclined to human nature and are associated with feelings and emotions. Caring is described as more of a feeling of care towards another individual (Bolderston et al. 2010:199). From this we can perceive that caring is beyond just a professional role and while radiographers might be meeting their professional responsibility of patient care, the manner in which they undertake this is very important in order for them to optimise the caring experience for the patient. With patient care, we still allude to the patient and not the person which is more aligned to the conventional understanding of care. Acknowledging this difference and gaining a good understanding of the two concepts is key to the development of a person-centred caring environment.

Ultimately, we want radiographers to know the "how" of person-centred care and not just the "what" of patient care. Higher education institutions offering a radiography qualification should look at evolving their curriculum to include a more comprehensive understanding of the concept of caring, person-centred care and caring science. Naidoo et al. (2020) created a model to facilitate the development of caring amongst radiography students. This model provides an illustration of how the educator and student can work together to foster a trusting relationship and begin a caring journey together from first year. The central concept outlined in this model is the "facilitation of a culture of caring". Furthermore the authors recommended authentic teaching methods such as simulation-based learning, role-playing exercises, reflective practice and peer discussions to promote the learning of caring (Naidoo et al. 2020). Clinical radiographers should also be cognisant of the role they play in student training, as students view them as role models so it is important that they echo the principles of person-centred care (Naidoo et al. 2020).

## Barriers to person-centred care in general radiography

Once a radiographer has completed their training qualifications, they are expected to meet the requirements of the profession. The dynamics within a general diagnostic radiography context proves different from other healthcare contexts. Advanced technology and limited patient interaction time creates unique caring expectations from patients within this environment. So, in order for us to achieve the ideal caring environment we must first acknowledge and understand some of the barriers that radiographers face on a daily basis. Below is a brief narrative of the several barriers which may pose challenging to person-centred care in radiography.

### Limited time with patient

General diagnostic radiography procedures, such as skeletal x-rays, comprise about 90% of most examinations that are carried out in the clinical environment (Hayre et al. 2016:245). However the duration of these examinations, depending on the body part/area being imaged, can range from 10 minutes to more than 30 minutes. In addition to this, a radiographer may only see that particular patient once in their healthcare journey. Therefore, comparing radiology imaging time to other healthcare professions, the time a radiographer has to interact with a patient is transient and very limited. Within this period they are expected to fully understand the patient history, gain the patients trust and provide an efficient service. This can be very challenging for radiographers especially if they are not equipped with the skills needed to overcome this.

Over the years, patient satisfaction has been often associated with reduced waiting times, therefore a reduction in the patient waiting times have been given priority in healthcare (Hayre et al. 2016:248). But this comes at a cost. When trying to reduce waiting times, patient interaction time is often sacrificed (Naidoo et al. 2018). Because of this, many diagnostic radiographers have acquired a "hurried" and "rushed" approach to imaging patients (Hayre et al. 2016:248). This is evidenced in radiography research studies which noted that radiographers are becoming disillusioned with caring and are focusing on patient throughput (Hayre et al. 2016:248).

Caring is a fundamental aspect of diagnostic radiography (Beyer & Diedericks 2010:22). Diagnostic radiographers work with terminally ill, mentally unstable, intoxicated, trauma and medical emergency patients, all of which require radiographers to make speedy and prompt decisions about their patients. At times this can be concerning as poor judgement, due to limited time and quick decision-making, may lead to poor communication or miscommunication and ultimately poor patient care (Strudwick 2016:54). Therefore, limited time spent with the patient is a crucial factor for caring and is a distinctive characteristic of diagnostic radiography, which cannot be disregarded. So the questions arising are: "how can a radiographer develop a trusting relationship with a patient in such a short time?" and "how can a radiographer adequately demonstrate that they care about a patient that they only just met and may not see again?"

### Workload challenges

Radiology services, especially general radiography, plays such a critical role within the healthcare system (Hussain, Salowa, Tedla, Saleh, Rizvi & Al-Rammah 2016:1374). Hence, the requested number of general radiology examinations, per year, persists to increase. The challenge, however, is that, while the number of patients needing radiological examinations increases, the number of radiographers available in most public health sector hospitals, especially in South Africa, continues to remain low. As a result, this uneven ratio proves demanding on radiographers as they must deliver a good service to every patient without compromising on person-centred care.

The greatest fear is that radiographers are becoming overpowered by these workload pressures and are perceived as robots in a factory, simply pushing through the production line and adopting the "in and out" culture (Hayre et al. 2016:248). In essence, less time with patients generally results in diminished caring levels of radiographers. While advanced technology has resulted in a reduction of the image acquisition times, which in principle helps with workload challenges and means that more patients can be imaged in a shorter time, this quick approach can often be viewed as radiographers distancing themselves from their patients (Hayre et al. 2016:248).

In modern day, patients have a much more enriched knowledge of health care services hence their expectations of their healthcare professional has changed. Patients want to be more involved in the decision making of their medical treatment and planning (Moller 2016:309). But in order to be able to make informed decisions, adequate time is needed by healthcare professionals to properly and satisfactorily explain procedures and outcomes in detail.

So, we must ask ourselves: "How can we afford patients this opportunity to ask questions and make informed decisions without neglecting service delivery or increasing patient waiting times?"

### Advanced technology

Technology has improved immensely over the years in radiography, and these advancements have enhanced image quality and reduced image acquisition times. However, this one-sided advancement has awarded priority to science and technology which has resulted in less importance given to the enhancement of the human element (Moller 2016:309). The downside of this is that diagnostic radiographers are now finding themselves spending even less time with the patient. The debate on the impact of technology on caring is ongoing and there is a persistent worry that technology is distancing healthcare professionals from their patients (Hayre et al. 2016:248).

A radiographer's main role might be to produce a good quality radiograph using advanced technology, but they should not lose sight of the person in the room with them (Bolderston & Ralph 2016:358).

There is no doubt that the advancement in technology has led to countless medical benefits – an example of which is the use of technology for assisting elderly patients with daily tasks such as testing blood sugar levels and reminding them to take medication on time, without the need for someone to physically be present (Pols & Moser 2009:160; Mistry et al. 2015). However, the technological advancements in radiography are intended to assist the radiographer in delivering a better service to the patient and not as a replacement of the radiographer. The love, compassion, understanding and sincerity of a human cannot be found in a machine (Pols & Moser 2009:160; Mistry et al. 2015). Sociologists believe that technology denies the patient their self-awareness and permits professionals to simply objectify patients (Reeves & Decker 2012:82). It is important to always remember that advances in technology are not always known to the patient and to them it might not make any difference to the procedure at hand. The question I pose is "In the machine world of radiography, how do we keep the human element alive?"

### Objectifying patients

When working in such a time-constrained and busy environment, radiographers usually focus on the examination at hand and refer to their patients by the body part being imaged (Reeves & Decker 2012:79; Strudwick 2016:51; Törnroos & Ahonen 2014:110). This is referred to as objectification of patients and can be seen as a way to distance oneself from the patient. However, researchers propose that professionals use objectification as a coping mechanism in order to not get too attached to patients (Reeves & Decker 2012:79). This helps radiographers deal with the job in front of them without getting too emotionally involved.

Diagnostic radiographers x-ray many trauma and medical emergency patients who are experiencing a lot of pain (Strudwick 2016:51; Reeves & Decker 2012:80) and because of this radiographers generally categorise their patients by body part in order to make quick decisions and work strategically (Strudwick 2016:51). Diagnostic radiographers habitually view image production as their long-term goal while caring for the patient is regarded as a short-term goal in order to achieve the x-ray (Reeves and Decker (2012:82).

Radiographers are in constant contact with patients so it is imperative for them to demonstrate caring attributes. These patients are human beings and are experiencing a number of different emotions and feelings. The question I would like you to think about here is "Can radiographers find a balance between producing high quality images and delivering high quality patient care?"

**Enablers for a person-centred caring environment**

Despite the distinctive environment and patient experience within radiology, the ability for a radiographer to be able to demonstrate caring is still very important and there are many ways that can be adopted by a radiographer to ensure a person-centred caring environment. At this point we have a fair understanding of the radiography environment, so let's look at some of the enablers for a person-centred caring. The following headings will review different caring approaches, which will be described and then an overview of how these can be used within a radiography setting will be provided. When reading this section, reflect on the earlier questions that were posed and see if you can find the answers embedded in these enablers.

*Ubuntu philosophy*

In a country like South Africa, there are a large number of spiritual beliefs which are passsed down from generation to generation. The ability to care is developed from our childhood and is grounded on various spiritual and cultural beliefs (Clouder 2005:505).

The African concept of caring emphasises the ideology of the community, village and family (Masango 2005). South Africa is rich in culture and diversity, and holding true to this description, is the philosophy of Ubuntu. Ubuntu is an ancient African philosophy and can be viewed from a caring sciences approach. Ubuntu is described as human kindness and humanity to others (Downing & Hastings-Tolsma 2016:215). There are two distinctive characteristics of Ubuntu: relationships between people and how those relationships could be conducted (Downing & Hastings-Tolsma 2016:216).

The philosophy of Ubuntu was initiated in different societies in Africa. "I am because of who we are", is such a powerful rendition of Ubuntu, which makes us understand that we are who we are because of others. This beautiful expression provides us with an insightful outlook to life and our profession. Radiographers are serving their communities, and this constitutes as part of their civic duty as individuals. This is such a compelling reason for radiographers to never forget that their patients are a part of them in some way or the other. Patients should be perceived as family, friends, neighbours or colleagues.

The philosophy of Ubuntu transcends continents, and the teachings of Ubuntu can be associated with many different cultures and traditions. Some of the values of Ubuntu are: communality, respect, dignity, value, acceptance, sharing, co-responsibility, humaneness, social justice, fairness, personhood, morality, group solidarity, compassion, joy, love, fulfilment and conciliation (Kroeze in Himonga, Taylor & Pope 2013). Recognising and understanding the philosophy of Ubuntu will allow radiographers to excel in delivering optimal patient care. Ubuntu can easily be reflected within a radiology environment by simply acknowledging the patients as being a part of who we are. This immediate awareness will transform the way in which we view patients and how we interact with them. Ubuntu allows us to look beyond skin colour, religion or gender. We have shared humanity and an interconnectedness that goes beyond blood relations.

Patients in general radiography are seldom known to radiographers due to their short and generally one-off visit to the department. If a radiographer embraces the Ubuntu philosophy, that brief encounter with a stranger will now become a brief encounter with a person you see as a father, mother, aunt, uncle, brother or sister. This belief will add a familiar association to an unknown face which will make it simpler to develop a relationship with the patient. When patients are viewed through the Ubuntu lens, caring science becomes more innate and sincere. Ubuntu, provides the backbone for a radiography culture change, which can bring about enhanced person-centred care.

Maya Angelou once said: "A bird doesn't sing because it has an answer, it sings because it has a song". To all radiographers, may Ubuntu be your song and may you sing it out loudly.

### Little gestures approach

The little gestures approach to caring science, while so humble, has a powerful impact on the patient. It is such a simple approach that can be implemented even if the patient's examination is only 10 minutes long. These little gestures include the power of touch, smiling at your patient, a friendly greeting of the patient or even merely providing the patient with a blanket (Bleiker et al. 2016:257; Hayre et al. 2016:245). These gestures show the patient that you are cognisant of them as people and that you care about their wellbeing. This display of empathy can be reassuring to patients and allows the radiographer to develop a connection with their patients (Bleiker et al. 2016:258).

Radiographers should keep in mind that, while innovative machines and technology assists the radiographer in improving their daily duties, patients are unaware of these benefits. The patient only sees their interaction with the healthcare professional (McMaster & Degiobbi 2016:298). For a patient, it is the blanket that was given while they waited or the warm smile that reassured them during the procedure, that makes a world of difference (McMaster and Degiobbi 2016:298). An example of how healthcare workers implemented a simple gesture approach to caring, was Canadian radiation therapists who made personalised immobilisation masks for their patients in order to take their minds off the treatment (McMaster & Degiobbi 2016:298).

Person-centred care involves a direct interaction with the patient and allows the patient the opportunity to ask questions, while the radiographer listens attentively (Smith and Topham 2016:375). While this might seem like a time-consuming pursuit, there are a number of listening responses which can be used while continuing with the examination. An example of this is Brown's five listening responses: nodding, pausing to look at the patient, casual remarks, echoing and mirroring an understanding of what was said (Hayre et al. 2016:249). These listening skills can undoubtedly help radiographers when time is limited (Hayre et al. 2016:249).

Providing compassionate care through "small gestures" and the "power of touch" does not require a large amount of time to undertake, however it has a meaningful influence on the patient (Bleiker et al. 2016:257–262). This approach is highly recommended to all radiographers and I believe that this can take us one step closer to achieving a sincere person-centred caring environment.

### Patient and family engagement through effective communication

The foundation of person-centred care is "working with" the patient rather than the "doing to" the patient. The pathway to achieve this concept is to engage with the patient and their families through effective communication. Effective communication is very important in person-centred care as this will help radiographers understand the needs of their patients. Effective communication will assist the radiographer in easing the emotional distress patients often experience and it will ensure cooperation and a good patient experience (Björkman et al. 2016:71).

Communication may be conveyed in different forms, namely speaking, texting, email, etc. A communication might be delivered to its recipient, however, for it to be considered effective, all parties involved in the communication must correctly understand the message transmitted. Effective communication must be:

- clear
- concise
- correct
- complete
- courteous
- constructive (Zeppetella 2012:11)

Robinson, Segal and Smith (2018) propose four simple steps that can easily be implemented by any individual when communicating with people. These steps are as follows (Robinson, Segal & Smith 2018):

1. Engaged listening

When communicating with people, we often focus on what we are saying rather than listening to the other person. Engaged listening is different from just hearing the person. Listening well means understanding the spoken words and the emotions of the speaker. There are a number of techniques that can be used to develop your listening skills. Below are a few techniques you could try:

a) Focus on the speaker: listen to their tone of voice, watch the body language and other nonverbal prompts.
b) Avoid interrupting or redirecting to your concerns: do not focus on your response. Be genuinely in the moment listening to the speaker.
c) Show your interest in what is said: simple gestures like nodding and smiling shows you're listening and interested.
d) Try to set aside judgement: set aside your judgement even if you do not like or agree with the speaker's point of view. If judgement is removed, even the most challenging conversation can lead to a sincere connection with someone.

## 2. Non-verbal communication

Often when people care about something, they communicate this using nonverbal actions. This includes facial expressions, body gestures, posture, eye contact, tone of voice, etc. When someone is speaking to you, do not cross your arms and try to maintain a neutral stance. Make eye contact with them and pay attention to your tone of voice, try not to be condescending or judgemental. Respond according to the context of the conversation.

## 3. Managing stress in the moment

Stressful conversations cause us to say things we often do not mean and later regret. Managing stress levels is important to ensure that one does not misinterpret the conversation. Some methods to manage stress in the moment:

a) Try to stall before responding; a stalling tactic may be asking the person to repeat the question. This will give you more time to think before responding.
b) Pause: take a moment to collect your thoughts and pause for a minute before answering.
c) Deliver your words clearly: how you speak is sometimes more important than what you say. Speak clearly and calmly.

## 4. Asserting yourself in a respectful way

Be assertive in an open and honest manner but not in a hostile and aggressive way. For example, learn to say "no" when there is a need. Feel empathy for a person's workload but still emphasise the need for the current task to be completed.

Creating effective communication spaces for the patient and radiographer will assist in elevating the level of care provided in a radiology department.

The techniques provided are easy to follow and, with practice, radiographers can become leaders in effective communication.

### Communicating with anxious patients

An everyday challenge when communicating with radiology patients is having the ability to communicate with patients that are anxious and fearful. This could be linked to them being in an unfamiliar environment or not being fully aware of what to expect during the radiology examination. Gustafson (2015) uses a simple guide to dealing with anxious patients:

1. Listen to your patient

Let your patient voice their concerns. Let them speak their minds and just listen. Sometimes all they need is to know that someone is listening to them and this shows you care.

2. Provide a full explanation

Explain the procedure in simple and easy to understand terms. Explain what you will be doing and why. Fear of the unknown can cause a lot of anxiety and a simple explanation removes their uneasiness.

3. Don't simply tell your patient to relax, rather show them how

When you tell a patient to calm down or relax they might get more agitated. So, rather provide them with techniques to calm down like breathing exercises.

4. Try using some form of humour

Use humour as a distraction technique by lightening the mood but be sincere in your gesture. Take the patient's mind off the procedure and ask them about their family or kids.

5. Be prepared for stressful situations

As an individual you know your triggers and how you would react to certain situations so take a moment to recognise those feelings and resolve them before you speak or act.

6. Show empathy

Do not be judgmental towards your patient. Be aware of your thoughts and feelings. Try to put yourself in your patient's shoes before you react.

A crucial part of interacting with people is learning how to communicate effectively. These methods and techniques provide a straightforward guide for radiographers and can certainly move the radiology department in the right direction to achieve person-centred care.

### Informed decision making

A critical part of person-centred care is allowing patients to make informed decisions and being a part of their health conversation. In recent years, patients have affixed more value to being acknowledged during medical examinations and procedures (Smith & Topham 2016:374). They want to be a part of the decision-making process and to be given adequate information regarding their treatment and medical procedures (Bolderston 2016:356). It is therefore critical for radiographers to provide patients with information regarding the general examination at hand and allow the patient the opportunity to ask questions. Your explanation should always use words that are easy to understand and avoid the use of many medical terms. This might confuse the patient even more. A practical example of allowing the patient to make informed decisions is to fully explain the benefits and risks of an x-ray examination and let the patient decide if this examination is suited to their health beliefs. If a patient refuses the examination, be respectful of their decision.

### Watson theory of human caring

Watson's theory of human caring describes Ten Carative Factors along with their processes and competencies (Watson Caring Science Institute 2018:3). The 10 carative factors are: (1) forming humanistic-altruistic value systems; (2) instilling faith-hope; (3) cultivating a sensitivity to self and others; (4) developing a helping-trust relationship; (5) promoting an expression of feelings; (6) using problem-solving for decision-making; (7) promoting teaching-learning; (8) promoting a supportive environment; (9) assisting with the gratification of human needs; and (10) allowing for existential-phenomenological forces (Watson Caring Science Institute 2018:3). While this was developed for the nursing profession, there are many aspects of this theory that can be adopted within radiography. I will explore the first four factors in relation to radiography.

The first factor of sustaining humanistic-altruistic values is extremely important in healthcare not just nursing alone. The healthcare system has a number of interrelated and interdependent disciplines working together to provide a service to the patient. A patient can easily be lost in the healthcare system and become just a number waiting for a service to be provided. The first carative factor is related to a demonstration of selfless concern and being kind and loving to the self and others. Therefore a radiographer should make a conscious effort to provide kindness and compassion to all patients. This will help to transform the way patients perceive healthcare as a whole.

The second factor is related to being genuinely present and listening to the patient. By doing this the professional will be able to instil faith and hope in the patient. The second carative factor of being authentically present is so crucial in ensuring the patient does not become objectified. The acknowledgement of the presence of an individual by showing that you are interested and listening to them can easily instil faith and hope for their wellbeing. It is critical for a radiographer to be authentically present with their patients when they are in the x-ray room. The patient should get their full attention and the radiographer should not be distracted, for example, by taking phone calls or responding to phone messages or even listening to music via ear pods or headphones during the examination. Just be present in the moment with the patient. This authentic presence will allow the radiographer to provide better care for the patient because the radiographer will be more observant of the patient's needs. This authentic presence will also increase the radiographer's response rate to patients in distress because they would quickly be able to ascertain if a patient's condition is deteriorating.

The third carative factor shifts focus slightly and encourages the healthcare professional to be sensitive to themselves and the patient. At times, the way in which we treat ourselves is reflected in our practice. Developing an awareness of the self and your own spiritual beliefs can in turn cultivate positive vibes to those around you. Patients are already in a vulnerable state and they look to their healthcare professional for help and support. This does not have to be just physical support but can be emotional as well. Radiographers are encouraged to look after their wellbeing by engaging in some form of physical or mental exercise like running, walking, weight lifting, yoga, or meditation. When you feel good about yourself it is easier to inculcate hope in others.

The next carative factor is developing and sustaining loving, trusting-caring relationships. This might seem challenging for radiographers, however using the little gestures approach and being guided by the principles of Ubuntu we can easily develop a trusting relationship without compromising patient waiting times. Speak to your patient and ask them probing questions to gain a better understanding of who they are and what their healthcare needs are. This form of showing interest in the patient will automatically help the radiographer to gain the patient's trust. Implementing these carative factors in radiography will enhance the patient experience and it provides a good foundation for radiographers to create a safe and comfortable space for their patients.

*Reflective practice*

Subconsciously, people have been reflecting their whole life: "thinking about and learning from past experiences to avoid things that did not work and to repeat things that did" (Koshy et al. 2017). "Reflection is the examination of personal thoughts and actions. For practitioners, this means focusing on how they interact with their colleagues and with the environment to obtain a clearer picture of their own behaviour" (Oluwatoyin 2015:28). Individuals gain perspective through reflection and understand themselves better. Reflection does not only mean focusing on our own experiences, we can learn from the experiences of others as well by placing ourselves in their situation (Koshy et al. 2017). There are two types of reflection: reflection-on-action, which is reflection after experiencing a situation or encounter, and reflection-in-action, which is about being reflective during an encounter (Oluwatoyin 2015).

A recommended model for radiographers to use to advance their reflective practice is Gibbs model of reflection (Lia 2016). This model is a practical method that can be used by radiographers as a way of reflecting on their experiences. Gibbs' reflective cycle has six stages:

1. Description
2. Feelings
3. Evaluation
4. Analysis
5. Conclusion
6. Action plan

Figure 6.1 below depicts each step of the model.

This model is effective in allowing individuals to:

- challenge their assumptions
- explore different/new ideas and approaches towards doing or thinking about things
- promote self-improvement (by identifying strengths and weaknesses and taking action to address them)
- link practice and theory (by combining doing or observing with thinking or applying knowledge)

If radiographers truly want to be caring practitioners then engaging in reflective practice is the key to unlocking their true potential.

## Conclusion

When we started this chapter, the development of a truly compassionate, dignified, kind, sincere and understanding radiography environment seemed eons away but after reviewing the various approaches to caring sciences, the reality of such a delightful person-centred caring environment is actually instantly achievable. The little gestures approach makes compassionate care doable in a busy and time constrained department. Jean Watson's theory

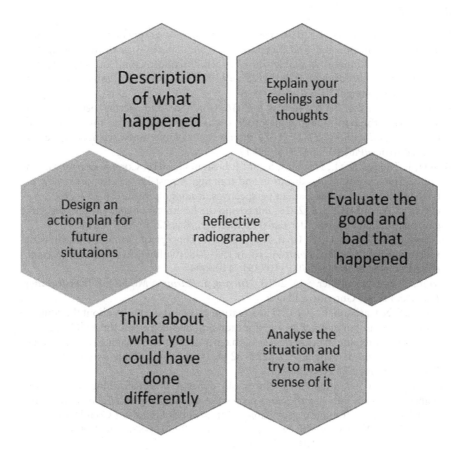

*Figure 6.1* Adaption of Gibbs' reflective cycle.

Source: Gibbs 1988.

of caring provides a beautiful backdrop for self-reflection and the development of radiographers who are sincerely caring. The concept of Ubuntu lays the perfect foundation for us to cultivate change and build meaningful relationships with our patients. It is now time for a new directive. May we break away from the stigma of being button pushers and move towards the association of being caring advocates. Let us be the smiling faces behind the big scary machines, the listening ear among the hundreds of busy healthcare workers and the healing power of touch that instils hope and faith in our patients. Sometimes all it takes is for one person to break away from the old habits and start a new culture of person-centred care. I would like to end this chapter with the words of Mahatma Gandhi: "You must not lose faith in humanity. Humanity is an ocean; if a few drops of the ocean are dirty, the ocean does not become dirty".

## References

Beyer, L. & Diedericks, P. 2010. The attitudes of radiographers towards patients in government hospitals in Bloemfontein. *The South African Radiographer*, 48(2):22–27. (Online). Available: http://web.B.Ebscohost.Com.Dutlib.Dut.Ac.Za/Ehost/Pdfviewer/Pdfviewer?Vid=1&Sid=93907d08-5fce-4648-9a04-2a2c43eb47b9%40sessionmgr120

Björkman, B., Enskär, K. & Nilsson, S. 2016. Children's and parents' perceptions of care during the peri-radiographic process when the child is seen for a suspected fracture. *Radiography*. 22(1):71–76. (Online). Available: www.sciencedirect.com/Science/Article/Pii/S1078817415000917

Bleiker, J., Knapp, K.M., Hopkins, S. and Johnston, G. 2016. Compassionate care in radiography recruitment, education and training: A post-Francis Report review of the current literature and patient perspectives. *Radiography*, 22(3): 257–262.

Bolderston, A. and Ralph, S. 2016. Improving the health care experiences of lesbian, gay, bisexual and transgender patients. *Radiography*, 22(3): e207–e211.

Bolderston, A., Lewis, D. & Chai, M. 2010. The concept of caring: Perceptions of radiation therapists. *Radiography*, 16(3):198–208. (Online). Available:www.sciencedirect.com/Science/Article/Pii/S10788174100004

Boykin, A. & Schoenhofer, S.O. 2013. *Nursing as Caring a Model for Transforming Practice*. Project Gutenberg Ebook

Carlsson, S. & Carlsson, E. 2013. "The situation and the uncertainty about the coming result scared me but interaction with the radiographers helped me through": A qualitative study on patients' experiences of magnetic resonance imaging examinations. *Journal of Clinical Nursing*, 22:3225–3234. (Online). Available: www.ncbi.Nlm.Nih.Gov/Pubmed/24118524

Clouder, L. 2005. Caring as a "threshold concept": Transforming students in higher education into health (care) professionals. *Teaching in Higher Education*, 10(4):505–517. (Online). Available: http://web.B.Ebscohost.Com.Dutlib.Dut.Ac.Za/Ehost/Pdfviewer/Pdfviewer?Vid=1&Sid=8499d015-88c4-4ebb-9a19-332b93a16e33%40sessionmgr101

Department of Labour South Africa. 1997. Government Gazette No. 18276. (Online). Available: www.gov.za/sites/default/files/gcis_document/201409/18276.pdf

Downing and Hastings-Tolsma. 2016. An Intergrative Review of Albertina Sisulu and Ubuntu: Relevance to caring and nursing. 21. 214–227. DOI: http://dx.doi.org/10.1016/j.hsag.2016.04.002

Gibbs, G. 1988. *Learning by doing: A Guide to teaching and learning methods*. Oxford: Oxford Further Education Unit

Gustafson, C. 2015. 7 ways to manage anxiety in patients, family members and yourself. (Online). Available: www.Travelnursing.Org/7-Ways-To-Provide-Exceptional-Patient-Care

Hayre, C.M., Blackman, S. & Eyden, A. 2016. Do general radiographic examinations resemble a person-centred environment? *Radiography*, 22:E245–E251. (Online). Available: https://doi.org/10.1016/j.radi.2016.07.001

Himonga, C., Taylor, M., & Pope, A. 2013. Reflections on Judicial Views of uBuntu. *Potchefstroom Electronic Law Journal*, 16(5), 369–430. https://doi.org/10.17159/1727-3781/2013/v16i5a2437

https://takestockpalmbeach.org/wp-content/uploads/2018/11/Communication-Skills.pdf

Hussain, A.M., Salowa, H., Tedla, J.S., Saleh, A.M., Rizvi, S.A. & Al-Rammah, T.Y. 2016. Radiography students' satisfaction during their practical and clinical training sessions at King Khalid University, Saudi Arabia: A cross-sectional study. *Biomedical Research*, 27(4):1374–1377. (Online). Available: http://0-Sea rch.Ebscohost.Com.Ujlink.Uj.Ac.Za/Login.Aspx?Direct=True&Db=A9h&An= 119485511&Site=Eds-Live&Scope=Site

Koshy, K., Limb, C., Gundogan, B., Whitehurst, K., Jafree, D.J. 2017. Reflective practice in health care and how to reflect effectively. *International Journal of Surgical Oncology* (N Y). Jul;2(6):e20. doi: 10.1097/IJ9.0000000000000020. Epub 2017 Jun 15. PMID: 29177215; PMCID: PMC5673148.

Lia, P. 2016. Using Gibbs Reflective Cycle in Course Work https://moodle.nptcgr oup.ac.uk/pluginfile.php/1113942/mod_resource/conte        nt/1/USING-GIBBS-REFLECTIVE-CYCLE-IN-COURSEWORK-DEC-2016- P-LIA.pdf

Masango, M. 2005. The African concept of caring for life. *HTS Theological studies*, 61(3). DOI: https://doi.org/10.4102/hts.v61i3.465

McMaster, N. & Degiobbi, J. 2016. It's the little things: Small gestures in patient care, big impact. *Journal of Medical Imaging and Radiation Sciences*, 47(4):298. (Online). Available: www.jmirs.org/article/S1939-8654(16)30123-0/pdf

Mistry, N. Keepanasseril, A. Wilczynski, N.L. Nieuwlaat, R. Ravall, M, Haynes, R.B. and the Patient Adherence Review Team. 2015. Technology-mediated interventions for enhancing medication adherence. *Journal of the American Medical Informatics Association*, 22(1) e177–e193. (Online). Available: https://doi.org/10.1093/jamia/ocu047

Moller, L. 2016. Radiography with the patient in the centre. *Journal of Radiology Nursing*, 35(4):309–314. (Online). Available: www.Sciencedirect.Com/Science/Arti cle/Pii/S1546084316300852

Munn, Z., Jordan, Z., Pearson, A., Murphy, F. & Pilkington, D. 2014. On their side: Focus group findings regarding the role of MRI. *Radiography*, 20(3):246–250. (Online). Available: www.sciencedirect.com/Science/Article/Pii/S107881741 400042x

Naidoo, K. Lawrence, H. Stein, C. 2020. A model to facilitate the teaching of caring to diagnostic radiography students: Original research. *Nurse Education Today*. Vol 86. DOI: https://doi-org.ezproxy.cput.ac.za/10.1016/j.nedt.2019.104316

Oluwatoyin, F.E. 2015. Reflective practice: Implication for nurses. *Journal of Nursing and Health Science*, 4(4): 28–33.

Paulson, D. 2004. Taking care of patients and caring for patients are not the same. *Aorn Journal*, 79(2):359–366. (Online). Available: www.sciencedirect.com/science/Article/Pii/S0001209206606121

Pols, J. & Moser, I. 2009. Cold technologies versus warm care? On affective and social relations with and through care technologies. *Alter, European Journal of Disability Research,* 3:159–178. (Online). Available: DOI: 10.1016/J.Alter.2009.01.00

Reeves, P.J. & Decker, S. 2012. Diagnostic radiography: A study in distancing. *Radiography*, 18(2):78–83. (Online). Available: https://Doi.Org/10.1016/J.Radi.2012.01.00

Robinson, L., Segal, J. & Smith, M. 2018. Effective communication improving communication skills in your work and personal relationships. (Online). Available: www.helpguide.org/articles/relationships-communication/effective-communication.htm

Smith, K. & Topham, C. 2016. Patient-centered care. *Journal of Medical Imaging and Radiation Sciences*, 47:373–375.

South African Government. 2018. Determine essential services. (Online). Available: www.gov.za/node/727484

South African Qualifications Authority (SAQA). 2014. 18 6. What is the South African Qualifications Authority? Available: www.saqa.org.za/about-saqa

Strudwick, R.M. 2016. Labelling patients. *Radiography*, 22(1):50–55. (Online). Available: www.sciencedirect.com/Science/Article/Pii/S1078817415000565

Törnroos, S. and Ahonen, S.M. 2014. Conception of man in diagnostic radiography research–A discourse analysis of research articles from the journal radiography. *Radiography*, 20(2): 107–111.

Watson Caring Science Institute. 2018. 10 carative processes. (Online). Available: www. Watsoncaringscience.org/Jean-Bio/Caring-Science-Theory/10-Caritasprocesses

Zeppetella, G. 2012. Communication skills. Palliative Care in Clinical Practice. DOI: 10.1007/978-1-4471-2843-4_2

# 7 Person-centred care in paediatric radiography

*Harry Bliss, Emma Rose and Niabh Kirk*

## Introduction

Patients and their families often express feelings of vulnerability and power-lessness in the face of illness and traversing healthcare processes. Feeling a lack of control in an unfamiliar environment can result in any patient experiencing distress or anxiety (Lerwick, 2016). For children, young people and accompanying guardians, attending the imaging department can be a distressing and anxious experience as this may be the place where the answer to their symptoms may be found (Bray et al., 2022). Depending on the child's development, they may be more likely to communicate their emotions through behaviour rather than words. Typical behavioural demonstrations of anxiety and fear include aggression, lack of cooperation and regression (Rodriguez et al., 2012). With transparency and consistent coordination of care these feelings can be put at ease (Ortiz, 2018).

Numerous different approaches have been suggested to meet patient needs and improve their experience. Paediatric patients differ from adult patients in their unique emotional needs, which can vary greatly depending upon their age, previous adverse childhood traumas and/or impact of any neurodiversity. Lerwick (2016) has proposed the CARE approach: Choices, Agenda, Resilience and Emotional Support when working with children.

Feeling a lack of control impacts the child or young person's experience. It is important children and young people are given as much choice and control as possible. In achieving this, the patient and guardian will feel empowered and trust will begin to build. Choices should be developmentally appropriate to ensure success with this approach. Identifying what choices are developmentally appropriate can be challenging and time-consuming initially. Examples of choice include allowing the child to explore, speaking to the child as well as the guardian, allowing the child to ask questions and offering appropriate explanations. When healthcare providers speak solely to the guardian about the child or young person, it does not provide the opportunity for exploring or asking questions and moves towards holding the person down for examinations without involving them. Choice is removed from the child or young person. Empowerment of the service user should be

DOI: 10.1201/9781003310143-9

a key goal of the healthcare provider. Empowerment is crucial in securing emotional safety for the person (Lerwick, 2016).

Agenda involves the healthcare provider explaining to the patient and guardian what to expect during their visit and what is required from them. Explanation and managing expectations is an important link in empowering patients and removes the unpredictable aspect of their visit. Agenda requires excellent communication skills with the child, young person and guardian. Allowing time for outlining the agenda and explaining each step will have a positive impact in trust and creating psychological safety for the child or young person and guardian.

Putting focus on the patient's resilience or strengths is a key component of empowering the patient and family. Healthcare providers can highlight a patient's resilience by asking the patient (and accompanying guardian) how they dealt with other struggles and allowing them the opportunity to provide the healthcare provider a better understanding of themselves and their situation. Discussing their strengths and resilience allows a positive conversation to occur between the healthcare provider, patient and guardian. Positivity is crucial and therefore it is important for the healthcare provider to avoid negative or critical conversation to occur. Questions can be rephrased to steer the question from negativity towards positivity. For a child or young person to hear about their positive qualities it can strengthen the relationship between the patient and guardian as well as reduce anxiety.

The child or young person and accompanying guardian will experience a range of emotions when attending the imaging department. Normalising emotions can be achieved by acknowledging feelings which are clearly evident and expressing wonder around emotions which appear vague or unspecific. By normalising the emotions the person is given the freedom to feel and express these emotions, which play a key role in their psychological safety. It is important the healthcare provider considers the emotions of both the child or young person and the accompanying adult. When acknowledging or discussing emotions it is important consideration is given to the tone and volume of the voice – soft tones will prove more successful in creating psychological safety for the person. By developing psychological safety for the child or young person and accompanying guardian, this will reduce anxiety and develop the relationship between them and the healthcare provider.

This chapter will now explore some of the concepts discussed above to gain a deeper understanding of patient-centred care for children and young people. Picker's principles of person care (Ortiz, 2018) will be used as a loose framework for this exploration.

## Provision of services

Picker's principles stipulate the importance of access, continuity, integration and coordination of care. There should be the processes and provisions in place for any child or young person to access high quality care no matter their location and to ensure smooth movement and communication of their needs and healthcare history from one clinical setting to another. During the research for this chapter, it has been identified that many policies and guidance documents from governments and professional bodies that form central standards for clinical departments to base local protocols upon are outdated – with some active guidance being written as far back as 2003. These policies therefore could not reflect the nuances of contemporary radiology, with regards to technology, societal attitudes and world events such as the Covid-19 pandemic. Although updated more regularly, public-facing procedures at hospital level are seen, but there are often inconsistencies from one location to another which would impact how children may experience consistent care if moving from one part of a country to another.

## The impact of poverty on children's health and well-being

As the fifth wealthiest country in the world, the UN Special Rapporteur (2022) has found the UK failing to comply with its obligations under the UN Convention of the Rights of the Child. This includes provisions to ensure children and families have enough to eat and a safe place to live. It is a fundamental principle of child development that children who are hungry, tired and worried are often disadvantaged and may not reach their full potential in adulthood (UNICEF, 2022).

The evidence linking poverty with ill-health is clear. This is evident even at birth, when the birthweights in the deprived areas are on average lower than in the richest, and children in disadvantaged families are more likely to die suddenly in infancy, to suffer acute infections, and to experience mental ill-health (RCPCH, 2022). Poor housing and lack of safe outdoor space also in part explains the worrying evidence that children in poverty are at significantly increased risk of injury and death from accidents, including in road accidents, fires, accidental drownings and accidental poisonings (RCPCH, 2022).

According to the RCPCH (2022) they also found that many parents in poverty worried that they would need to choose between their children's health needs and the immediate stress of trying to keep a roof over their heads. For example, they might lose their job if they take a day off work, or have their benefits sanctioned if they cancel an appointment at the jobcentre. No clear data has been found to determine the correlation between poverty of cancelling or missing appointments, identifying a potential avenue of research. All missed hospital appointments have the potential to directly result in adverse management of a patient's medical condition.

Poverty within families has also been linked to overcrowding within the household, which can make the care of children with long-term health issues, learning difficulties or neuro diversity more challenging (RCPCH, 2022).

Although a wider range of appointment times are now being offered by imaging departments, more can be done to offer appointments to families that coincide with school holidays and out of standard working hours, therefore reducing the strain on families from poorer socioeconomic backgrounds.

## Specialist services

Although variations of healthcare provision change throughout the world, most systems can be divided into primary, secondary and tertiary care. Primary care is the first point of contact for patients in need of healthcare, such as General Practitioners or family doctors, dentists and pharmacists. Secondary care represents traditional hospital and community care, from elective procedures to urgent care such as treatment for a fracture via the emergency department. Whereas tertiary care involves specialised treatment such as neurosurgery, transplants and secure forensic mental health services. Unfortunately interconnection between these services may be disruptive. An example of this, especially within the paediatric population, is the insertion of central venous access which may occur with a tertiary centre and subsequently may be used by community primary care teams but there is a lack of communication with regards to the acceptable use of the device and transfer of associated radiographic images and reports (NHS Providers, 2022).

There are wide variations in the access to specialist paediatric imaging throughout the world. Most specialist services are concentrated within dedicated children's hospitals and major teaching hospitals. However, the majority of routine, emergency and trauma imaging takes place in district or non-specialist hospital settings. As a result there are variable levels of local expertise and support from specialist units to smaller hospital imaging departments. Despite this, children rarely breach the governmental targets, such as those for cancer and Accident & Emergency and there are proportionally fewer complaints about paediatric imaging services when compared to those of adults (Department of Health, 2010).

There are currently 27 specialised NHS children's hospitals spread across the UK, with reduced access to these services in Wales, north Scotland and the south west of England (British Association of Paediatric Surgeons, 2022). Specialised services (within tertiary care) are designed to support patients with a range of rare or complex conditions. This may involve access to treatment options or specialist referrers that are not available in district hospitals. Specialist centres often launch cutting-edge innovation, and support pioneering clinical practice (NHS England, 2022). Standards to form a network approach to paediatric radiology provision were laid out in the UK through the "Delivering quality imaging services for children" by the Department of Health (2010). Unfortunately, according to Halliday (2016), the aims of this piece of work are largely unmet, which can put vulnerable children at risk. Equipment and facilities suitable for children of all ages, ranging from premature infants to teenagers is needed in modern healthcare. These are often different to the resources required to support adult services. Therefore, imaging departments need to have access to child-focused infrastructures, specific to the age of the children in their care. This would include appropriate procedures to support the nuances relating to radiation protection and safeguarding, which are of great concern for this age group. It is vital that children are considered in their own right, with imaging techniques and protocols focused to their needs rather than using techniques based on the idea of imaging small adults.

For example, access to interventional radiology (IR) services can be variable in England, ranging from smaller hospitals with no service provision, to a small number of centres that provide a 24/7 service in larger cities. Although most hospitals offer a service during normal working hours, many of the radiologists providing IR services may not be trained specifically in children's interventional radiology. This directly impacts children accessing services, and therefore influences patient outcomes and satisfaction. Even at specialist paediatric hospitals the interventional radiology service is limited, with many utilising adult interventional radiologists and equipment (Department of Health, 2010).

## Specialist staff

Paediatric radiology is repeatedly the smallest radiology sub-specialties provided by imaging departments. Historically recruitment for both radiology and radiography trainees is low.

To combat this, all radiology and radiography training schemes in the United Kingdom now include core paediatric training as part of their curriculum. Currently paediatric radiography is not recognised as an extended role for registered radiographers and has no recognition within Agenda for Change. Paediatric radiography has no formal career structure and thus there is little incentive for radiographers to specialise in this area. As a result, few dedicated paediatric radiographers exist outside specialist children's units. This is being combated by the Society of Radiographers and Association of Paediatric Radiographers with varying success.

A special interest group (SIG) is a collection of individuals from a specific or mixed professional backgrounds with a joint interest in the progression and development of their field. Such groups have been found to be highly effective in increasing knowledge and understanding with their peers by organising learning events and advising upon development of policy (Louw, Turner & Wolvaardt, 2018). In the UK, groups such the British Society of Paediatric Radiology and the Association of Paediatric Radiographers, are available to give professional bodies, radiology managers and practitioners a point of contact to share clinical expertise on children and young people imaging.

Variation in practice within radiology is a common issue. This is partly due to a repeated need for training staff that move from one clinical site to another and patient satisfaction implications when experiences vary from one location to another. By standardising protocols nationally or even internationally this may be resolved. However, to make this process successful within any imaging department, clear leadership is required to articulate the importance of protocol standardisation in improving the quality of patient care (Venkataraman et al., 2019).

In order to qualify and register, all radiographers must have obtained training and experience in working with children (HCPC, 2013). Although according to Foster and Clark (2019) most newly qualified radiographers consider their preparation for paediatric imaging adequate, there is a notable feeling that additional specialist training would be helpful once in clinical practice. This has been identified as far back as the work of Hardy (2000), who identified post graduate education in this area of radiographic practice as lacking and in need of further educational support. With this additional training a better understanding of children and their families is hoped to be achieved, and with it improved patient experience and outcomes.

## Dedicated children's space

According to Brems and Rasmussen (2018), it is vital that any child attending a healthcare appointment has access to a space which has been furnished and designed in such a way that all feel safe and welcome. Although their time spent in an imaging department may be short, the need when a child comes for any medical imaging is unchanged. With paediatric patients and their families often spending extended periods of time waiting for their imaging studies, consideration must be made that those waiting in radiology departments are likely to become bored, depressed and anxious, which may negatively affect their overall experience of seeking any form of medical attention in the future (Qi et al., 2021). With children accounting for a small proportion of non-specialist centres' workload, often reduced resources are allocated. Specialist resources can be in short supply and of uneven quality compared to what is provided to adults. Studies in China, for example by Qi et al. (2021), found that the waiting space in children's clinics failed to meet key patient needs in areas such as mother and infant rooms, children's play areas and drinking water facilities, and there are widespread problems with the creation of natural environments, such as views of natural scenery from windows and indoor green plants.

By optimising the design of children's waiting spaces in radiology departments, with a focus on functional layout, flow organisation, supporting facilities and environmental details, healthcare providers can improve the overall satisfaction of both paediatric patients and their families (Qi et al., 2021). Where possible, spaces within radiology rooms should be adapted to be more welcoming for children. Colourful lights or television screens (as demonstrated in Figure 7.1) can be cost-effective methods of improving the ambience and friendliness of the clinical space for both paediatric patients and adults.

*Figure 7.1* Colourful lights and projector screens improving the ambience of a nuclear medicine SPECT/CT scanner.

Source: Picture provided by Great Ormond Street Hospital for Children, London.

## Transition from paediatric to adult services

As patients move through their teenage years, the need to transition into the adult services becomes inevitable. Preventing adolescents becoming lost in this transfer between paediatric and adult health services remains a major challenge in all countries. Transition is the process of preparation for this transfer. This includes the initial planning, the transfer itself and the support provided throughout, including that provided in adult care. Ideally, transition should be a purposeful and planned process. In reality, however, the process is often poorly actioned with frequent changes and unlearnt lessons. According to a recent Royal College of Physicians (RCP) report, only about half of adolescents and young adults with chronic disease receive any preparation for this transfer of healthcare (Willis & McDonagh, 2018).

## Communication

Picker's principles of care highlights the need for effective communication to provide patients and families with all the information required to ease their experience of healthcare. This includes both written and verbal communication.

In the imaging department, many radiology appointments will provide a written information leaflet to prepare patients for their upcoming procedure or scan. It is important that leaflets are sent for parents, and also in a way that the child will understand. This could be through pictures for children and simplified language for young adults.

For both parents and children, it is important to consider how written information can be accessible for all. This might be making the information available in different languages or in braille. In addition, children and young people are very familiar with video-based information, and it can be beneficial to explore sources such as YouTube. A video of the department and machines that patients can easily access at home provides a beneficial tool for preparing the expectations of the child.

The breadth of age within paediatric patients provides a challenge with verbal communication, which must be adapted based on the patient. The level of communication ability can be difficult to establish in a short interaction and may require adjustment of your language as you understand the child during the imaging or procedure. Establishing a rapport with a child can make a huge difference to the success or failure of a scan or procedure. A three-point identity check, required for all imaging studies, will likely be overwhelming for most children, but it can be broken into segments to allow the child to feel a valued part of the interaction. Asking "when is your birthday" instead of "date of birth" is an example of tailoring your communication to allow the child to be invested in their care.

Other services exist in the hospital which can be utilised to benefit a patient's interaction with the imaging department. Speech and Language Therapists (SLTs) can make recommendations of the best tools to communicate with patients, including non-verbal alternatives such as "Talking Mats" (Talking Mats, 2022) and other pictorial cues, symbols or signs. Some families will benefit greatly from the hospital's multi-faith chaplaincy. Whilst the chaplaincy cannot provide any medical advice or guidance, they can provide a safe space for patients and families to discuss options, especially in line with their faith or culture.

The role of play cannot be undermined. It can be a useful tool in daily practice and preparing children for invasive or uncomfortable imaging procedures. Play refers to structured activities designed with focus on the health condition, psychosocial and cognitive development of the child (Koukourikos et al. 2015). Play therapists play a crucial role in ensuring maximum benefit from play within the hospital setting and ensuring relevance to the child's personal experience.

Play can help contribute to the healthcare provider's understanding of the impact attending the imaging department can have on the child. At the same time it can contribute to the emotional development of the child and their confidence. It can also be beneficial in reducing anxiety and familiarising the child and guardian with imaging procedures. This is all crucial in empowerment of the family and turning a visit to the imaging department from a negative experience into a positive experience.

## Hospital passport

An established tool used within the adult sector especially within dementia services, is the use of a hospital passport. This system has been implemented in many specialist centres but is relatively new to paediatrics in general hospital settings. The hospital passport is a method to communicate information across the different services a patient will encounter whilst under the care of a medical team. Within paediatrics the aim therefore would be to enable the voice of the child and their family to be heard. Such passports would need to contain important information, such as previous medical history, medications and baseline health informatics of the child. But more importantly it would also contain information about how a child prefers to communicate; how they express pain and anxiety, their preferences etc. This is especially important with children who have long-term health care issues or those with neurodiversity such as autism.

To use this tool successfully the healthcare practitioner would read the hospital passport prior to every interaction with the patient so that they may build a rapport and put into place any required modifications. By doing this before every interaction, this would allow any updates to the patient's file to be made and represent the child at that moment rather than patient preferences at an earlier stage of their development. This system may also be beneficial to families whose first language is not English, and may struggle to find the right words to explain their child's needs. The hospital passport for a child would vary from those of adults as it may include pictures, and important information from services outside of the hospital including the child's school or social worker (Academy of Fabulous Stuff, 2022).

## Involvement of family and care-givers

Picker's principles discuss the impact of friends and families involvement and respecting patient's values and preferences. The balance and inclusion of these two values for paediatric patients can be challenging. A key factor in this is assessing the maturity and understanding of the child or young person. A psychologist, Piaget, formed a theory of cognitive development in the 1950s which described four main stages as seen in Table 7.1 (Pakpahan & Sariba, 2022).

Children will not be able to appreciate the value of a healthcare intervention until they reach formal operations stage, where they can visualise the future. For example, it is difficult for young children to understand that the discomfort of a cannula is needed in order to feel better overall. However,

*Table 7.1* Four main stages of cognitive development

| Age | Piaget's descriptor | Description |
| --- | --- | --- |
| 0–2 years | Sensorimotor | Understanding the world through sense and touch |
| 2–7 years | Pre-operational | Thoughts are driven by what a child sees in front of them |
| 7–11 years | Concrete operational | Learns rules and has some rational thinking but still tied to concrete objects |
| 12+ years | Formal Operations | Has logical thinking and reasoning. Can conceptualise the future |

in all children above the age of 2, it is important to take into account some of their preferences so that behaviour is improved. Up to 60% of children may show signs of changes in behaviour in the two weeks following surgery including bed-wetting, sleep disruption and withdrawal (Kotiniemi et al., 1997). To minimise these adverse behavioural issues post-hospitalisation, it is important to prepare children and respect their wishes as much as is practical. Studies have shown that children age 7 and upwards actively wish to receive comprehensive information about surgeries or intervention, particularly how much pain they may be in afterwards (Fortier et al., 2009). In line with Piaget's theory, pre-adolescent children are particularly interested in the environment they would be in – aligning the rational thinking to the concrete objects. It is also important to respect the viewpoint of children who do not wish to know anything. Given the option, most young children would prefer to not have the healthcare intervention and so balance must be struck between respecting the patient's values and preferences whilst also being able to provide healthcare in line with needs and care-giver wishes.

Alongside respecting preferences, is the issue of consent and capacity. In the United Kingdom, the assessment of capacity is termed Gillick competence, stemming from a legal case in 1986. If a young person can demonstrate that they understand the risks, benefits and long-term consequences of a medical intervention or procedure (for example, having an x-ray), then they are deemed to be Gillick competent. However, this must be assessed each time, and may differ. A young person may have capacity to consent for a routine x-ray, but perhaps not an invasive complex surgery. Any doubts of capacity or competence should be referred to specialist teams within the hospital for guidance.

In English law, a patient is presumed to have capacity to consent at 16, but this can be overridden by care-givers with parental responsibility (GMC, 2018). Confidentiality is also important, unless the disclosure is required by law, or the patient does not have the capacity or maturity.

## Family and care-giver needs

As Piaget's model shows, there is a clear need for involvement of family and friends for paediatric patients that will differ from the support structure required for adult patients. Babies and toddlers rely solely on their care-givers, for their basic needs such as food and comfort, and for all their decision-making. Younger children will need their care-givers to guide them and make decisions on their behalf as, whilst they can demonstrate preferences, they do not have the cognitive ability to make the decision themselves. Older children will also require support and assistance in their decision-making, particularly for more involved treatments or surgeries.

It is important to consider cultural and religious beliefs of the family when undertaking healthcare interventions for children. Religion and spirituality can provide great comfort and support for families in difficult situations, and although not always possible, it is imperative to respect traditions and beliefs where conceivable.

Needing to decide a healthcare decision for someone else can take a toll on the care-giver. Support services should be in place to allow the care-givers time and space to reflect on any decisions that need to be made.

## Physical comfort

Physical comfort is an important dimension of patient-centred care for children and young people. Physical comfort can directly impact experience for the patient, family and health care provider. The level of physical comfort experienced by the patient will impact their level of anxiety. As each age group has different needs, physical comfort needs may differ from patient to patient. It is important the healthcare provider considers the factors which influence physical comfort for the child or young person when attending the Imaging department.

It is firstly important to consider patient presentation when attending the Imaging department. The child or young person may be in pain or feeling unwell. They may not understand their symptoms. They may require assistance in performing daily tasks where previously they did not require assistance, which may be distressing or confusing for them.

Maintaining privacy is also crucial in achieving physical comfort for the child or young person, an aspect sometimes forgotten in paediatric imaging. Considerations such as offering a patient an appropriately sized gown so they maintain their dignity can be highly effective in building a relationship of trust and respect between the healthcare professional and patient.

Within the imaging examination room, achieving physical comfort can sometimes become challenging. The comfort of the imaging room will impact service user experience, such as room temperature and physical comfort of the equipment required for the examination. Examination rooms are clinical areas and as a result tend to be clinical in appearance. Simple measures can be implemented to improve this. Where possible, involving play specialists can be beneficial in reviewing the physical environment and identifying steps which can be taken to improve physical comfort throughout the environment. Other factors to consider include whether the child or young person may be required to remain still and in a particular position for a prolonged period of time (Bray et al., 2022). Particular aspects of imaging examinations can impact physical comfort such as intravenous cannulation, catheterisation, and insertion of feeding tubes.

The role of parent or guardian should be considered in securing the patient's physical comfort. The patient may seek physical comfort from the accompanying guardian. Depending on the requirements of the imaging examination this may not be possible, or may require some creative thinking to achieve this safely. To improve physical comfort, distraction techniques can be more beneficial in the younger age groups (Dastgheyb et al., 2018) . As the age of the child or young person increases, distraction techniques become less impactful and comforting positions/pharmaceutical comfort (analgesia) becomes more beneficial. Topical anaesthetic can be useful in maintaining physical comfort for the patient. Storyboards can also be beneficial in preparing the patient.

The impact of noises and lighting on patient comfort cannot be underestimated (Bosch-Alcaraz et al., 2018; Bray et al., 2022). For examination rooms using ionizing radiation, warning signs and lighting may cause anxiety for the child or young person. Limited comprehension due to age may add to this distress. It is important to consider the particulars of each imaging modality. For example, MRI can be an extremely noisy environment and therefore may be particularly challenging to those with auditory sensory sensitivities.

Sensory sensitivities, or sensory processing difficulties can also play a vital role. With sensory sensitivities the person may experience sensory inputs more or less intensely than others. Sensory sensitivities can be identified in children with other conditions, but can also be recognised in children with no known medical conditions. Children and young people with sensory sensitivities or sensory processing difficulties can find the hospital environment an overwhelming experience. The child or young person may be overly responsive to sounds, sights or touch, or under responsive to sights, touch or movements. A change in the person's routine can also have an impact on their experience and behaviours during their visit to the Imaging department. Studies have demonstrated the positive impact of educating healthcare staff, appropriate communication strategies, parental collaboration and changes within the physical environment (Wood et al., 2019). Gupta et al. (2019) also demonstrated the role of the sensory pathway within a tertiary hospital for patients with sensory processing difficulties. The key components of this sensory pathway include staff training, sensory toolkits and storyboards, collaboration with allied professionals and parental involvement. Sensory toolkits can be particularly beneficial within an imaging department. They typically compose of noise-cancelling headphones, fidget toys, light spinners, weight devices and blankets. Storyboards can also be a simple yet useful tool within the Imaging department. Storyboards are used to visually describe a process or procedure in a precise, step-by-step manner. They can be adjusted to meet the requirements of the patient.

## Conclusion

In conclusion, working with children and young people is a challenging but rewarding part of radiography practice. Patient-centred care for paediatric patients is complex due to the wide range of ages and understanding of the children and young people. Radiographers providing care for paediatric patients must also consider the care-givers, the physical environment and possible consequences to behaviour that a poor experience in healthcare can have on a developing child. To assist radiographers, the following points are suggested using the acronym "C.H.I.L.D.R.E.N."

### C is for Calm

It is vital to maintain a calm environment for all those present in the radiology room. This may be done by assessing and meeting the needs of patients, caregivers and staff. This might be appreciating that a child's pain management had not been optimised, or that the child is hungry or tired and may require a break from the radiological examination to meet these needs and try again.

### H is for Help

The medical imaging of children can be challenging. As radiographers this should not be done alone. Always ask for assistance where possible. This may be from fellow radiographers, associate clinical staff or the child's caregiver(s), to assist in the safe holding of the child during the examination.

### I is for don't Intimidate

Radiological examinations can be intimidating experiences for children and their caregivers. It is vital as radiographers we do all we can to make patients and their families feel as comfortable as possible. This may include: maintaining a clear welcoming environment; lowering ourselves to the child's eye level when speaking with them; allowing autonomy in the child's movements wherever possible; and keeping the child and their families informed at all times during their journey through their health care encounter.

### L is for Language

Child language development progresses through the years with variations in a child's understanding and ability to express themselves. Radiographers must always keep this in mind, adapting accordingly. This may include knowing when and when is not appropriate to ask questions of the child to determine the accuracy of clinical information, to considering the terminology being used in front of child and their families to avoid unnecessary stress within their visit to the clinical setting.

### D is for Distractions

A wide range of distraction techniques are available to radiographers to use during radiological examinations and procedures. The availability of smartphones and tablets has given clinical staff the opportunity to let the child watch their favourite film or TV show during a challenging time, to ease anxiety. Although infection control restraints may sometimes apply, radiographers should consider where possible the child's favourite cuddly toy may be able to stay by their owners side during different medical imaging experiences.

### R is for Rewards

Rewards for children such as stickers and certificates have been commonplace in healthcare settings for many years. Ensuring a good supply of rewards is vital to any department that cares for children. An important lesson for some professionals is to never offer a reward that is not in your power to give. For example, offering a child a trip to their favourite fast food restaurant or to be able to go straight home after their radiological examination etc., is an unreasonable reward to offer by a radiographer and would put pressure on caregivers who might not be able to action what has been offered.

### E is for Engage

The skill to build a rapport with a patient in the short period of time of a radiological examination is key to any radiographer's practice. Doubly so when working with children. Not only must a radiographer engage with the patient (the child) determining a gateway conversation starter such as a favourite cartoon character or sports team, but in addition must build an accord with the caregiver as well. By engaging with both child and caregiver, the radiographer will often ease the anxieties of both parties, allowing a smoother journey through the imaging department.

### N is for Needs

As the child being examined may vary in age from new born to 17 years old, a large range of communication levels may be encountered. Our patient may not always be able to express their needs, which may make the imaging process more pleasant for all involved. As radiographers it is important to be observant of the patient and communicate well with the caregivers to identify any unmet needs the child may have. Something as simple as changing a baby's nappy, giving them some food/drink or checking that appropriate pain relief has been given, maybe the difference between a successful examination and needing to stop imaging midpoint due to a distressed child.

## References

Academy of Fabulous Stuff (2022) Children's Hospital Passport. Available at: https://fabnhsstuff.net/fab-stuff/childrens-hospital-passport

Bosch-Alcaraz, A., Falcó-Pegueroles, A. and Jordan, I., 2018. A literature review of comfort in the paediatric critical care patient. *Journal of Clinical Nursing*, 27(13–14), pp.2546–2557.

Bray, L., Booth, L., Gray, V., Maden, M., Thompson, J. and Saron, H., 2022. Interventions and methods to prepare, educate or familiarise children and young people for radiological procedures: a scoping review. *Insights into Imaging*, 13(1), pp.1–33.

Brems, C., & Rasmussen, C. H., 2018. *A Comprehensive Guide to Child Psychotherapy and Counseling: Fourth Edition*. Waveland Press.

British Association of Paediatric Surgeons (2022) Map of Specialty Paediatric Surgical Services for the UK. Available at: www.baps.org.uk/patients/map-of-specialty-paediatric-surgical-services-for-the-uk/

Dastgheyb, S., Fishlock, K., Daskalakis, C., Kessel, J. and Rosen, P., 2018. Evaluating comfort measures for commonly performed painful procedures in pediatric patients. *Journal of Pain Research*, 11, p.1383.

Department of Health (2010) Delivering quality imaging services for children. A report from the National Imaging Board. London: DoH.

Fortier, M.A., Chorney, J.M., Rony, R.Y.Z., Perret-Karimi, D., Rinehart, J.B., Camilon, F.S. and Kain, Z.N., 2009. Children's desire for perioperative information. *Anesthesia and Analgesia*, 109(4), p.1085.

Foster, L.P. and Clark, K.R., 2019. Pediatric radiography education: a survey of recent graduates' experiences. *Radiologic Technology*, 91(1), pp.18–26.

Gupta, N., Brown, C., Deneke, J., Maha, J. and Kong, M., 2019. Utilization of a novel pathway in a tertiary pediatric hospital to meet the sensory needs of acutely ill pediatric patients. *Frontiers in Pediatrics*, 7, p.367.

Halliday, K., Drinkwater, K. and Howlett, D.C., 2016. Evaluation of paediatric radiology services in hospitals in the UK. *Clinical Radiology*, 71(12), pp.1263–1267.

Hardy, M., 2000. Paediatric radiography: is there a need for postgraduate education?. *Radiography*, 6(1), pp.27–34.

HCPC (2013) The standards of proficiency for radiographers. Available at: www.hcpc-uk.org/standards/standards-of-proficiency/radiographers/

Kotiniemi, L.H., Ryhänen, P.T. and Moilanen, I.K., 1997. Behavioural changes in children following day-case surgery: a 4-week follow-up of 551 children. *Anaesthesia*, 52(10), pp.970–976.

Koukourikos, K., Tzeha, L., Pantelidou, P. and Tsaloglidou, A., 2015. The importance of play during hospitalization of children. *Materia Socio-medica*, 27(6), p.438.

Lerwick, J.L., 2016. Minimizing pediatric healthcare-induced anxiety and trauma. *World Journal of Clinical Pediatrics*, 5(2), p.143.

Louw, A., Turner, A. and Wolvaardt, L., 2018. A case study of the use of a special interest group to enhance interest in public health among undergraduate health science students. *Public Health Reviews*, 39(1), pp.1–12.

NHS England (2022) Specialised services. Available at: www.england.nhs.uk/commissioning/spec-services/

NHS Providers (2022) The NHS provider sector. Available at: https://nhsproviders.org/topics/delivery-and-performance/the-nhs-provider-sector

Ortiz, M.R., 2018. Patient-centered care: Nursing knowledge and policy. *Nursing Science Quarterly*, 31(3), pp.291–295.

Pakpahan, F.H. and Saragih, M., 2022. Theory of cognitive development By Jean Piaget. *Journal of Applied Linguistics*, 2(2), pp.55–60.

Qi, Y., Yan, Y., Lau, S.S. and Tao, Y., 2021. Evidence-based design for waiting space environment of pediatric clinics – Three hospitals in Shenzhen as case studies. *International Journal of Environmental Research and Public Health*, 18(22), p.11804.

Rodriguez CM, Clough V, Gowda AS, Tucker MC., 2012. Multimethod assessment of children's distress during noninvasive outpatient medical procedures: child and parent attitudes and factors. *J Pediatr Psychol*, 37, pp.557–566.

Royal College of Paediatrics and Child Health, 2022. Child health inequalities driven by child poverty in the UK - position statement. www.rcpch.ac.uk/resources/child-health-inequalities-position-statement

Stogiannos, N., Carlier, S., Harvey-Lloyd, J.M., Brammer, A., Nugent, B., Cleaver, K., McNulty, J.P., Dos Reis, C.S. and Malamateniou, C., 2022. A systematic review of person-centred adjustments to facilitate magnetic resonance imaging for autistic patients without the use of sedation or anaesthesia. *Autism*, 26(4), pp.782–797.

Talking Mats (2022) For Health and Social Care. Available at: www.talkingmats.com/talking-mats-in-action/for-health-and-social-care/

UNICEF (2022) Unicef UK responds to UN Special Rapporteur's UK poverty report. Available at: www.unicef.org.uk/press-releases/unicef-uk-responds-to-un-special-rapporteurs-uk-poverty-report/amp/

Venkataraman, V., Browning, T., Pedrosa, I., Abbara, S., Fetzer, D., Toomay, S. and Peshock, R.M., 2019. Implementing shared, standardized imaging protocols to improve cross-enterprise workflow and quality. *Journal of Digital Imaging*, 32(5), pp.880–887.

Willis, E.R. and McDonagh, J.E., 2018. Transition from children's to adults' services for young people using health or social care services (NICE Guideline NG43). *Archives of Disease in Childhood-Education and Practice*, 103(5), pp.253–256.

Wood, E.B., Halverson, A., Harrison, G. and Rosenkranz, A., 2019. Creating a sensory-friendly pediatric emergency department. *Journal of Emergency Nursing*, 45(4), pp.415–424.

# 8  Person-centred care in DXA and secondary fracture prevention

*Jill Griffin and Karen Knapp*

## Person centred care in DXA services

Dual energy X-ray absorptiometry (DXA) equipment is used to generate measurements of bone mineral density (BMD) and in the UK is regulated under the Ionising Radiation (Medical Exposure) Regulations (IR(ME)R) 2017 in the same way as other x-ray examinations. It is, however, unlike other imaging procedures in that the diagnosis is not informed by the *image* but the BMD *measurements*. There are therefore greater requirements for accuracy, precision, and minimisation of precision errors to ensure the measurements and their application inform appropriate care for the individual patient based on their unique clinical and lifestyle risks associated with bone health. It is for these reasons that DXA should be regarded as a specialist technique that relies wholly upon the knowledge, skills and competency of the operator and reporting practitioner.

A large percentage of the population attending for DXA scans are older men and women, though there are younger people who also require DXA, including oncology populations, particularly, or those with breast and prostate cancer those with other secondary causes of osteoporosis. It is therefore essential that patient care in DXA is able to adapt to a range of patients from paediatric patients to older adults and support and those with impaired cognition and comorbidities.

## Pre-scan information and patient preparation

Patients are frequently concerned about coming for a DXA scan and often associate "scan" with going inside a "tunnel". Optimising person-centred care therefore starts before the person enters the department. Providing a clear letter with not only the time, but date and where to attend, coupled with explaining the procedure is essential. Pictures of the DXA scanner and department can help patients to feel more comfortable about coming for their scan. If particular clothing is required, for example avoiding zips, metal and buttons, then this should be included in the letter, but it should also be considered that not all people have the right clothing and that assurance of the ability to change into a gown if needed should also be included.

DOI: 10.1201/9781003310143-10

Patients attending osteoporosis services and DXA scans have frequently had a fracture or multiple fractures, often as a result of low or minimal trauma and this has been reported to have a negative impact on their quality of life (Tarride et al., 2016). Many people who have sustained low trauma fractures live in fear of having further fractures *"on a daily basis and we live in constant fear of sustaining another fracture"* and many find scanners uncomfortable (Arnold et al., 2018).

Patients' with fragility fractures have described their needs as (ROS, APPG, 2023):

- to live without fear of fracture;
- to be able to carry out their activities of daily living;
- to live a normal and enriched life;
- to understand their fracture risk;
- to feel informed how they can live well with advice and support.

### A patient-friendly waiting room

Creating a person-centred environment in the department is essential and co-creation with users of a service is an important step to making the environment as person-centred as possible; this can include the information patients receive prior to their scan, the physical environment and the format of the results they receive. Many patients with vertebral fractures struggle to sit comfortably in chairs, so consideration of chairs that provide support, arms to assist with sitting to standing and vice-versa, and offer comfort for those with tender spines. When planning a new department or updating a department, including patients in the design should be considered essential.

Infographics and posters are an appropriate addition to waiting areas to help convey information to patients awaiting their scan. There is evidence to suggest that the co-creation of such resources with patients is a powerful way to ensure they are appropriate; these can also help patients in communicating with healthcare professionals (Piil et al., 2023).

### Preparing for the scan

The bone density measurement achieved from a DXA scan informs clinical recommendations in combination with clinical risk factors. As such, DXA scanning in clinical practice must revolve around the individual patient and at every point in the clinical pathway. This is so that reliable and accurate measurements that inform diagnoses are made. clinical interpretation of the measurements consider the individuals' medical and personal context, and the clinical recommendations made in the DXA report are based on all these things. Additionally, when acquiring the scan, this is achieved in collaboration with the patient to optimise choices of anatomical scan site, scan mode and positioning, taking into account the individual's needs, medical history, body habitus, capacity and capability.

## The questionnaire

It is best practice to make use of a patient questionnaire at the DXA appointment to support discussions with the individual patient and take a focused history for their individual risk of fragility fracture. The questionnaire may also be used as a prompt and method of recording statutory radiation protection information such as confirmation of identity and inclusive pregnancy policies. It also provides the opportunity to discuss with and understand the patients' needs and goals for the examination in terms of bone health and any onward referral recommendations, such as to a falls service for those reporting falls. Providing the questionnaire along with the appointment letter allows the patient to discuss their answers with their family or carers if needed, which can be particularly useful for those with cognitive impairment or language barriers.

For those with language barriers, facilitation of a translator, family member or carer who can translate is important to enable the examination to progress with the patient feeling comfortable and welcome, along with being able to answer the questions regarding clinical risk factors. Likewise, for those with hearing impairment, a sign language translator may be appropriate if the patient requests or needs one.

Patients with additional needs such as those with learning disabilities, neurodiversity and children should be appropriately supported before they attend. This could include the use of longer scan slots, visits to the department prior to the day of their scan and the use of play specialists, where appropriate, to assist the smooth running of the scan, to ensure a positive patient experience.

The information gained from the questionnaire is essential and the focused history enables the clinician interpreting and reporting the DXA measurements, to be able to consider factors contributing to low bone mass or increase facture risk, so that reporting advice regarding further investigations, supplementations and treatments are tailored for that patient's risk rather than a single factor when applying the diagnostic thresholds. The inclusion of the questions relating to osteoporosis treatments is also good practice. Osteoporosis treatments are often taken incorrectly, are reported to have poor adherence (Fry, 2019) and gaining information relating to how often, when, how and the total duration of treatment is essential to underpin appropriate reports (Edwards S., 2023). The process of discussing this with the patient enables guidance to be given, for example if a patient describes taking an oral bisphosphonate in the morning with a cup of tea, then the radiographer can provide guidance on the correct way to take the medication at this point and also alert the reporter or referrer, to potential reasons for any treatment failure identified on the scans.

## The scan

The DXA scan requires patients to be measured and weighed prior to their scan so that this can be input into the scanner to help guide the scan mode used. For some patient groups, this can be a daunting prospect. For those with eating disorders, blinded weight checks may be the norm with their clinicians, so care discussing whether they want to know or be told their weight is important (Forbush et al., 2015). There are also key indicators of potential vertebral fractures in patients with height loss and it is therefore useful to ask the patient their height at the age of 21 years or as a young adult on their questionnaire, so that height loss can be assessed. A height loss of four or more centimetres can be an indicator of vertebral fracture and can indicate the need for a vertebral fracture assessment (VFA) scan to be undertaken in addition to the standard DXA scan in these patients (Schousboe et al., 2008).

The weight of the patient is also essential to assess whether they exceed the scanner weight limit. Some scanners have high weight limits of around 250kg, while others are less than half of this weight. It is essential that the radiographer or operator knows the weight limit of the scanner they are using. Conversations with patients who exceed the weight limit can be difficult and there are different options available in this situation. It may be possible to refer the patient to another centre where they have a scanner with a higher weight limit. Alternatively, the forearm can be scanned with the patient sat by the scanner and this can provide some information on their bone density, although this is a poorer predictor of fracture risk than either the spine or hip bone density measurement (Leslie et al., 2007; Marshall et al., 1996). Considering a holistic clinical picture of the patient may enable a treatment decision to be made based on their clinical risk factors, particularly if they are over 75 years of age. It would be ill advised to request the patient to lose weight and return once they are within the weight limit of the scanner. This could delay treatment and result in subsequent fractures if they have a high fracture risk.

It is important that non-ambulatory patients can access DXA scans. Wheelchair users are particularly at risk of lower limb fractures due to disuse osteopenia and as such require bone assessment and appropriate treatment if required (Knapp et al., 2010; Smith and Carroll, 2011). While some wheelchair users can transfer easily with minimal assistance, others will not be able to, and a hoist and safe working practices for hoisting should be available in departments for these individuals. Excellent communication skills are required when using a hoist to reassure the patient, particularly if they are not used to being transferred using a hoist.

The DXA scanner couch can be uncomfortable for some patients and it's essential to remember that those with a kyphosis resulting from their vertebral fractures cannot lie flat. Therefore sufficient pillows at hand to ensure their head is supported is essential (Arnold et al., 2018). Accurate positioning is key in DXA to ensure that scans are precise and diagnostic. This means elevating the patient's legs onto a foam block for the lumbar spine scan and internally rotating the leg for the hip scan. Both positions can cause discomfort for some and it's essential that good communication is maintained throughout the positioning and the scan. Looking at the patient's face as they are being positioned can check for signs of discomfort. This is also key because some patients are very stoic and will avoid articulating their discomfort. If the patient is unable to maintain the optimum positions, it is essential that this is recorded for the reporter of the scan and for future follow-up scans so that the positioning can be as close to the original as possible to reduce precision errors (Knapp et al., 2012).

The scan analysis is an essential aspect of DXA and radiographers must not only focus on the technical features of this, but also review the image for indicators of pathology. Vertebral fractures can often be seen as vertebrae with a reduced height on the lumbar spine DXA and if a vertebral fracture is suspected, it is essential to follow this up. A vertebral fracture is a strong predictor of future vertebral and hip fracture and as such must be identified so that appropriate treatments can be commenced (Schousboe et al., 2019).

A vertebral fracture is so important that the department should have a protocol to undertake vertebral fracture assessment (VFA) scans on those who meet a protocol based on probability of risk, in addition to those where there is a significant clinical suspicion (Schousboe et al., 2008). If a new vertebral fracture is identified on the low resolution VFA scan, it is critical that further imaging is undertaken to confirm and characterise the fracture (Knapp et al., 2018). In the ideal person-centred setting, a direct referral for a radiograph should be made at the time of the scan, with the patient able to undergo the imaging while they are still on site. This reduces the need for additional hospital visits, but also means a definitive report with treatment recommendations, which includes any confirmed vertebral fractures. However, this requires the scanning radiographer to be educated as and delegated as a referrer and to be given sufficient time in the examination process to identify any possible vertebral fractures and be able to explain to the patient why they need a radiograph. This should not be taken lightly, because informing a patient that they may have a vertebral fracture, which needs further investigation, is an unexpected and alarming diagnosis for many. Patients may believe that vertebral fractures are related to paralysis and a lot of anxiety can be created in any discussion around this area, so the radiographer must have appropriate education, training and time to have these discussions in a meaningful way with the patient.

### Advice and guidance

During the scan, the radiographer remains in the room with the patient and this provides an opportunity for conversation. Many patients will ask questions about the best way to protect their bones, whether their children should be concerned about osteoporosis, their diet and a range of other queries. It is essential that DXA radiographers are educated in the pathophysiology of osteoporosis, the clinical risk factors and basic details on dietary calcium intake and Vitamin D so that they can support patients in their lifestyle choices. Having leaflets to hand or websites that patients can access for further information is essential. Weight-bearing exercise is important for bone density, but it's important that patients are advised regarding appropriate exercises for those suffering from osteoporosis. The Royal Osteoporosis Society has published guidance in their "strong, steady and straight" document and patient can be signposted to this (Brooke-Wavell et al., 2022). There is evidence that small bouts of exercise of 2–3 minutes can be beneficial for bone in those who undertake this habitually (Stiles et al., 2017), but further research is required to explore how this can impact on those taking up exercise to protect their bones.

During the examination, there is an opportunity for radiographers to discuss with patients how they are taking their medication. Oral bisphosphonates need to be taken on an empty stomach (most people do this first thing in the morning) with a large glass of tap water. Nothing else must be taken with the oral bisphosphonate for 30 minutes following taking the medication and patients must remain upright during this period to reduce the potential of upper gastro-intestinal problems such as oesophagitis, a common adverse effect resulting from oral bisphosphonates (Pazianas and Abrahamsen, 2011; Poole and Compston, 2012). Some patients will disclose taking their medication incorrectly, for example, with a cup of tea in the morning, and this provides an opportunity for the radiographer to discuss the correct method for oral administration.

### Just a number – a patient experience

My journey navigating through the GP and hospital services for diagnosis and treatment of osteoporosis.

Following two GP visits and two hospital visits at the accident and emergency department due to severe pain from coughing due to flu, I managed to request and was approved for a CT scan. I needed to get to the bottom of why I was having so much pain. The visits to these healthcare centres caused me anxiety and this played on my mind.

The CT scan showed a healing left sixth rib fracture probably due to coughing. This led me to search the internet as it seemed to me something was not quite right. The search led me to NHS and Royal Osteoporosis Society pages and I found out that I probably have osteoporosis as a fractured bone without trauma is most probably one of the reasons for osteoporosis.

I made an appointment with my GP for two months later and discussed this with my GP. He said: "You probably have osteoporosis with your recent fracture and age". I then requested a DXA scan. I was not given any information by GP at this point and came out feeling deflated. I needed to know about the DXA scan as this worried me. Would I need to undress? Would I need to be Nil by Mouth and questions kept coming into my mind with no answers.

At this point, while waiting for a DXA scan, I stopped all my activities. I was scared that I would break my bones and this would impact on my activities of daily living. The walks I enjoyed came to a standstill and so did my shopping trips and work around the house. I lived with this fear of what osteoporosis was and what my prognosis would be.

A month later, I went to the hospital to keep my DXA scan appointment. I was given instructions to lie down and told the machine would pick up images. While having this done, my mind kept thinking – what information if any will I get from the scan? After the DXA scan all I was told was that the results would be sent to the GP. I asked what DXA scan measured and I was told the bones and that was the end of my session. It really was not a good session in terms of giving me more information about DXA scan and my mental anxiety started to accelerate. I feel that the radiographers need to be trained to support the patients coming through their clinic in answering any queries appropriately. To support patients in alleviating anxiety associated with this diagnosis.

Another month went by and I received a letter from GP informing me that I had osteoporosis and weak bones and I needed to make an appointment. A month later, the first available GP appointment was given to me. During this interim period, I realised that I now have osteoporosis. The swimming sessions had to stop I thought to myself. Questions like: "What can I do to prevent fractures?" "Will my bones get better?" "Will I have to take medications for a long time?". "Will I be offered another DXA scan?" Unfortunately none of these questions were answered except being told at this appointment that I have osteoporosis and to start Alendronic acid weekly. No discussion was forthcoming and I felt that I was just a number in a queue of patients waiting to be seen.

I needed a short talk about my condition with an opportunity to ask questions but this was not given. I left the GP surgery feeling this is not good. Where do I get these information from and my several telephone calls with ROS managed to answer most of my queries. I felt I was on my own on this journey with no support whatsoever. This is a lonely journey for me. The word osteoporosis never left my mind and I was constantly thinking day and night about the diagnosis and treatment.

## Conclusion

My anxiety levels kept going up and this affected the way I conducted my work. Every time I had to pick up the phone to call my GP, go to see him or call the hospital, I would force myself to remain calm. Inside I was petrified not knowing what sort of reception I would receive and whether my queries would be answered appropriately.

## Red flags

There are important red flags which patients might disclose relating to their anti-resorptive bone medication, which need urgent flagging with the person who is reporting the scan, so that follow-up can occur. Bone sparing therapies such as bisphosphonates and Denosumab are associated with both atypical femoral fractures and osteonecrosis of the jaw (Querrer et al., 2021p Rudran et al., 2021), but both of these are rare events. However, it is essential if a patient discloses lateral thigh, groin or hip pain for three or more weeks that this is investigated with bilateral femoral and pelvis radiographs. Some scanners have the ability to undertake opportunistic screening for atypical femoral fracture using long-leg DXA scans, which can identify any "beaking" in the lateral cortex, which is a feature of incomplete atypical femoral fracture (Smith et al., 2023). Further imaging is still required to provide a higher quality image for treatment planning, which may include prophylactic orthopaedic surgery or conservative treatment (Shim et al., 2023; Rudran et al., 2021; Knapp et al., 2018).

Discussions around poor dental health, problems with gums or teeth should be taken seriously. While osteonecrosis of the jaw is a rare side effect associated primarily with oncology-related bisphosphonates, where higher doses are prescribed, any issues raised by a patient needs mentioning in their report so that appropriate dental investigations and care can be sought (Wan et al., 2020).

## Safeguarding and Elder abuse

Patients attending DXA services include those who are elderly, frail and vulnerable. It is important that the radiographer is aware of the potential of elder abuse and neglect because these are frequently overlooked. Elder abuse can be physical, sexual, emotional or financial (Switzer and Michienzi, 2012). Elder abuse has been reported in 13% of patients presenting to the emergency department and is more likely to be linked to skull and rib fractures than femur, foot or ankle fractures (Gardezi et al., 2022). There is a positive correlation between abuse of the elderly and their activities of daily living index (Aslan and Erci, 2020) and in those with cognitive impairment and/or a history of domestic violence (Switzer and Michienzi, 2012; Kavak and Özdemir, 2019). Patients, when away from their carers, may disclose abuse to the radiographer and it is essential that any such disclosure is documented verbatim and referred on using the internal safeguarding processes within the healthcare setting to raise these concerns.

## Concluding the examination

Concluding the examination is as important as the rest of the process leading up to this point. If any glasses or hearing aids have been removed (for total then these should be given back to the patient before they move from the scanner couch. Patients may need assistance sitting up and getting off the bed and it's important that the radiographer assists within their manual moving and handling limits and ensures they do not put themselves at risk in the process. As many of the patients are elderly, they may suffer from blood pressure changes moving from laying down to a vertical position and it's important to allow them time to sit on the edge of the bed if they feel lightheaded. Some patients may need assistance with putting their shoes back on after the examination if they have been required to remove them.

Clear instructions regarding when and how to get their results are essential. If the results will take three weeks to report, it is essential that the patient is told this so that they are not making wasted appointments with their doctor for their results. Equally, if a patient has a clinic appointment soon after the scan, it is essential to know this so that the results can be reported in time for the appointment. This would usually be achieved by flagging the urgency with the reporter through internal processes.

If advice has been provided for the patient during the scan consultation, then it can be useful to provide leaflets or for those who have the internet available, websites where they can look up further information or for an aid memoir regarding what the radiographer has advised. The Roayl Osteoporsis Society has a suite of information resources in differnt formats to support patietns understanding of diagnosis, treatment and lifestyle.

Some centres provide education sessions for people living with osteoporosis and this multi-disciplinary approach can be useful in helping patients manage their own health and adjust to their new diagnosis. Having professionals to discuss treatment options, diet, exercise, diagnosis and scan intervals, and falls prevention is helpful. Osteoporosis support groups also exist in many places and in the UK, the Royal Osteoporosis Society helps support their network of support groups. These are a valuable meeting space for people suffering from osteoporosis to meet others who are living with osteoporosis and to develop a peer support network. Signposting to the local support group if there is one, is an important role for the radiographer, particularly for patients who indicate they would like to make contact with others.

While most settings do not give results at the time of the scan, there are some who do and this needs to be done in a suitable setting with appropriately trained staff who can give the patients all of the information they require about managing their disease.

### Following the examination

Unlike many other imaging modalities, DXA requires careful analysis and it is essential that this is performed in line with scanner manufacturers' and local guidelines to ensure optimum scan quality for reporting. Any notable aspects from the scan or conversations with the patient need to be included in the information for the reporter to ensure that any red flags are followed up and all the information required to optimise the report is available to them.

Some patients attending DXA services will be on immunosuppressant medications and it is essential that they have a safe environment for their scan. Some of these patients will be more vulnerable to viruses and infections, so it's essential that good infection prevention and control is in place with handwashing and cleaning to reduce the risk of hospital-acquired infections. Many viruses are airborne and staff should be respectful if these patients request to wear masks and for staff to mask to ensure they feel safe for their scan (Lu et al., 2023).

In conclusion, DXA provides the opportunity for person-centred care through not only optimising the scan experience for the patient, but also in exploring their wider clinical history and listening to any concerns they have. The DXA radiographer needs to be well educated to provide advice to the patient on top of providing an excellent scan experience.

## References

Arnold, J., Ewings, P., Handel, L., Langd On, M., Powell, H., Rhydderch-Evans, Z., Stone, W., Woodland, S., Hoade, L. & Bleiker, J. 2018. Fragile: Please handle with care. *Radiography*, 24, S9–S10.

Aslan, H. & Erci, B. 2020. The incidence and influencing factors of elder abuse and neglect. *Journal of Public Health*, 28, 525–533.

Brooke-Wavell, K., Skelton, D. A., Barker, K. L., Clark, E. M., De Biase, S., Arnold, S., Paskins, Z., Robinson, K. R., Lewis, R. M. & Tobias, J. H. 2022. Strong, steady and straight: UK consensus statement on physical activity and exercise for osteoporosis. *British Journal of Sports Medicine*, 56, 837–846.

Edwards S, S. C., Knight K, Knapp K. M. 2023. Improving understanding of patient treatment duration and adherence using a structured pro-forma during dual-x-ray absorptiometry scan visits. *Journal of Bone and Mineral Research Plus*, In Press.

Forbush, K. T., Richardson, J. H. & Bohrer, B. K. 2015. Clinicians' practices regarding blind versus open weighing among patients with eating disorders. *International Journal of Eating Disorders*, 48, 905–911.

Fry, K. R. 2019. Why Hospitals Need Service Design: Challenges and Methods for Successful Implementation of Change in Hospitals. *Service design and service thinking in healthcare and hospital management: Theory, concepts, practice*, 377–399.

Gardezi, M., Moore, H. G., Rubin, L. E. & Grauer, J. N. 2022. Predictors of Physical Abuse in Elder Patients With Fracture. *JAAOS Global Research & Reviews*, 6, 1–8.

Kavak, R. P. & Özdemir, M. 2019. Radiological appearance of physical elder abuse. *European Geriatric Medicine*, 10, 871–878.

Knapp, K. M., Rowlands, A. V., Welsman, J. R. & Macleod, K. M. 2010. Prolonged unilateral disuse osteopenia 14 years post external fixator removal: A case history and critical review. *Case Reports in Medicine*, 2010:629020. doi: 10.1155/2010/629020. Epub 2010 Apr 21. PMID: 20445732; PMCID: PMC2858376.

Knapp, K. M., Welsman, J. R., Hopkins, S. J., Fogelman, I. & Blake, G. M. 2012. Obesity increases precision errors in dual-energy X-ray absorptiometry measurements. *Journal of Clinical Densitometry*, 15, 315–319.

Knapp, K. M., Meertens, R. & Seymour, R. 2018. Imaging and opportunistic identification of fractures.

Leslie, W. D., Tsang, J. F., Caetano, P. A., Lix, L. M. & Program, M. B. D. 2007. Effectiveness of bone density measurement for predicting osteoporotic fractures in clinical practice. *Journal of Clinical Endocrinology & Metabolism*, 92, 77–81.

Lu, Y., Okpani, A., Mcleod, C., Grant, J. & Yassi, A. 2023. Masking strategy to protect healthcare workers from COVID-19: An umbrella meta-analysis. *Infection, Disease & Health* Aug;28(3), 226–238. doi: 10.1016/j.idh.2023.01.004. Epub 2023 Feb 16. PMID: 36863978; PMCID: PMC9932689.

Marshall, D., Johnell, O. & Wedel, H. 1996. Meta-analysis of how well measures of bone mineral density predict occurrence of osteoporotic fractures. *Bmj*, 312, 1254–1259.

Pazianas, M. & Abrahamsen, B. 2011. Safety of bisphosphonates. *Bone*, 49, 103–110.

Piil, K., Pedersen, P., Gyldenvang, H. H., Elsborg, A. J., Skaarup, A. B., Starklint, M., Kjølsen, T. & Pappot, H. 2023. The development of medical infographics to raise symptom awareness and promote communication to patients with cancer: A co-creation study. *PEC Innovation*, 2, 100146.

Poole, K. E. & Compston, J. E. 2012. Bisphosphonates in the treatment of osteoporosis. *Bmj*, 344, 1–3.

Querrer, R., Ferrare, N., Melo, N., Stefani, C. M., Dos Reis, P. E. D., Mesquita, C. R. M., Borges, G. A., Leite, A. F. & Figueiredo, P. T. 2021. Differences between bisphosphonate-related and denosumab-related osteonecrosis of the jaws: a systematic review. *Supportive Care in Cancer*, 29, 2811–2820.

Royal Osteoporosis Society. 2023. All Party Parliamentary Report. https://strwebprd media.blob.core.windows.net/media/31tbj2dt/appg-on-osteoporosis-and-bone-health-fls-inquiry-inquiry-report-2021.pdf

Rudran, B., Super, J., Jandoo, R., Babu, V., Nathan, S., Ibrahim, E. & Wiik, A. V. 2021. Current concepts in the management of bisphosphonate associated atypical femoral fractures. *World Journal of Orthopedics*, 12, 660.

Schousboe, J. T., Vokes, T., Broy, S. B., Ferrar, L., Mckiernan, F., Roux, C. & Binkley, N. 2008. Vertebral fracture assessment: the 2007 ISCD official positions. *Journal of Clinical Densitometry*, 11, 92–108.

Schousboe, J. T., Lix, L. M., Morin, S. N., Derkatch, S., Bryanton, M., Alhrbi, M. & Leslie, W. D. 2019. Prevalent vertebral fracture on bone density lateral spine (VFA) images in routine clinical practice predict incident fractures. *Bone*, 121, 72–79.

Shim, B.-J., Won, H., Kim, S.-Y. & Baek, S.-H. 2023. Surgical strategy of the treatment of atypical femoral fractures. *World Journal of Orthopedics*, 14, 302.

Smith, D., Knapp, K., Wright, C. & Hollick, R. 2023. Dual energy X-ray absorptiometry (DXA) extended femur scans to support opportunistic screening for incomplete atypical femoral fractures: A short term in-vivo precision study. *Journal of Clinical Densitometry*, 26, 101352.

Smith, É. & Carroll, Á. 2011. Bone mineral density in adults disabled through acquired neurological conditions: a review. *Journal of Clinical Densitometry*, 14, 85–94.

Stiles, V. H., Metcalf, B. S., Knapp, K. M. & Rowlands, A. V. 2017. A small amount of precisely measured high-intensity habitual physical activity predicts bone health in pre-and post-menopausal women in UK Biobank. *International Journal of Epidemiology*, 46, 1847–1856.

Switzer, J. A. & Michienzi, A. E. 2012. Elder abuse: an update on prevalence, identification, and reporting for the orthopaedic surgeon. *JAAOS-Journal of the American Academy of Orthopaedic Surgeons*, 20, 788–794.

Tarride, J.-E., Burke, N., Leslie, W. D., Morin, S. N., Adachi, J. D., Papaioannou, A., Bessette, L., Brown, J. P., Pericleous, L. & Muratov, S. 2016. Loss of health related quality of life following low-trauma fractures in the elderly. *BMC Geriatrics*, 16, 1–11.

Wan, J. T., Sheeley, D. M., Somerman, M. J. & Lee, J. S. 2020. Mitigating osteonecrosis of the jaw (ONJ) through preventive dental care and understanding of risk factors. *Bone Research*, 8, 14.

# Section three

# Person-centred care in emergency settings

# 9 Beyond general radiography

## Patient-centred care in the specialty environment of the emergency department

*Michael Fuller and Lynne Ingram*

## Introduction

It is easy for hospital staff to overlook a patient's apprehension and uncertainty when entering a hospital. Some patients' only experience of a hospital might have been when a friend or relative was unwell or died, and their subsequent hospital visits can be tainted by this negative memory. A visit to hospital can trigger patients to relive unrelated traumatic events. For other patients a hospital emergency department may only have been experienced through the medium of television: the "Grey's Anatomy effect" (Serrone et al., 2018). The reality shock of a real emergency department can present a jarring experience.

## The patient in the emergency department

### Meeting the patient's basic needs

Person-centred care is focused on meeting the patient's all-round needs. Maslow (1943) proposed that the most basic of these needs are physiological needs such as breathing, food, water, sleep, and homeostasis.

It is common for patients in airconditioned buildings to feel cold (Robinson and Benton, 2002). Asking a patient if they are feeling cold is addressing one of the patient's common physiological needs. If your department has a blanket warmer, make good use of it. The smile on a patient's face after you provide them with a warm blanket is an indication of how important this simple act can be. It is more effective to pull back the existing patient blanket, lay the warm blanket over the patient, and then put the existing blanket back over the top of the warm blanket to keep the warmth in. Providing a warm blanket can relax muscles and reduce pain as well as providing a feeling of tangible comfort during a stressful time.

Another level of human need that is especially relevant in a hospital setting is a sense of safety and security. A patient who presents to the emergency department is often experiencing health and wellbeing concerns and, less obviously, may also be concerned about the potential impact of their illness

DOI: 10.1201/9781003310143-12

or injury on their employment and their family's financial security. An elderly patient may be facing the possibility of moving from independent living to nursing home care: this can be a daunting prospect. A patient's imaging in the emergency department is the point at which the extent of an injury or illness and the potential impact on the patient and their families' lives is going to be revealed. It is entirely expected and reasonable for the patient to be experiencing high levels of stress and they may not be focused on what you are doing and saying. The emergency department is different to other areas of patient care because the patients often do not already have a confirmed diagnosis.

### The gentle art of appropriate patient empathy while multitasking

It is a critical personal quality and practice for radiographers to empathise at the appropriate level with the patient. You will find that you naturally empathise at a higher level with patients who remind you of your partner, child, or parent. These key age groups will change as you progress through life. It is an important part of patient care to find the balance between concern for the patient's welfare and focus on the technical task; too great a focus on one may be at the expense of the other. Radiographers working in the emergency department will engage in multitasking and experience rapid task-switching and interruptions. These are challenges that will inevitably require some level of mastery. Student radiographers may find this difficult as they focus on learning the technical aspects of radiography.

### The stress of waiting and uncertainty

Distress for patients can increase significantly if there are delays in being triaged, examined, diagnosed, and treated in the emergency department. Waiting in a busy emergency department waiting area with other sick and distressed patients can be a miserable experience. It can be comforting for patients to verbalise the stress of waiting with you and for you to acknowledge and discuss their experience.

*A patient satisfaction survey was undertaken in a hospital waiting room to gauge the feelings of patients waiting to be seen. One patient described the waiting area as a "pit of misery".*

If a patient has been waiting for a long time, it is preferable to say "thank you for waiting" rather than "sorry for the wait". The choice of wording emphasises that their patience is genuinely appreciated. The phrase "sorry for the wait" also implies that you were at fault, which is normally not the case.

Patients may be seen quickly by the doctor or nurse on arrival in the emergency department then wait for a much longer period to be called for their X-ray examination. During this period, the patient's anxiety and list of questions grows and can be blurted out to the radiographer. When patients are waiting long periods to be seen they will, of course, be very focused on their own injury or illness and may not be aware of the presence of other patients with more urgent needs. Patients waiting for a protracted period can seem to develop a *bubble* around themselves, and the longer they wait, the more impenetrable the bubble becomes. Keeping this in mind will make it easier to be understanding and validating of the patient's experience.

### The distracted patient

A common legacy of a patient's unplanned visit to the emergency department is a high degree of mental distraction. When you experience a patient doing something that might seem odd or unusual, it is worth considering the level of mental distraction and stress they are experiencing. Their day may have started like any other day and taken a sudden and unexpected turn for the worse. A chest X-ray examination might be a very common task for the radiographer, but the patient may be facing the realisation that they have had a heart attack and should not be expected to behave as if it is a normal day in their life.

*The duck pond at the front entrance of my hospital was removed because of the number of distracted patients accidentally falling into it.*

### Patient advocacy, patient safety, the multidisciplinary team and situational awareness

The practice of medicine has evolved into a somewhat fragmented system characterised by increased role specialisation. We see evidence of this in medical imaging with the development of ultrasound imaging, computed tomography (CT) scanning, positron emission tomography (PET), magnetic resonance imaging (MRI) and nuclear medicine. It is easy for radiographers to be focused on the technology of their specialisation at the expense of the *big picture*. Patients are sometimes described as being on a *journey* (although they don't always like to hear this jingoistic terminology) and the radiographer is part of the team that takes them on that journey.

There are times when it is particularly important for a radiographer to step back and consider their role in the context of the roles and activities of the staff around them: this is referred to as *situational awareness* (Calder et al., 2018). This is an important consideration in the resuscitation/emergency room where multiple staff with different roles are undertaking different activities on a critically ill or injured patient (often blocking your access to the patient). The team leader may ask you to step in and "take a chest X-ray". There may be occasions when you are watching the activities of the other staff and it appears that a chest X-ray is the next most important activity. When this happens, check with the team leader then continue with a high level of awareness of what is going on around you. Be prepared to stop what you are doing and step back if needed. This is a very fluid environment where you are required to act, react, and adapt quickly.

Patient advocacy is important. This refers to your role as an advocate and protector of the patient in a more holistic sense. This is reflected in the phrase "if you see something, say something". There is no longer a place for the notion of *I'm just the radiographer*. Caring for the patient is your business, and you may be required to step outside the traditional confines of your role in order to fulfil this need. A holistic approach to your role can be particularly important if you work in an isolated or remote location. Staff in isolated and remote areas may be required to engage in activities and patient advocacy that would not normally be seen as part of their role in a larger metropolitan healthcare facility.

### Technology, efficiency and patient care

With an ever-increasing workload, radiographers may be torn between the need for speed and patient care. As a guiding principle, an unacceptable quality of imaging and patient care can rarely be justified by a high workload. In the battle between speed and effectiveness, effectiveness should take priority. It should also be remembered that the "small touches" (such as a kind word) are often the first casualty of a busy shift in the Emergency Department.

### Non-verbal communication

It is very easy to overlook the impact of your facial expressions on a patient's level of anxiety. You may be unaware that your facial expression belies what you are saying. Your words may convey one sentiment ("everything is ok") and your facial expression may suggest the opposite. An ill-timed grimace or look of concern may completely contradict your words and ultimately be the greater influence on your patient's level of stress. In the same way that an injured child looks to a parent's face for cues (does this warrant tears?), the patient will look at the radiographer's face for cues. This is an aspect of non-verbal communication that can occur in the emergency department more than other areas of imaging.

### Judging patients

It is common to make judgements about a person by categorising them in some manner based on their appearance or behaviour. Patients commonly arrive at the emergency department unexpectedly. They may have an unpleasant body odour, ungroomed hair or bad breath. They may be under-dressed (pyjamas, swimwear, gym clothes) or overdressed (formal evening wear). The patient may be uncomfortable and embarrassed by their appearance despite the distress of their trauma or illness. Some patients are already so fearful of being judged (even before arriving at the hospital) that it impacts the way they seek care. Your professionalism can reassure the patient that their health and wellbeing is paramount.

A female patient was referred for lower limb radiography revealing an unstable displaced fracture of her lower leg. During the examination the patient engaged in friendly conversation with the radiographer where she confided that she insisted her partner drive her home before presenting to the emergency department so she could shave her legs. She described hanging her fractured lower leg out of the passenger door while her partner shaved her leg.

*It can be particularly challenging to undertake an examination on a patient who is, for example, engaging in culturally unacceptable behaviour. Your behaviour should always incorporate your best possible professional practice.*

### Paediatric patients

The X-ray room is an intimidating environment for paediatric patients. To get a sense of what the child is experiencing, it is a useful exercise to lay on the X-ray table and have a colleague move the X-ray tube over you. You are aiming for a quick and efficient examination to reduce the child's stress. It is helpful to have the X-ray room fully prepared with the exposure set and the X-ray tube and cassette in position.

An additional consideration with paediatric patients is the potential for a lasting distressing memory of an X-ray examination. It is reasonable to consider the possible long-term negative memory of an X-ray examination, particularly when the child is extremely distressed. Where the cost-benefit considerations suggest an examination will have a marginal benefit, it is worth questioning the merits of the examination. Is it reasonable to undertake an extremely stressful X-ray examination in cases where a treatable diagnosis is unlikely? If you are not comfortable raising this with the referrer, seek the advice of a senior colleague.

A paediatric chest X-ray performed in the erect position is both potentially technically better and less threatening than a supine chest technique.

Child-friendly décor can also help to reduce stress in paediatric waiting areas and X-ray rooms. Commissioning an artist to paint a mural on the X-ray room wall can help to reduce the child's fear of the room. Drawing the patient's attention to elements of the décor (e.g., "Can you count the number of fish on the wall?") can further distract them from their fears.

### Care of the elderly

Elderly patients are more likely to be frail, unsteady, slower to respond, and afraid of falling. It is also more likely that elderly patients will have difficulty hearing the radiographer's instructions, especially over background noise. Short and simple patient instructions in a clear, slow, and audible voice will help to get your message heard.

Elderly patients can have very fragile skin which is subject to skin tears. They can tear their skin with what might be considered normal activities of life. These patients are commonly referred to as having *paper-thin* skin. The radiographer in the emergency department should take special care to avoid patient skin tears, particularly when the patient has alerted them to this risk.

### Cultural competency

It is important for health staff to develop cultural awareness, particularly for commonly encountered cultural and religious groups. Gadsden et al. (2019) when looking at Australian Aboriginal presentations to emergency departments noted that Aboriginal people can feel culturally unsafe when using mainstream health services. Taylor et al. (2009, p551) quoted an Aboriginal health worker as follows:

> the Aboriginal patient is probably a little bit shy, a bit overwhelmed, spends a lot of time looking at the floor ... and not making eye contact, and the staff take that the wrong way.

Australian Aboriginal people are one of the cultural groups that tend to refrain from making eye contact and this should not be interpreted as rudeness or disinterest. Other cultural groups may tend to respond positively to every question rather than risk appearing to be disagreeable or rude. The tendency to answer "yes" to all questions can also be associated with language difficulties. Some cultural groups pride themselves on independence and may not want mobility assistance when it is appropriate for safety reasons.

You should respect the personal space of the patient when communicating. Some cultures prefer you to be close when communicating and others will find this uncomfortable. If you are invading the personal space of the patient, the body language and non-verbal cues provided by the patient will indicate you are standing too close to them. Some radiographic procedures can cause the patient to feel exposed and vulnerable. These patients may feel more comfortable with a radiographer who presents as the same gender identity as themselves. This may have a cultural or religious basis, but may be neither.

Whilst cultural considerations are important for radiographers, it is equally important to avoid behaviour based on stereotyping. Talking to the patient will help avoid misunderstandings and avoid a perception that the radiographer lacked sensitivity or was not willing to accommodate cultural differences.

### Patients with an intellectual disability

You are more likely to be successful when imaging patients with an intellectual disability if they are accompanied by someone who is familiar with their abilities and behaviours. The patient will likely have a history, rapport and trust relationship with their carer which can be critical in achieving good positioning. Patients with intellectual disabilities may have very specific behaviour triggers which will likely be known to the parent or carer.

It is best practice to engage with both the carer and the patient. Speaking to the carer instead of the patient may seem like a way of expediting the examination, but to do so is likely to offend the patient and the carer. Patients with disabilities are every bit as individual as any other patient you may come across. It is important to remember that these patients also deserve your best. The extra moments taken to perform your role in an informed manner will help you gain the best cooperation that the patient is capable of.

### Detained patients

A patient who is in police custody may be aggressive, agitated, and distressed. Equally, they may be resigned to their circumstances. Ask the accompanying police officer for guidance before the examination (this is an example of inter-professional communication). The police officer may require the patient to remain restrained (shackled) during the examination.

The patient's distress and agitation may become heightened under questioning. It is generally not appropriate to ask the patient about the circumstances of their injury. This limitation can be challenging when trying to establish more detailed information about their mechanism of injury. Be mindful that in some circumstances the patient may be motivated to lie about their presentation or to exaggerate their injury and symptoms.

## Before the examination

### Justification and the cost–benefit question

For every X-ray examination, it is essential that radiographers consider whether it is justified prior to commencing the examination. The justification considerations, criteria, and process will not be the same for all countries, regions, or even for all health care facilities within the same region. For example, if your department does not allow for radiography which is unlikely to change the course of clinical management (such as rib or nasal bone radiography), the radiographer should ask the referrer if there are unusual or compelling circumstances that justify the X-ray examination in contravention of a specific local policy, protocol, or clinical pathway.

The radiographer is usually the professional who is licenced by the state to expose the patient to ionising radiation. In some countries, governments have passed legislation and enacted regulations (including licencing and/or registration of radiographers) that regulate how this is done and who can expose a patient to medical ionising radiation. The decision to expose a patient to ionising radiation has an implied cost–benefit assessment that should come down on the side of benefit; that is to say, with every exposure to ionising radiation there is an implied assumption that the potential benefits of the examination are greater than the potential cost (ICRP, 2000). The radiographer provides an additional level of safety and should *say something* if the potential patient benefit of an examination does not appear to be greater than the potential patient cost.

If the patient has arrived at your facility with imaging from another site, it would be reasonable to ask if the imaging needs to be repeated. Of course, there may be grounds for repeating the X-ray imaging. These measures are all taken to meet the broad principle of ALARA: to keep the radiation dose to the patient *as low as reasonably achievable.*

### Checking previous imaging

Checking previous imaging in the emergency department might be something which occurs before, during, or after the examination. Checking before the examination is important if the patient's clinical information states that the patient has re-injured an old trauma or has presented with exacerbation of existing disease. For example, it is useful to review old imaging before an examination if the patient has presented with exacerbation of obstructive airways disease to establish if the patient has *long lungs*. Radiographers often ask the patient to let the radiographer know about their long lungs before future chest X-ray examinations. It is also noteworthy that checking old imaging before an examination may be important in terms of justification: has the examination already been performed? Checking old imaging during an examination could be warranted if it is unclear whether an unusual appearance is acute, old, normal anatomical variant, or congenital/developmental. Checking old imaging at the end of an examination can be a useful review and education exercise.

### The importance of the patient conversation

It can be beneficial to ask the patient before starting an examination "why did you come to the emergency department today?" or "how are you today?", or a similar open and engaging question. This can start a conversation in which the patient has an opportunity to share their most pressing concerns. For example, the patient may be experiencing severe pain, they may be feeling cold, or worried about their urgent need to go to the toilet, or they may be concerned for the welfare of their child left in the care of a neighbour. You may find the patient has concerns about the radiation dose from the examination, employment implications of their injury, or legal implications if they were injured in a car accident. Regardless of the nature of the patient's concerns, the conversation with the patient is an important part of person-centred care. These discussions demonstrate that you are placing the patient at the centre of the examination. Note that this is largely a listening exercise;

don't interrupt the patient while they are still providing a context to the point they wish to make. Your gentle handling of this conversation may help the patient to feel less alone. The patient may convey personal information in the privacy of the X-ray room that may be withheld in a corridor.

A brief history from the patient will guide your approach to the examination in terms of what anatomy to image and how you are going to approach the examination. For example, there will be a significant difference in your approach when a patient is experiencing shoulder pain from tendon calcification compared with a patient who has a neck of humerus fracture. Even within the narrow consideration of trauma to the shoulder, the patient could have an acromioclavicular joint injury, gleno-humeral dislocation, or clavicle fracture to name but a few shoulder pathologies. Some injuries will be suspected because they are associated with a specific mechanism of injury and age-group. An injury may be suspected because of a deformity, soft tissue appearance, level of pain, restricted range of joint movement, or some other associated feature. Beware the distracting injury; the patient may not indicate pain at the site of a fracture because the pain is masked by another more painful injury. This is seen, for example, in patients with base of fifth metatarsal fractures where the ankle ligament injury pain masks the pain at the fracture site.

### The radiography team

It is important to consider whether the resources needed for an examination are available before calling for the patient. For example, a chest X-ray examination on a semi-conscious patient in the emergency department will usually require more than one radiographer. If the patient presents with additional imaging challenges, the number of staff required to complete the examination safely and efficiently may be more than two. It is also worth considering delaying the examination until you have enough staff present to carry out the procedure efficiently, effectively, and safely for you and safely for your patient. An efficient and effective radiographic team will be seen positively by the patient.

### Working with radiography students

It is common for radiographers to be accompanied by radiography students. Students will participate to varying degrees in examinations depending on their degree of training, experience, and confidence. The student is the responsibility of the qualified radiographer. It is important to note that the student is not only learning the technical aspects of the role, but is also being socialised into the culture of the role. If you are displaying genuine empathy and care towards the patient, the student will be more likely to behave in the same manner. Students will potentially emulate and internalise attitudes and behaviours towards patients by observing them in qualified radiographers. It is important to remember that you are a role model for students on clinical placement.

### Prepare the X-ray room

It is both professional and efficient to have the X-ray room prepared for the examination before bringing the patient into the room. If there are two radiographers involved in an examination, one can be preparing the room while the other collects the patient.

### Introducing your team and patient identification

It is good practice to introduce yourself (and your colleagues and student radiographer) to the patient at your first point of contact. If you adopt this practice, it will become a routine starting point for every patient interaction to say "Hello, my name is...". This should be followed by positively identifying the patient. Current recommendations endorsed by the Society and College of Radiographers (United Kingdom) suggest a minimum of three points of identification (e.g. full name, date of birth and address). It is very poor practice to identify a patient using a phrase like "is your name John Smith?"

You must also confirm that you are imaging the correct anatomy on the correct side (RCR, 2020). It is best practice to ask the patient "which arm are we going to X-ray today?" rather than "I am going to X-ray your left arm". Don't assume, for example, that the patient's arm in a sling is the injured arm (the patient may have an old and a new injury). Each radiographic examination should be considered in the context of the patient's presentation. If the patient's presentation does not match the information provided by the referrer, the discrepancy should be clarified. The discrepancy can be due to the referrer selecting the wrong patient name, the patient's symptoms may have changed, or there may simply be a laterality error.

### Timeframe

It can be comforting for the patient and relatives to be provided with a timeframe for the examination. The patient will appreciate you stating that "X-raying your shoulder should take about ten minutes". If the patient is expecting the X-ray examination to be painful, the timeframe provides an expected endpoint. Most radiographers have fallen into the trap of advising the patient that "this is the last X-ray". It can be a loss-of-trust moment when this turns out to be incorrect.

### Collecting the patient

Collecting the patient and transporting them to the X-ray room is sometimes undertaken by the radiographer. It is important to introduce yourself to the patient, positively identify the patient, and ask them if it is a convenient time for their X-ray. The mode of transport will be negotiated with the patient whilst considering any previous mobility assessment and clinical information offered by the referring clinician. A few important notes are as follows:

- Ask the patient about their level of pain on a scale of 0 to 10 (of increasing pain) and ask whether they have had pain relief. If they are waiting for pain relief medication, it is usually kinder to delay the examination.
- Take careful note of the patient's attachments and the level of oxygen being administered to the patient. You may need to ask the nurse or referrer if the patient can be without supplemental oxygen during the transfer to the X-ray room.
- If the patient is lying on an inflatable mattress, make sure it is on a "travel" setting to ensure it remains inflated.
- Be careful to ensure that the patient's fingers will not be injured when going through doorways and narrow passages.
- Ask the patient if they want a relative or support person to come with them.
- Reconnect any attachments you disconnected when returning the patient to the cubicle. It is particularly important to hand the patient the nurse call buzzer and ensure they understand when and how to use it.

### Who should accompany the patient?

It is usually in a child's best interest for a parent or guardian to accompany the child during the examination. It can be useful to ask the child which parent they would like to come with them (this is often followed with some light humour!). If one of the parents is highly distressed, the other parent will usually be a better choice. The parent or guardian can provide comfort and security for the child during the examination. The X-ray room and equipment can be particularly confronting and threatening to a child. A patient with a disability may be more comfortable and compliant with a carer or family member present. A patient who has been detained may need to be accompanied by a law enforcement officer.

### Transporting patients

Patients typically arrive at the door of the X-ray room in a bed, wheelchair, or walking. The examination can be encumbered if a patient with a high level of immobility arrives in a wheelchair; a bed-to-bed transfer is usually easier than a wheelchair-to-bed transfer. Typically, a patient with a neck of femur fracture should arrive in a bed, a patient with a radial head fracture may need a wheelchair, and patient with a finger foreign body can usually walk into the room. If a patient is gripping the bed handrails for security, it is important to ask them to bring their "hands in" when pushing the bed through a doorway.

A patient can also be injured when the bed rails are lowered: ensure their limbs will not be impacted when the bed rails are lowered or raised.

### Pain management, patient care and compliance

Pain management is a significant consideration in the emergency department. Radiographers often need to move the patient during examinations more than any of the other staff in the emergency department, and should be thinking about minimising patient pain and discomfort. It is good practice to have an early conversation with the patient regarding their level of pain. There is a useful 0 to 10 pain scale (0 = no pain, 10 = worst possible pain) that can be explained to the patient to obtain an approximate indication of the patient's level of pain (Bijur, 2003).

One of the important imaging choices for the radiographer is whether to perform the examination on the patient's bed or on the X-ray table. This is an on-balance consideration seeking the best way of answering the clinical question with the least possible discomfort to the patient.

Informed consent should include discussing the expected pain that will be caused by transferring and positioning the patient during the examination. For example, if a hip fracture is suspected and the patient will need a slide-transfer to the X-ray table, informing the patient that the slide will be painful is facilitating informed consent. You should advise the patient that the radiographers will minimise their pain as much as possible and you are seeking to gain consent for this movement. In this scenario, ask the patient if they have had pain relief recently and whether they would like further pain relief before attempting the slide transfer onto the X-ray table. This will be beneficial in maintaining patient trust and comfort, and improving image quality.

Some patient beds will have a space for an X-ray cassette under the mattress. This bed design can avoid the need to transfer the patient on to the X-ray table and avoid the associated patient discomfort and pain. Alternatively, some wall bucky designs allow for the bucky to be lowered into a horizontal position. The patient can then be positioned over the bucky on their bed. This approach requires a radiolucent patient bed which is usually designed for this purpose.

### During the examination

### Maintaining patient dignity

Patients experience a heightened level of vulnerability while they are in your care. It is important to be mindful of the need to maintain the patient's modesty and dignity. This can be as simple as advising the patient that they can keep their underclothes on when changing or offering to tie up their gaping hospital gown. Bariatric patients may require two patient gowns; one applied to the patient's front and one to their back.

Certain radiographic procedures and positions may present an affront to patient modesty (e.g. the horizontal ray lateral hip position). For these types of examinations, it is important to communicate with the patient what you are doing and make every effort to cover the patient appropriately without introducing image artifacts.

### When patients need to use the toilet/bathroom

Patients may not realise they need the toilet until they start to move. If a patient asks to use the toilet, the key questions are *when, how and where*?

When: how urgent is their need?
How: this question relates to their mobility and fluid balance monitoring.
Where: direct the patient to the nearest toilet.

If in doubt, ask the nursing or medical staff for advice, particularly regarding fluid balance monitoring. If the patient's fluid balance is being monitored, the nurse will want to know the volume of urine the patient has passed. Finally, some patients may ask to use the toilet habitually: this may be behavioural rather than a genuine need.

### Unexpected patient behaviours

When you observe patients behaving in an unusual way, it is worth considering that you may be seeing their particular *illness behaviour* (Mechanic, 1986). There is no right or wrong way to experience or express illness and pain. Some cultural groups openly demonstrate pain and others suppress the overt demonstration of pain and distress. Patients may exhibit behaviour that is uncharacteristically withdrawn, or at the other extreme, they may appear aggressive or engage in awkward humour. Illness behaviours can be culturally defined, but are invariably tempered by a patient's personality traits, life experience, and age.

Some children may not demonstrate the level of pain that you would expect to accompany their injury. This can be associated with conditions like autistic spectrum disorder (Allely, 2013) or the child may simply have a high threshold for pain. A high pain threshold is more commonly seen in elderly patients and in patients with Alzheimer's disease (Benedetti et al., 1999). Patients can be highly distracted, and may seem oddly unable to follow simple directions. A patient might lay prone on the X-ray table when asked to lie on their back. It would be easy to categorise these patients as unintelligent rather than acknowledging that their *brain fog* reflects their stress level.

*The patient's connections: Lines, tubes, leads, and monitoring*

Patients can present with a variety of attachments which need monitoring during the examination. For example, it is common for patients to present for X-ray examinations with an intravenous line(s) attached. With the potential for patient movement during the examination, the lines should be monitored at all times. When appropriate, if a nurse is present with the patient, it may be useful to disconnect the intravenous line before starting the examination. When positioning the patient, ensure the intravenous line does not become inadvertently dislodged: this is a relatively common occurrence. If the intravenous line access is dislodged during the examination, ensure you have appropriate personal protective equipment (PPE) then apply pressure to stop the bleeding, raise the limb, and advise the referring practitioner or nurse. The patient's intravenous access cannulation point may not be obvious if a drip is not connected. The radiographer should monitor the intravenous cannula if it is at risk of dislodgement during patient movements.

The patient may have a variety of attached drains. It is important to be aware of what these drains are and any special requirements. An underwater sealed drain (UWSD) needs particularly careful attention. This three-chamber device provides a gentle continuous pressure to remove air or fluid from the patient's pleural space. It is important that the UWSD remains below the level of the patient's chest to avoid backflow of fluid into the patient's pleural cavity. Ensure that it is not dislodged on doorframes, accidentally kicked, or otherwise damaged. You may find the nursing staff have clamped the UWSD tube prior to the X-ray examination as a risk mitigation measure.

A patient with a urinary catheter may have a collection bag attached to the side of the bed. Care should be taken to ensure that this bag is not dragging on the ground where it could be caught under a wheel of the bed. The bag may need to be placed on the patient's bed temporarily when lowering the bed side rail. Ensure the bag moves with the patient during a bed-to-bed transfer and is returned to a position below the level of the patient's urinary bladder as soon as possible.

It is common for patients to present with an oxygen mask (Hudson mask) or nasal tubing (nasal "specs"). It is important to continue the oxygen therapy during and after the examination. Take note of the oxygen flow setting before disconnecting the tubing from the oxygen bottle and reconnecting to the wall oxygen outlet. The oxygen flow rate should be set by adjusting the flowmeter until the *middle of the ball* is aligned to the correct number on the scale.

*Emergency protocols, procedures, and processes*

Your emergency department will have protocols and facilities for patient-related emergencies. You must be aware of how to call for assistance in an emergency. The emergency bell may be needed if a patient has suddenly deteriorated, fallen, or become violent. If there is a call button on the wall of the X-ray room, you should be aware of its position and when to use it.

## Communication

Verbal and written communication permeate every moment of our working day. We spend so much time communicating with our colleagues and patients that it doesn't always register with us as a foundational aspect of patient care. Person-centred care and communication are very closely linked. The overwhelming nature of the unfamiliar technology is threatening and dehumanising for the patient: your communication skills and practices will make a difference.

Typically during an X-ray examination, there is a very limited timeframe to establish rapport, trust, and the confidence of your patient. Communication with patients begins with an introduction. It is worthwhile establishing the habit of starting every examination with "hello, my name is ... and I will be taking your chest X-ray". This introduction is usually followed by positive identification of the patient and a clear and succinct explanation of what you are going to do. During the examination, the patient will appreciate an ongoing explanation of what you are going to do (and why). This affords them a degree of comfort, control, reduction in anxiety, and an opportunity for the patient to continue to provide informed consent. Patients can be particularly stressed by your deflection of what they consider to be a legitimate question; this can look dishonest, rushed, or uncaring and insincere. A patient's experience in the emergency department may be more influenced by how you made them feel than what you did.

It is important to develop adaptive communication skills. The way you talk to your colleague will be different to the way you speak to other members of the multidisciplinary team (MDT), and different again to communicating with an elderly or paediatric patient. You might say to the patient that you are going to "X-ray their lower back", but this would be unprofessional language when speaking to a colleague or other health care professional.

Patients generally respond well to light conversation during an examination. Communication should ideally strike a balance between directions, technical communication, and light-hearted polite conversation that will make the patient aware that they are being cared for by a person. This balance will be directed by the patient's response. The patient may engage in conversation or may even take the conversation to light banter or humour. There is a subtle verbal ping-pong where the conversation bounces back and forth until you establish a mutually comfortable level of communication. Comments such as "You did that particularly well" can act as an important supportive element to the patient's experience. While you are focused on the technical skills of your role, this is the *other* important skill. You should be mindful that the patient will be more aware of your communication skills than your technical skills. Developing a rapport with the patient is likely to enhance patient cooperation; the patient might be more willing to push the boundaries of their abilities to help you optimise image quality.

### Pitch, intonation, tone and inflection

The qualities of your voice can influence the patient's perception of you as much as your actual spoken words. While it may be appropriate with some patients to use a firm tone, it is generally the case that the qualities of your voice should reflect empathy and concern rather than authority. The patient should perceive that they are a partner in the imaging process rather than subject to a technical process with routine directions.

### Negotiated problem solving

A significant part of patient communication involves problem solving. This is often a discussion between the radiographer and the patient regarding the best approach to an examination. For example, the patient may be asked "can you lie flat on your back on the table?". The patient may respond with "I can't lie on my back". What follows is a discussion covering varied approaches to mitigate the difficulties associated with the patient lying supine on the X-ray table. A compromise position can usually be achieved through consideration of alternatives. Radiography can be described as *the art of continuous compromise* and listening carefully to the patient is likely to help you arrive at the best negotiated compromise.

### The "extra view"

Performing an extra view might be routine for the radiographer but could be interpreted by the patient as an indication that there is a problem. If the patient asks "is there something wrong?", the answer should be neither dismissive nor misleading. If there is consideration of pathology versus a normal anatomical variant, this should be conveyed to the patient. It is also reasonable to convey to the patient that the extra view will provide extra information for the doctor, facilitating a more accurate diagnosis.

### Radiographer teamwork

Patients will quickly pick up on a lack of teamwork, a lack of shared purpose, and poor camaraderie between the members of the radiography team. This is not to suggest that the radiographers cannot discuss varied approaches to an examination in front of the patient, but the tone should be cooperative and purposeful rather than confrontational.

*Technical terms*

The use of technical terms by radiographers can be so pervasive in their daily communication that they can overlook that the patient is unaware of their meaning. There is usually a non-technical word or phrase that can be employed in preference to a technical term. For example, "please turn and stand with your left side against the board" is easy for the patient to understand, whereas "please stand in a lateral position" will be unintelligible and confusing for the patient. Equally, it would be preferable to advise the patient that you will "take the picture after you have taken a big breath in" rather than "on full inspiration". Simple short sentences are more likely to be understood and remembered than long complex instructions.

A common error when giving expiration chest X-ray breathing instructions is to ask the patient to "breathe in, breathe out, and hold your breath" – this wording is unclear. It is better to explain to the patient that you want to take a picture with their breath out then say "breathe in, breathe out, and stop".

*Humour in medicine*

Patients will often set the tone and communication style for an examination. For example, if they are choosing to engage in humour or light-hearted banter, encourage them by mirroring that choice. A light-hearted quip can help patients to cope with a stressful moment. Their decision to join in or deflect your humour will provide you with a direction to take the conversation. Alternatively, if they are seeking strong demonstration of professional ability, let them see your technical grasp of the procedure.

Some patients put up a barrier to distress in the form of jocularity. If the patient appears to be excessively or inappropriately jovial, it may be a coping behaviour concealing their true level of distress and anxiety.

*Being called away from an examination*

The smooth flow of an X-ray examination can be suddenly interrupted when you are called to another task. For example, there may be an urgent need to attend to another examination or the operating theatre. It is sound practice to explain to the patient that you are urgently needed elsewhere and to introduce the radiographer taking over the examination. You must provide sufficient handover information to your colleague to ensure the successful completion of the procedure.

*Communication with the elderly*

It is expected that X-ray examinations of elderly patients will become more frequent as the age profiles of populations change. Elderly patients usually require simple, audible, non-threatening, and clear communication. Elderly patients can need more time to process instructions and take longer to adopt the positions required for particular views. Use short sentences and don't use technical jargon. Elderly patients often like some light-hearted conversation at the end of an examination. In the context of a busy emergency department, this can be perceived as a threat to efficiency, but a minute spent listening to a patient's story will be appreciated and should be seen as an integral part of the examination.

Elderly patients are often in fear of falling and their fear may be particularly evident when they are in an unfamiliar environment and feeling that they are not in control. It is important to reassure elderly patients that they are not going to fall, particularly during transfers.

### Communicating with children

Effective communication and care with paediatric patients requires a general awareness of a child's perceptions and needs and how these tend to change as they get older. In general, the younger the child the more they will suffer separation anxiety if separated from the parent or carer. Engage the child in conversation at their level from the first contact. Don't mislead the child by saying "this won't hurt" then cause them pain. This will result in a loss of any trust that had developed, resulting in difficulty with subsequent views. A child will develop a more logical rather than reactive thought process as they get older and may react well to explanations such as "this is like having a photograph of your arm".

The parents, guardians, or carer will often volunteer information about the child's normal range of behaviours. For example, they may say "this is unusual behaviour for her" or "this is how he behaved when he broke his arm previously". An older child or adolescent patient may take pride in the fact that they can cope with an X-ray examination without a parent in the room.

It is appropriate to assess the state of the child and likelihood of compliance before commencing an examination. It may be prudent to stop an examination if the child becomes frantic, traumatised or at risk of injury. A discussion with the parent or guardian is usually the best approach. The referrer may re-consider the cost–benefit of the examination and, in some cases, delay or cancel the examination. Alternatively, some level of sedation for the child may facilitate a safer, kinder, and more effective examination.

### Patients with an intellectual disability

Communicating with patients with intellectual disabilities needs to be tailored to the individual as much as possible. You may have received information from the patient's carer or relative that will allow you to use effective speech and non-verbal actions that can be understood by your patient. While timeliness is important, don't rush – a little extra time spent may help the patient to feel safe and cared for. Use short and clear sentences that simply describe what you are going to do. Reiterate the information at each step before proceeding. It may not be apparent that the patient is understanding more than they can verbalise. You will often need to get your cues from the patient's carer or family, but it is appropriate to address initial questions to the patient. This allows the patient to understand that you are requesting their cooperation during the examination.

Emphasise to the patient that they are safe and you are there to help. Allow time for each thought to process and take a moment to let them finish what they are doing. Use informative yet simple phrases that paint a picture of what steps will occur giving a general timeframe and context to your actions. For example: "First, we'll go to the X-ray department, move your bed next to our X-ray camera and then we'll take pictures of your sore foot. After the pictures are finished, we will come back here to your cubicle in about 15 minutes time".

When addressing the carer, ask them if they are happy to accompany the patient to the department and remain either in the room with the patient or within the patient's line of sight in the control room. Ask the carer if there is anything pertinent that will help with the examination. Many patients will have personal information folders that may be helpful with planning. A commonly used aid to avoid heightening the patient's anxiety is the use of noise-cancelling headphones. Guidance for this would be obtained from the patient's carer or relative.

### Communicating with patients who have a mental illness

When responding to a patient who has a mental illness, don't correct errors in reality or apply contradictory logic to your responses: the patient may be experiencing a different reality to yours. Focus on the task at hand and make your comments practical and task-specific. Don't stare or make strong eye contact. This can be very confronting to some patients and may act as a trigger for physical reactions. Look at a mutually acceptable object like the X-ray table and use non-verbal actions and simple words to convey your care and concern.

While directions to the patient should be straightforward, avoid more assertive language when trying to correct an action or direct the patient. The more assertive the language that you use, the more confronting the experience for mental health patients. Also consider the possibility of co-existing intellectual disabilities (such as Autism Spectrum Disorder) that may hinder the patient's ability to process a stronger delivery of language. While you are responsible for keeping both you and your patient safe in your environment, don't stop sudden patient movements. Your patient may grab your hand when you are about to touch them; this is acceptable as long as the action is attempting to stop you rather than hurt you. You should always assume that the patient is in *fight* or *flight* mode, meaning that an action intended as a mild gesture by you may be misconstrued as a threatening move to someone less able to read your intent. It can also be useful to point out that the light is dimmed within the x-ray room to help you see the light from the x-ray camera in order to line the patient up for their picture. If the low light is an issue for your patient, turn up the lighting for this examination. Some mental health patients may also have specific trigger words that need to be avoided.

### Addressing patients appropriately (sweetie, love, hun, etc)

It is generally appropriate to refer to patients by their first or preferred name. It should be considered whether referring to patients as "sweetie" or "love" is appropriate and professional. It is noteworthy that some regional cultural practices consider this form of address appropriate.

It is also important that you do not misgender a patient. Dolan et al. (2020, p150) noted that "misgendering within the health care system can significantly affect the mental and physical health of transgender … individuals and can negatively impact future engagement with the health care system". Misgendering a patient can increase their feelings of distress, and prevent the building of rapport and trust.

### Inter-professional communication

Care of patients in the emergency department is a team endeavour. There will invariably be occasions where there are questions, discussion points, information, and feedback for other members of the emergency department team. Effective communication with other members of the multi-disciplinary team will involve a level of confidence in communication, a level of communication adaptability, and a level of shared knowledge and common understanding. The development of effective interprofessional communication requires practice. Developing a knowledge of the roles and goals of the professionals around you will enhance your working relationships and facilitate a more collaborative approach to patient care.

### Immobilisation and positioning aids

Radiographers in the emergency department commonly make use of immobilisation devices. Given that you might be restraining the free movement of the patient with an immobilisation device, consideration should be given to the accepted local practices as well as the legal and policy environment in which you are working. There is an unclear distinction between an immobilisation device and a restraint, particularly in relation to paediatric patients (Hardy and Armitage, 2002). It may be unclear whether your immobilisation of a patient is acceptable or could be seen as coercive. Good communication with the patient and parents before using a positioning aid or immobilisation device can help to reduce patient anxiety and misunderstanding.

*Routine views and adaptive radiography as patient care*

It is best practice to establish a set of radiographic routine views for every body part as a starting point for every X-ray examination. Patients don't always injure themselves in a manner that neatly fits the conventional radiographic anatomical regions. Trauma radiography requires an adaptive approach. Collimate the X-ray beam to include the anatomy of interest. Focus on the anatomy to be imaged in lieu of the conventional centre points. There is an element of creativity in trauma radiography that starts with the question: "how can I produce the highest quality X-ray images that achieve the best balance between answering the clinical question; minimising patient discomfort; ensuring patient safety and minimising patient radiation dose?" This often involves working around the patient's limitations and working from first principles. For example, if the patient has an unstable wrist fracture, PA and lateral views can be achieved with little or no movement of the patient; simply angle the X-ray tube and image receptor to the patient's anatomy rather than moving the patient's anatomy to suit the X-ray tube position. This is the art and practice of trauma radiography and is a specific skill. The trauma radiographer is involved in continuous problem solving and this is the challenge that trauma radiographers find at the centre of their role and job satisfaction.

It is often valuable to take a *scout* overview image including as much of the symptomatic anatomy as possible. If this scout view can be achieved with minimal patient movement, the pathology may be revealed and the approach for the remainder of the examination may become clearer.

Follow-up radiography of a known injury requires a different approach. The clinical question is around the progress of an injury and the success or otherwise of a particular treatment. The collimated X-ray field can usually be restricted to the treated pathology. An exception can occur when there is a suspected missed pathology on the initial imaging. These considerations may seem very technical in nature, but they are also patient care considerations; a patient in severe pain should not suffer the rigors of unnecessary imaging.

There can be a case for prematurely ending an X-ray examination when the clinical question has been answered and it is clear that the patient will need a CT examination. Care must be taken when there are multiple pathologies or a base-line set of X-ray images will be needed for treatment planning or follow-up imaging.

*A patient presented for facial bone radiography following an assault. The radiographer asked the patient about her vision and she indicated that she was experiencing diplopia (double vision). An orbital blowout fracture was suspected and was subsequently demonstrated on the initial occipito-mental image. Knowing that this patient would be having a CT examination to assess inferior rectus muscle trapping, the radiographer did not perform any additional views. The radiographer discussed the case with the referring doctor and it was agreed that no further X-ray imaging was warranted.*

*The alteration to the routine facial bone views was documented by the radiographer.*

## Normal anatomical variants

Normal anatomical variants are anatomical features that are unusual, but not specifically abnormal or pathological. The radiographic features of these anatomical variants can be confusing and may be mistaken for pathologies. It is useful to have a textbook of normal anatomical variants available such as Keat's (2013) atlas of normal anatomical variants. It is undesirable for the patient to have additional imaging of a suspected pathology which is actually a documented normal anatomical variant. Alternatively, if you have access to a radiologist, a discussion of a potential normal anatomical variant can be both time-saving and a useful learning experience.

## Patient care in multi-trauma: The minor injury trap

A patient who experiences life or limb-threatening injuries presents specific challenges to the radiographer and the broader multi-disciplinary trauma team. The trauma team are initially focused on life and limb-saving diagnosis and treatment, however the patient's minor injury revealed during the secondary survey may present the greatest threat to the patient returning to their previous life.

*Multi-trauma patients admitted to the intensive care unit (ICU) have follow-up appointments with the Trauma Coordinator Nurse. These follow-up appointments are an audit tool that provide an opportunity for staff to review the long-term effects on the patient of the care that was provided. One such patient reported the long-term impact of their multi-trauma was associated with a missed thumb fracture on their dominant side: this injury prevented the patient returning to their former occupation.*

## Patient care for specific injuries and presentations

### Neck of femur fracture

Neck of femur fractures are a very common injury and are becoming more common in aging populations (Veronese and Maggi, 2018). It is appropriate to ask hip and pelvis trauma patients about their level of pain and whether they have had pain relief prior to imaging. If it is known that the patient is going to be receiving pain relief, it is preferable to work with the referrer or nurse to conduct the examination when the pain medication is at maximum effectiveness (consider the elimination half-life of the pain medication).

*An elderly female patient was referred for additional views of a neck of femur fracture. On presenting to the X-ray room, it was clear that the patient was in severe pain. The radiographer considered whether the examination should be delayed until the patient received a higher level of pain relief. On discussion with the referring clinician, it was revealed that the patient had not received a planned femoral nerve block because she had been transported to X-ray. The patient was returned to the referrer for the nerve block and then further views were taken with little discomfort to the patient.*

Care should be taken to ensure the patient's legs are lifted and not dragged during the transfer to the X-ray table. Also, there may be an opportunity to change the patient's bed linen and to organise a hospital bed if the patient is shown to have a proximal femur fracture. It is noteworthy that some patients will be referred for hip radiography for neck of femur fracture despite being able to mobilise. Be mindful that on rare occasions patients will have a stable neck of femur fracture and will mobilise.

### Radial head fracture

Intracapsular radial head fractures are usually associated with a significant elbow joint effusion. The patient's fracture and associated distended elbow joint capsule are notoriously painful, and the patient can be at risk of vaso-vagal syncope. When in doubt, transport the patient to the X-ray room in a wheelchair or, in extreme cases, on a bed.

### Paediatric supracondylar fracture

Paediatric supracondylar fractures are commonly associated with falls from play equipment. These fractures can be extremely painful and are distressing for the child and parents. Discussing pain relief before the examination may be warranted and *adaptive radiography* (working around the patient) can reduce the patient's pain and distress.

### Orthopantomogram (OPG) for facial injuries

It is worth considering that patients may not have seen their face following facial trauma. Any equipment that has a mirror or highly reflective surface (such as an OPG machine) may invoke a severe reaction from the patient. Ensure that the patient is an appropriate candidate for an OPG as they will be standing unaided for about five minutes. If the patient is unable to stand safely and hold themselves in position, a tall chair without a tall backrest may be used.

### Finger injuries in young children

Children are prone to getting their fingers caught in closing doors. The parents may be more distressed than the child, and you may find yourself caring for the parents as well as the child.

*Knee radiography*

The patient with a knee injury may be reluctant to flex their knee. The clinical information may refer to the patient having a "locked knee", but this will need clarification with the patient. The patient's knee may be mechanically locked with a loose body, or the patient may be unwilling to flex their knee because of pain or fear of pain. When positioning the patient's knee for a horizontal ray lateral or skyline view, it may be easier to raise the whole leg then flex the knee. This allows a knee bend with less-threatening language. Appropriate gentle persuasion may result in the patient adopting a position that they thought was not possible.

*Forearm radiography*

A *mixed view* forearm position with a PA wrist and lateral elbow is often more easily achieved in a trauma setting where movement of radial or ulnar fractures will cause significant pain. The orthogonal view to this (lateral wrist and AP elbow) can then be achieved by raising the patient's wrist in the air with torso and shoulder movement, keeping their elbow on the X-ray table, and then instructing the patient to lower their wrist to the table (in the lateral position). Straighten the patient's elbow as much as their pain tolerance will allow. This approach avoids supination and pronation associated movement at the fracture site.

*The inebriated patient*

Patients whose behaviour is affected by alcohol or drugs require special consideration by the radiographer. These patients are less likely to comply with instructions and more likely to be at risk of falling. It is prudent and reasonable to discuss the patient's condition and compliance with the referrer and delay the examination if, on balance, it is in the interests of the patient to do so. It is noteworthy that a patient who appears to be influenced by alcohol or drugs may be under the influence of neither. Patients can appear inebriated when their behaviour is actually the result of an injury or illness. This is especially true of patients with low blood sugar levels and patients with head trauma.

**Radiography in the resuscitation/emergency room**

Radiography in the resuscitation or emergency room is a specific skill. The radiographer is initially required to provide quick answers to a few specific life-threatening clinical questions. An initial chest X-ray image will determine if the patient has a pneumothorax that requires immediate treatment. The chest X-ray image will also convey important information such as line and tube positions, and evidence of intrathoracic bleeding, spinal trauma, and rib fractures. It is likely that you will be directed to perform a chest X-ray soon after the patient's arrival in the room.

The radiographer should carefully monitor what is happening to the patient before *stepping in* to start imaging. Situational awareness is critical in the emergency room and is a skill that is developed with experience. If you are not experienced in *reading the room* you should follow the direction of the team leader closely.

## After the examination

### The diagnosis

After the examination (or at any time during the examination) the patient, relative, or carer may ask for a simple diagnosis: "is it broken?" Patients may have a compelling need to hear a diagnosis in cases where they are concerned about the potential treatment (e.g. operative treatment) or where the diagnosis is going to have a significant impact on their lives. Whilst policies on radiographers providing the diagnosis directly to the patient will vary between service providers and institutions (and may be strictly forbidden), it is important to provide empathetic and accurate information. The patient may be comforted as much by a timeframe as a specific diagnosis. For example, you could reply with "the images are available immediately for the doctor to examine and you should receive a diagnosis soon". The potential scenarios and policies are varied, but you should be guided by the need to avoid communicating a misdiagnosis as well as the need to provide empathetic support, and indicate a probable time-frame.

### Flagging significant pathology

There is a compelling argument suggesting the radiographer's duty of care to the patient should include flagging significant and urgent X-ray findings with the referrer. If a patient has a pathology such as a pneumothorax or pneumoperitoneum demonstrated on an X-ray image, the referrer will want to know as a soon as possible. In Australia, Preliminary Image Evaluation (PIE) is emerging as best practice in this area (Brown et al., 2019).

### Patient attachments and external devices

It is important to be aware of and care for the patient's attachments and devices such as drainage bags which may be connected to the sides of the bed. During transportation of the patient in a bed or wheelchair, these attachments are at risk of becoming disconnected or pulled from the patient. You should be just as aware of these attachments upon returning your patient as you were upon collecting them.

### Connections: lines, tubes, leads, and monitoring

When the patient is returned to their cubicle in the emergency department, all connections including ECG leads, blood pressure cuff, oxygen tubing, and oxygen saturation monitoring devices need to re-connected. Importantly, a patient who was receiving oxygen therapy will need to be reconnected to wall (mains) oxygen. Of particular note is the patient's call bell; it can be very distressing and potentially dangerous if the patient has not been given the nurse call-bell button when returned to their cubicle.

### Patient handover

When an examination is completed, it is important to hand over relevant information to the team managing the patient. This could include reporting such matters as: a change in the patient's level of consciousness; a vomiting episode; a change in the patient's behaviour; or any other significant finding or event. The patient may also have a fluid-balance order in place. If the patient voided prior to returning to the cubicle the volume of urine will need to be recorded in the patient's notes.

### The final patient interaction

There are a few parting gestures that can be made to conclude the patient examination. A fresh blanket from a blanket warmer can be a great comfort to a patient. A smile and a parting comment such as "I hope everything goes well" marks a pleasant end to an examination and a reassuring hand on the patient's shoulder can be a powerful gesture of support (Osmun et al., 2000).

### Caring for yourself is caring for your patient

It is very easy to focus on caring for the patient at the expense of oneself. This broad principle could be applied to manual handling, overwork and "burn out", radiation protection (occupational exposure), and psychological well-being. If you have experienced a traumatic event in the emergency department, it can be valuable to take some time out immediately after the event to reflect with colleagues. If you find that you are continuing to replay the experience in your mind or you are having ongoing sleep disturbance, seek counselling (many hospitals offer employee assistance programs) and discuss your experiences with your line manager.

### Reflection and shared learning

A particularly complex, difficult, or stressful examination should be followed by reflection. It is good practice to consider what went well and, in retrospect, what could have been done differently. If an examination was particularly challenging, an informal shared discussion with colleagues can be beneficial.

*Peer review*

Peer review is one of the hallmarks of professional practice. In the setting of a multi-radiographer department, it is useful to conduct case review meetings with colleagues in which there is discussion and reflection on specific examinations. This can cover both technical and patient care aspects of an examination.

## Conclusion

There are many areas of skill required to produce X-ray images of a high standard in the Emergency Department while providing a caring and supportive experience for the patient. Trauma radiography should be considered as an imaging sub-specialty appealing to radiographers that possess a skill mix specific to this area. Emergency department radiographers need to continually add creative methods and approaches to their list of "tips and tricks" as they observe or learn from others. Radiographers always need to be listening for more points that resonate with them. This includes the element of reflection as part of continuing professional development. Patient cooperation is paramount. An understanding of patient behaviours is helpful when eliciting the required patient response.

Communication is key. Imaging request/order interpretation skills are used as often as image interpretation skills. Assuming responsibility to follow up on findings is a key facet of professional development. A successful trauma radiographer is a multi-skilled, multi-talented individual whose skill should be recognised during professional role discussions. The gap in knowledge between patients and radiographers is getting wider. Equally, so is the gap in knowledge with referrers who may struggle to keep up to date with new and emerging trends and technologies in medical imaging. Radiographers need to help bridge the gap. It is helpful to have a clear perception of your role including your duty of care to your patients and your duty of care to yourself. Rise to the challenge and you and your patient will reap the benefits.

## References

Allely CS. Pain sensitivity and observer perception of pain in individuals with autistic spectrum disorder. *Sci World J.* 2013 Jun 13;2013:916178.

Benedetti Fabrizio, Vighetti Sergio, Ricco Claudia, Lagna Elisabetta, Bergamasco Bruno, Pinessi Lorenzo, Rainero Innocenzo. Pain threshold and tolerance in Alzheimer's disease. *Pain* 80(1):p 377–382, March 1, 1999.

Bijur PE, Latimer CT, Gallagher EJ. Validation of a verbally administered numerical rating scale of acute pain for use in the emergency department. *Acad Emerg Med.* 2003 Apr;10(4):390–2.

Brown C, Neep MJ, Pozzias E, McPhail SM. Reducing risk in the emergency department: a 12-month prospective longitudinal study of radiographer preliminary image evaluations. *J Med Radiat Sci.* 2019 Sep;66(3):154–162.

Calder LA, Bhandari A, Mastoras G, Day K, Momtahan K, Falconer M, Weitzman B, Sohmer B, Cwinn AA, Hamstra SJ, Parush A. Healthcare providers' perceptions of a situational awareness display for emergency department resuscitation: a simulation qualitative study. *Int J Qual Health Care.* 2018 Feb 1;30(1):16–22.

Dolan IJ, Strauss P, Winter S, Lin A. Misgendering and experiences of stigma in health care settings for transgender people. *Med J Aust.* 2020 Mar;212(4):150–151.e1.

Gadsden T, Wilson G, Totterdell J. *et al.* Can a continuous quality improvement program create culturally safe emergency departments for Aboriginal people in Australia? A multiple baseline study. *BMC Health Serv Res* **19**, 222 (2019).

Hardy M, Armitage G. The child's right to consent to x-ray and imaging investigations: issues of restraint and immobilization from a multidisciplinary perspective. *J Child Health Care.* 2002 Jun;6(2):107–19.

Keats TE., & Anderson MW.. *Atlas of normal roentgen variants that may simulate disease* (9th ed.), 2013. Elsevier/Saunders.

Maslow AH. "A Theory of Human Motivation". In *Psychol Rev.* 1943;*50* (4), 430–437. Washington, DC: American Psychological Association.

Mechanic D. The concept of illness behaviour: Culture, situation and personal predisposition. *Psychol Med.* 1986;16(1), 1–7. p1.

Osmun WE, Brown JB, Stewart M, Graham S. Patients' attitudes to comforting touch in family practice. *Can Fam Physician.* 2000 Dec;46:2411–6.

RCR. The Royal College of Radiologists . IR(ME)R: Implications for clinical practice in diagnostic imaging, interventional radiology and diagnostic nuclear medicine. June, 2020. P39, 59.

Robinson S, Benton G. Warmed blankets: an intervention to promote comfort for elderly hospitalized patients. *Geriatr Nurs.* 2002 Nov-Dec;23(6):320–3. doi: 10.1067/mgn.2002.130273. PMID: 12494005.

Serrone RO, Weinberg JA, Goslar PW., et al. Grey's Anatomy effect: Television portrayal of patients with trauma may cultivate unrealistic patient and family expectations after injury. *Trauma Surg Acute Care Open.* 2018;3:e000137.

Taylor Kate P., et al. "Exploring the impact of an Aboriginal Health Worker on hospitalised Aboriginal experiences: lessons from cardiology". *Aust Health Rev.* 2009;33(4):549–557.

Veronese N, Maggi S. Epidemiology and social costs of hip fracture. *Injury.* 2018 Aug;49(8):1458–1460.

# Person-centred care in fluoroscopy and angiography

Section Four

Person-centred care
in fluoroscopy and
angiography

# 10 Patient care in the screening room

*Elio Arruzza*

## Introduction

Since Thomas Edison invented the first fluoroscope in 1896, a year after x-rays were discovered by Wilhelm Röntgen, fluoroscopy has maintained a crucial presence within the medical imaging landscape. Though newer imaging technologies have superseded its use in some imaging applications, dependence on fluoroscopy persists. Between February 2017 and February 2018, over 1 million fluoroscopic procedures were performed in the UK alone (NHS 2018).

Fluoroscopy is a technique which produces dynamic or continuous x-ray images. After an x-ray passes through the body, an image is produced and displayed on a monitor, so that the motion of internal body structures, or a medical instrument can be seen within the body. These images are observed in real time, but can be stored for viewing after the examination. Fluoroscopic procedures can therefore be used to diagnose pathology directly at the time and place of imaging, or to perform interventional medical procedures.

Fluoroscopy is generally performed in three primary settings: 1) a dedicated fluoroscopic/screening x-ray rooms within a radiology department; 2) within a surgical theatre using a mobile c-arm machine; and 3) within an angiography suite using a stationary c-arm machine. The two latter environments use fluoroscopy for surgical guidance in orthopedic, neurologic and urologic surgeries, as well as coronary and cerebral interventions by specialized physicians. This chapter will focus on the former environment, where radiographers are generally accredited to practice upon graduation, without further formal training. Common applications of fluoroscopy within a dedicated fluoroscopic/screening x-ray room include:

- Gastroenterology: These examinations assess the gastrointestinal tract. 'Swallows' and 'meals' assess the upper GI tract whilst an 'enema' assesses the lower GI tract. Defecography/proctography assesses abnormalities of the pelvic floor.
- Myelography: A study where contrast material is injected to evaluate the spinal cord, nerve roots and spinal lining (meninges).

DOI: 10.1201/9781003310143-14

- Hysterosalpingography: Commonly known as an HSG, this imaging helps diagnose disorders of the female reproductive system.
- Urology: A plethora of procedures which can be further divided into urethrography (urethra), cystography (bladder), cystourethrography (bladder and urethra) and pyelography (ureter and renal pelvis).
- Sialography: Contrast is injected into the salivary ducts to check for blockages and other pathologies.
- Placement of catheters and tubes: PICCs (peripherally inserted central catheter) and NGTs (Nasogastric tube) for example.

This chapter will provide a step-by-step description of the role that the radiographer plays in the care of patients within the screening room.

## Booking an appointment requiring fluoroscopic screening

Compared to the act of scheduling a general x-ray examination, booking a screening examination typically requires a greater knowledge of crucial preparatory activities and timing considerations. According to Australian national guidelines (MRPBA 2019), it is a radiographer's responsibility to identify and communicate patient preparation requirements. This is especially vital in cases where technologists are not reliant on administration staff, and are directly responsible for booking and scheduling screening examinations.

Patients requiring a screening examination can often present with vague symptomatology, or ambiguous clinical details written on their imaging referral. They may also require multiple imaging modalities to examine the pathology queried. The technologist should make enquiries about these examinations, not only to improve workflow within the department, but more importantly, to question the value and need for further risk from radiation or medical intervention. Today, some screening procedures have been superseded by more accurate, timely, and less invasive procedures using ultrasound, CT, and MRI. Demonstrating a sound knowledge of the clinical indicators which suggest more appropriate use of an alternative imaging modality, is essential for modern radiographers where technology and diagnostic protocols are ever-changing.

Scheduling recommendations, which consider the busy nature of the radiology department as well as simultaneously acting in the best interest of the patient, include:

- Patients requiring fasting should have their examinations performed in the morning. This is to ensure the patient does not experience an excessive duration without eating. This is especially crucial for high-risk patients such as the elderly, obese and pediatrics.
- Examinations requiring contrast should only be booked if a registered medical practitioner is guaranteed to be on-site. Though the potential for

serious adverse effects from contrast administration is small, a qualified medical doctor is required to be nearby in the unlikely event.

- If the patient requires multiple examinations, consideration should be made to whether contrast use will affect the examination performed after.

### Prior to the examination

#### *The room*

When a screening room is untidy, it can resemble that of a surgical theater. This can increase patient anxiety. It can also demonstrate a lack of respect and care, potentially conveying a message that the professionals facilitating the patient's procedure are either incompetent or ignorant to care about their working environment, and further, their patients.

Further to comprehensively understanding the purposes and processes pertaining to a range of fluoroscopic examinations, the technologist must be able to frame and tailor their communication for their patient. Often, examinations requiring screening are more complex and less comprehensible by patients, than an x-ray for a queried fracture for example. Sometimes, the referring practitioner will not explain the procedure to their patient, meaning the patient will often arrive at the radiology department unbeknownst to the procedure they are undertaking. The role of the radiographer in introducing themselves and their role, obtaining informed consent, explaining the process and providing reassurance is central to patient satisfaction and a successful examination.

It can be argued that the most important responsibility for the technologist is to develop a harmonious practitioner relationship with the patient and their caregivers. This is often influenced by the patient's first impression of healthcare staff, implying a respectful introduction may go a long way. Radiographers should introduce themselves and validate patient information. The technologist should also briefly describe their role in the patient's examination.

Though the screening radiographer often undertakes an auxiliary care role during the procedure, it is crucial that technologists fully understand the procedure to be performed and are aware of its risks and benefits. Retaining this knowledge can be challenging, given that it is naturally more difficult to understand a process which is being observed (i.e. the radiographer) rather than actively performed (i.e. the doctor or nurse).

It is good practice and legislation in many countries, that the technologist delivers information regarding the benefits and risks of the procedure involved. These risks should include those brought on by radiation, as well as those resulting from the interventional aspect of the examination. The latter is often reiterated by the medical practitioner administering the intervention (i.e. radiologist), but it is best practice if the patient is provided this information earlier. This allows the patient time to compute the information and formulate questions for the radiologist.

The radiographer's role also involves supporting and promoting the rights and interests of patients. Furthermore, it is important that the radiographer pauses to take in the thoughts and views of the patient at every possible occasion. The radiographer's explanation should highlight that the imaging can provide crucial information that will permit for deliverance of the most appropriate care. No imaging examination should be conducted where the risk outweighs the benefits, and the patient needs to be aware of this. The radiographer, at every possible opportunity, should verify that the patient has understood the information provided.

Explicit consent from patients is necessary to undertake screening procedures as they may involve substantial risk. All procedures require thorough explanation of procedures, though some may necessitate written signature by the patient and a witness. The radiographer is often the staff member responsible for obtaining informed consent. A patient must be competent in order to sign and understand the procedure in their own language, so medically trained interpreters should be used if the technologist's speech cannot be understood. The radiographer needs to be aware of the procedure which requires such consent and for ensuring documentation is fulfilled prior to conducting the examination. These procedures will differ by site and legislation, but generally, written consent is necessary in cases where anesthesia or sedation is utilized, any invasive or surgical procedures are performed, or where administration of high-risk medication is required. Verbal informed consent is generally reserved for purely diagnostic procedures with minimal invasiveness or intervention, provided that full records are documented. Signing a consent form does not eliminate the patient's right to withdraw from the examination once it has started.

Screening procedures are often the most invasive procedures within the radiology department. Given patients are required to sacrifice some amount of dignity in pursuit of diagnosis and a positive health outcome, cultural considerations may have a large bearing on the success of the examination and patient satisfaction. Many cultures hold modesty as a highly sensitive aspect of one's life.

### Principles of communication

For the radiographer, Dutton & Ryan (2019) emphasized key concepts of therapeutic communication including:

1) Responding to the underlying message
2) Reflecting the main idea
3) Exploring
4) Validating

Examples pertaining to these aspects have been tailored for use in screening room procedures.

### Responding to the underlying message

When a patient questions the value or necessity of an imaging procedure, the patient may find it helpful if the radiographer's response acknowledges their views:

> You are feeling discouraged because you don't feel the procedures are doing any good.

### Reflecting the main idea

Reflecting the main idea evolves the previous technique by exploring the matter, and involves the patient in decision-making. Once again when a patient questions the value or necessity of an imaging procedure:

> Do you feel that you should refuse this procedure?

### Validating

When essential information is required or being conveyed, the radiographer should look to validate the patient's understanding, thoughts or experience:

> Are you feeling pain in your arm when this band is tightened?

### Exploring

When a patient briefly describes a thought or experience that may influence their outcome, the radiographer should probe further:

> Have you been walking around by yourself in the ward?

Using some of the examples provided in the introduction, the following speech proformas can be used to guide the radiographer's explanation of the examination. In no way should this act as a strict protocol for informing the patient, but rather a template which should be altered based on local preferences and the patient is being directed to.

### Barium swallow

In this examination, you will stand in front of an x-ray camera and the doctor will ask you to swallow a liquid which shows up bright on the x-ray, highlighting structures internally and providing a more accurate picture. This fluid is called 'barium'. Barium is a dry, white, chalky powder that is mixed with water to make a thick, milkshake-like drink. Unfortunately, the liquid does not taste too pleasant, but it should not give you any pain upon swallowing. You may be asked to hold the liquid in your mouth and then swallow when asked. As you swallow the barium, the radiologist will take single pictures, a series of X-rays, or a video (fluoroscopy) to observe the barium moving through your throat. The radiologist will view the liquid as you swallow and take images. You may be given some additional liquid, which is used to produce gas within your stomach. This gas makes your stomach easier to see, but may prompt you to burp. Please try to refrain from burping until the conclusion of the examination. The table on which you are standing may also be laid down so your internal organs can be assessed when lying down. In this position, you may be asked to turn onto your abdomen or side and drink some more barium through a straw.

### Small bowel series

You will be given some liquid barium to drink, and after a short wait, you will lie on an x-ray table for the first x-ray of your abdomen. Barium is a dry, white, chalky powder that is mixed with water to make a thick, milkshake-like drink. Unfortunately, the liquid does not taste too pleasant, but it should not give you any pain upon swallowing. It is used to highlight structures internally and provide a more accurate picture. You may be asked to lie on your right side between x-rays to assist in moving the barium through you quicker. The length of time it takes for the liquid to pass through the entire small bowel may take anywhere from 45 minutes to a few hours. X-ray images of the abdomen are repeated every 20–40 minutes as the barium moves through your small bowel. Usually, you will lie on the table for the first part of the test. After this, you may be encouraged to drink more barium and walk around. You may be given a cup of tea and something to eat if it is taking a long time for the barium to pass through your system. It is important to wait until the barium reaches the end of your small bowel before the examination is finished. Once the large bowel is reached, the radiologist will use more x-rays to look more closely at the small bowel. Sometimes, the radiologist may need to gently press on your abdomen to separate sections of the bowel in the pictures.

### Barium enema

In this procedure, you will lie on an x-ray table for a primary abdominal x-ray image. The x-ray will be used to make sure the bowel preparation has cleared your bowel. A short, flexible, lubricated tube will be inserted into your rectum. Some tape will be placed on the skin to secure the tube in place.

Sometimes, a small balloon may be inflated to prevent the tube from falling out. Barium liquid is passed through the tube and into the bowel. This may give you some urgency to use your bowel, which you must resist. Once the bowel has been adequately filled, it is partly drained out. A small amount of air is then introduced to the bowel using the same tube. You will be asked to move on the table into a variety of positions while x-rays are taken. The radiologist will make sure that your bowel is well coated with the mixture. The air will make you feel pressure and a feeling like you need to use your bowels again, but it is important to try to hold the liquid and air inside. There may be some discomfort during the procedure, like a belly ache. As soon as the x-rays have been taken, the air will be released in a controlled way. The tube will be removed, and you will be able to go to the toilet. A bowel motion in the toilet will provide some immediate relief.

## PICC

PICC stands for 'Peripherally inserted central catheter'. In this procedure, you will lie on the x-ray table with your arm out to the side. An ultrasound machine is used to identify a suitable vein in your upper arm to insert the PICC. This skin is then prepared as a sterile field with antiseptic on the skin and sterile drapes placed on your upper body. A local anesthetic is administered by a radiologist or specially trained nurse and a needle or short catheter placed in the vein. A small incision will be made, and the longer catheter advanced. You should not feel any sensation of the catheter inside your chest. An x-ray is taken to confirm the correct position of the catheter tip. Occasionally, x-ray contrast may be injected into the catheter to show the veins. If it is not possible to advance the catheter on one side, it may be necessary to use the other arm.

### Dacryocystogram

This examination is used to investigate your tear canals. Soon, you will lie on an x-ray table with a small cushion under your head. The radiologist will place a small cannula into the tear duct opening of your lower eyelid. A bright light or torch and magnifying equipment may be needed as the ducts are very small. You are asked to close your eye once the cannula is in place. A small amount of tape around the cannula tubing will hold the cannula in place against your cheek. You need to remain still when the contrast is injected as the ducts are difficult to see on x-ray if you move. It takes a few seconds to take the pictures. The dye will pass into the back of your mouth if the ducts are functioning. Sometimes if the ducts are blocked the dye may well near your eye. This is safe and can be wiped away with a tissue by myself or the radiologist. This process may then be repeated on the other eye for comparison.

## Sialogram

This examination is used to image your glands which make saliva. You will be asked to lie flat on an x-ray table with a small cushion under your head. The radiologist will place a small cannula into the salivary duct of concern by asking you to open your mouth widely. A bright light or torch and magnifying equipment may be needed to see the ducts, as they are quite small. Sometimes, a small amount of lemon juice is squirted in your mouth to open the duct and make the placement of the cannula much easier. You are asked to close your mouth once the cannula is in place. A small amount of tape around the cannula tubing will hold the cannula in place against your cheek. You will be asked to remain still when the contrast is injected as the ducts are difficult to see on x-ray if you move. It takes a few seconds to take the necessary pictures. The dye will pass into the duct, and you may feel a little pressure. You may get a small amount of dye in your mouth, but this is safe.

## HSG

For this procedure, you will lie on your back on the x-ray table, positioned with your knees raised and bent. You will be assisted by a female nurse and a radiologist; nurse and radiographer will be present for the procedure. A speculum is inserted into the vagina to allow the radiologist to visualize the cervix. A small catheter is inserted into the uterus through the cervix, and dye is injected. The x-ray machine is used to take pictures of the dye in the uterus and fallopian tubes. At the time of injection, you may experience mild discomfort and cramping similar to period pain. Once the test is done, the catheter and speculum are removed.

### Retrograde urethrogram/cystogram

You will lie on your back on an x-ray table. If you already have a urinary catheter, it will be used for the test. A new catheter is not needed. If you do not already have a urinary catheter, a specialist nurse or radiologist will need to put one in. To insert a urinary catheter, the nurse or radiologist will begin by carefully cleaning the genital area with a specific antiseptic wash. A sterile drape (a piece of cloth) is used to cover your groin and surrounding area. A small amount of anesthetic gel is applied to the urethral opening to minimize discomfort. A urinary catheter is then inserted in the tip of the urethra until it reaches the bladder. A small balloon is inflated to hold it in place. The catheter bag will be removed, and a syringe or drip-line that is filled with x-ray dye is connected. The radiologist will fill the bladder until you start to feel full. X-rays are taken as the bladder is filled. Once there is sufficient dye in the bladder, the radiologist or myself will ask you to move into various positions to assess your bladder on x-ray at different angles. Once the x-rays are taken, the dye is drained via the syringe or drip bottle. The catheter bag is then reconnected.

### Micturating cystourethrogram

You will lie on your back on an X-ray table. The radiologist will carefully clean the genital area with a specific antiseptic wash. A sterile drape (a piece of cloth) is used to cover your groin and surrounding area. A small amount of anesthetic gel is applied to the urethral opening to minimize discomfort. A urinary catheter is then inserted in the tip of the urethra until it reaches the bladder. A small balloon is inflated to hold it in place. The bladder is filled with x-ray dye until you feel full. Once there is sufficient dye in the bladder, the urethral catheter is removed. The table will be raised by the radiographer until you are in a standing position. You will be handed a urine bottle that will allow you to urinate (void) while x-rays are taken. X-rays are taken as the dye passes out the bladder through the urethra. It is common to have some difficulty voiding on command. Our experienced and patient staff will support you as needed.

These explanations are provided courtesy of Jones Radiology (2022).

### During the procedure

The screening room is unique in that it demands a multifaceted approach whereby the radiographer is required to attend to the patient, the doctor and the equipment simultaneously. This can feel like a lot to juggle, which is why experienced radiographers are often rostered in this role. After explaining the procedure to the patient, the radiographer's role involves assembling the necessary equipment and supplies, offering support to the physician if required, completing the technical radiation aspects of the procedure and ensuring the patient feels comfortable. During the examination, the radiographer should observe the patient for any change in appearance that might indicate a change in status. Listening to any verbal or nonverbal complaint and providing reassurance is important. Though during the examination, interaction with the patient is mostly facilitated by the physician, auxiliary role of the technologist can contribute to the satisfaction and comfort of the patient extensively. Checking on the patient periodically with simple remarks such as 'Feeling okay?' or 'Doing well' can go a long way. Offering comforting items such as pillows or blankets may be valued.

Examinations requiring contrast media are prevalent in the screening room. The radiographer must understand the properties of contrast agents and be able to alert of any symptoms of reaction, monitor vital signs effectively, assist with drug and contrast administration, and participate in correct surgical aseptic technique. They must take all precautions to maintain infection control.

When transporting or handling patients with drainage bags, the technologist must ensure their bag remains lower than the level of the urinary bladder to ensure flow of urine. Tension of the catheter should be avoided.

## Interaction with the radiologist

A key aspect in the screening room is communication. The need for rapid examinations can exacerbate personal differences between coworkers. Though they do not feature care of the patient directly, the outcome of conversations contributes to both workers' ability to prioritize the patient. Being patient, considerate, validating communication and maintaining a positive attitude can go a long way in ensuring the patient remains the priority. In this environment, all staff members work together as an interdependent team meaning each must depend on the other for safe and successful diagnostic and treatment outcomes.

The screening room can be a stressful environment for all involved. Stress in these environments is often caused when the examination is not proceeding as envisioned, when the status of the patient is negatively impacted, or where disagreement occurs between colleagues. Strategies which alleviate personal stress are often individualized; however, some general strategies to minimize stress resultant from such situations in the screening room specifically include:

- Using a quieter voice, speaking slowly and clearly
- Being non-judgmental
- Not allowing a colleague or patient's inappropriate behavior prompt a similar response
- Requesting an answer when the patient or colleague's understanding is unknown

## Interaction with the patient

Stressful situations are often felt by the patient, which can lead to behaviors which would otherwise not characterize the patient's personality. Most patients have been informed of a potential illness and the procedure they are undergoing may potentially confirm it. Many are also in continual physical pain. The atmosphere within the room can cause anxiety; often the room will resemble a foreign environment, with medical equipment and surgical drapes surrounding them. In pandemic affected times, healthcare members will be masked and hidden behind various lead and personal protective gowns and screens.

In situations where a patient becomes aggressive or agitated:

- Seek help from others before the issue escalates and do not handle the situation alone.

- Be pleasantly firm whilst explaining that your role is to provide them with care.
- Never let an individual get between yourself and the room exit.

## After the procedure

After the examination is complete as determined by the radiologist, it may be the radiographer's role to farewell the patient. The patient should be thanked for their cooperation and provided a description of the timeframe and pathway related to their healthcare. This may include providing answers relating to the availability of the examination's results, and how these results can be obtained. The patient should be escorted out of the department if necessary.

What the patient does after their examination can be just as important as what the patient does before. Furthermore, health professionals' final interaction with a patient can determine their pathway ahead; if a patient leaves the technologist's care feeling confused or misunderstood, their continued care may not be fulfilled. Instead, patients who leave feeling cared for are more likely to continue their diagnostic/treatment pathway and let health professionals influence their health, and their loved one's health, in the future.

Patient care encompasses the messaging that a radiographer must convey in order to ensure the patient cares for themselves after the procedure. If the department's policy is to leave aftercare entirely to the patient, radiographers must be explicit, transparent and complete with their instructions.

For patients who have undergone bowel images with barium, constipation or bowel obstruction in severe cases can occur due to barium's tendency to clump. It is important to instruct the patient to increase their intake of water as well as high-bulk, high-fiber foods. A cathartic fluid such as milk of magnesia may also be recommended. In the case of enema, a significant amount of contrast and air can remain in the bowel so the patient should be instructed to be near the toilet for 4–6 hours post-procedure.

Patients with recently implanted PICC lines should be advised to avoid getting it wet, particularly by hot water, and maintaining the clear dressing laid upon it. If leakage is noted, nursing staff should be alarmed as it may need splitting or replacing. If soreness of bruising occurs, paracetamol can be suggested and an ice pack to provide relief.

## Documentation

Some form of data entry post-procedure is often required from radiographers, whether this solely involves recording the radiation dose output for each examination, or whether their role extends to documenting drug/implant administration. Medical recording will once again vary by institution, and practitioners must be aware of their site's requirements. Recording of data unique to the radiology department includes reporting the use of contrast media, reactions to contrast media, treatments administered and changes in patient status experienced whilst in the practitioner's care.

Radiology departments are sometimes not operated by the same institution where inpatients are cared for. Differences in the reporting practices between the two departments can therefore produce inconsistencies in how patient records are documented. The radiographer should therefore be aware of the guidelines which govern correct documentation in their department, being complete and transparent in any case to minimize miscommunication. Personnel should avoid erasing data if deletion is required or a mistake is made; a line should be drawn through the error, and be initialed and dated.

## Summary

Fluoroscopic procedures within the screening room require proper patient care measures. These examinations are often invasive and require consent and cooperation to optimize patient outcomes. Scheduling these examinations and developing a harmonious radiographer–patient relationship from the outset contributes to their success. Communication differs across the numerous procedures performed and should be tailored to the patient's personal and cultural needs. Radiographers must attend to their obligation to reduce dose for the patient. Aftercare and documentation conclude the healthcare process.

## References

Jones Radiology 2022, *Diagnostic Fluoroscopy (X-Ray)*, https://jonesradiology.com.au

NHS 2018, *Diagnostic Imaging Dataset Statistical Release*, www.england.nhs.uk/statistics/wp-content/uploads/sites/2/2018/06/Provisional-Monthly-Diagnostic-Imaging-Dataset-Statistics-2018-06-21.pdf

Medical Radiation Practitioner Board of Australia (MRPBA) 2019, 'Professional Capabilities for Medical Radiation Practice', MRPBA, www.medicalradiationpracticeboard.gov.au/Registration-Standards/Professional-Capabilities.aspx

# 11 Patient care in interventional radiology and angiography

*Robert Whiteman*

## Introduction

The advantages and desire to use minimally invasive techniques, such as those used in Interventional Radiology (IR), over more traditional open methods of surgery, has steadily increased over recent times for many reasons.

For example, reduced recovery times and length of stay, reduced complication and blood loss rates and potential to be more readily available (Royal College of Radiologists, 2019).

Due to this, the likelihood of a patient encountering IR is higher than before and ever increasing, hence there is a greater need for an awareness and understanding of patient care especially pre- and post-procedure, which is what this section will mainly focus on.

## Patient preparation

Aside from the usual pre-theatre preparations, there are some specific considerations for IR. The institution in which you are working along with the intervention the patient will undergo will largely dictate this; however, there will be some common themes (Mahnken et al., 2021). Please note: a text entitled CIRSE Clinical Practice Manual is an excellent reference text for this subject and it is recommended you read it in conjunction with this section.

### Blood testing

There are several reasons for the interventionalist to be aware of a patient's blood results prior to commencing a case and to provide a baseline prior to intervention. This is done to minimise risks of the procedure and intra/post procedure complications, allow correction of abnormal results or postpone/cancellation of the case if it is deemed not to be safe, or finally allow risk stratification if the case is deemed to be essential. Tests often include but not limited to:

DOI: 10.1201/9781003310143-15

- Full Blood Count primarily to capture haemoglobin and platelet levels. These may well be low in a bleeding patient, for example at baseline, however will improve after an embolisation procedure to stop the haemorrhage, or equally can be used as a marker of bleeding in the aftercare environment if their baseline was normal (RCR, 2014).
- Urea and Electrolytes is of particular importance in assessing creatinine and estimated glomerular filtration rate (eGFR) especially in the context of intravascular use of iodinated contrast media and acute kidney injuries (van der Molen, 2018).
- Clotting to assess prothrombin time (PT), international normalised ratio (INR) and activated partial thromboplastin time (APTT).
  The ranges of what is acceptable from a risk perspective will be dependent on the intended case. For example, a puncture of a vascular structure such as a liver or kidney will require a lower INR than that of a superficial drain of a collection or ascites. Each centre will have a different classification of these procedures, so check your local guidelines.
- Group and save/crossmatch (dependent on the clinical scenario) samples are also recommended for patients undergoing arterial and organ punctures or biopsies in the event of complications either during or post procedure

### Imaging

In order for a patient to be considered for an IR procedure they will have already undergone necessary imaging as part of their assessment process by the parent team, whether that be CT, MRI or US.

There may however be additional imaging that has been requested prior to the actual case starting by an assisting team, for example a chest x-ray as part of an anaesthetic review, and so it is important to ensure this has been completed before the patient leaves the ward/admission area.

### Pre-procedural medications

The intended procedure will heavily dictate what medications need to be stopped prior to the case, for example antiplatelet/anticoagulants to reduce the risk of bleeding during or in the recovery phase. The type of medication and the urgency of the case will be evaluated between the interventionalist and the referring parent team to ascertain the urgency and then the benefit vs. the risks related to their medications.

Alternatively, medications such as antibiotics may need to be given or prepared in good time prior to the case to prevent infections (Venkatesan, 2010), or pain relief to ensure the patient is comfortable especially in the context of cases that seek to deliberately stop blood supply to tissue, such as fibroid embolisation.

In the setting of life-threatening bleeding, it is likely the patient will be continually receiving fluid resuscitation with the addition of blood products such as blood, platelets and fresh frozen plasma to prevent further deterioration (NICE, 2007).

### Consent

Historically, if there was a need for written informed consent this was likely to have been done immediately before the procedure in the Radiology department. However, there is now very much a move towards following a two-step approach to such consent, which is done away from the IR department then confirmed by the interventionalist prior to commencing the procedure (Prashar, 2021).

### Intra-operative considerations

In the pre-stage the IR team, in conjunction with the parent team and anaesthetics if appropriate, will also assess if the patient needs any additional pain relief, sedation or more formal anaesthetic to be arranged to be given during the procedure (Kessel and Robertson, 2016).

Similar to other inter-operative procedures, the patient's vital signs will be continually monitored to closely assess for changes in their condition that may need to be addressed.

This is of particular importance in IR when performing procedures that may result in blood vessel rupture such as angioplasty and stenting, but also those procedures that are likely to cause vagal stimulation, which may result in profound hypotension and bradycardia.

### Post-operative care

This portion of a patient's care can be subdivided into general post-surgical care that you are likely already familiar with, and then IR specific.

Vital signs still need to be monitored closely to check for changes, especially relating to respiratory rate changes (which are often early signs of a deteriorating patient), pulse and blood pressure, and recorded appropriately.

Using a scoring chart such as NEWS2 can aid early detection of a deteriorating patient which will then need to be escalated appropriately via your local protocols, often using a structured approach such as the ISBAR format or similar (Resuscitation Council UK, 2021).

In the situation of embolisation (deliberately stopping blood to an area or organ) in certain cases such as uterine fibroids, as that lesion begins to atrophy it can cause huge amounts of pain, and so the patient may have been given a patient controlled analgesic device (PCA). (Chan et al., 2021).

However, it is of course important to monitor the effectiveness of such devices; equally, pain may also be an indication of an evolving complication similar to other post-operative procedures and so you must act accordingly to seek medical or senior assistance.

*IR specific – puncture sites*

Due to the minimally invasive nature of IR there need to be some additional observations carried out on these patients that will differ from the usual post-operative surgical patient.

Principally this relates to the access site used, and awareness of the structures beneath that have been punctured, and so checking this site will need to be added into the usual routine observation task.

The interventionalist will also specify in relation to vascular punctures specific instructions with regard to bedrest as this may well be for several hours to prevent large volume but difficult to detect bleeding which ultimately can lead to death. It is important to have an appreciation of what the baseline was for the puncture site in order to be able to quickly recognise and alert the medical team to complications that may arise.

*IR specific – indwelling lines and catheters*

Often patients will return from IR with a drain that is in a deep structure such as a kidney or liver, a vascular access device such as a renal dialysis line or a feeding tube in their nose or directly through their abdomen into their gastrointestinal system.

Particular care needs to be afforded to these devices both in terms of good nursing care to keep the area clean, dry and with appropriate dressings, but also to guard against them being inadvertently displaced during manual handling manoeuvres.

The patient may well be dependent on it to prevent further deterioration and pain in cases of nephrostomy and liver drains; the procedure to place it may have been of great technical difficulty, and furthermore may need a specific team or interventionalist to replace it, all of which impacts on a patient's care (Royal College of Radiologists, 2014).

## Conclusion

Although patient care for those undergoing IR procedures is broadly similar to many other patients undergoing traditional surgical intervention. In many respects there are also some very specific considerations that need diligent attention to prevent delays prior to the case.

Likewise, there are aspects of their aftercare that you must be aware of to prevent but also recognise deterioration and misplacement of indwelling devices.

## References

Chan, P et al., 2021, August. Managing postembolization syndrome–related pain after uterine fibroid embolization. In *Seminars in Interventional Radiology* 38(3), pp. 382–387. New York.

Kessel, D. and Robertson, I., 2016. *Interventional radiology: A survival guide.* Elsevier Health Sciences.

Mahnken, A et al., 2021. CIRSE clinical practice manual. *Cardiovascular and Interventional Radiology*, 44, pp. 1323–1353.

National Institute for Clinical Excellence, 2007. Acutely ill adults in hospital: recognising and responding to deterioration. *NICE Guidelines no July*, pp.1–30.

Prashar, A et al .,2021. Informed consent in interventional radiology–are we doing enough?. *The British Journal of Radiology*, 94(1122), pp. 20201368.

Resuscitation Council UK, 2021. *Adult Advanced Life Support Guidelines 2021.* London, RCUK.

The Royal College of Radiologists UK, 2019. *Provision of Interventional Radiology Services*, 2nd Edition, London, RCR.

The Royal College of Radiologists, 2014. *Guidelines for Nursing Care in Interventional Radiology: The Roles of The Registered Nurse and Nursing Support.* London: RCR.

van der Molen, A.J., et al., 2018. *Post-Contrast Acute Kidney Injury–Part 1: Definition, Clinical Features, Incidence, Role of Contrast Medium and Risk Factors: Recommendations for Updated Esur Contrast Medium Safety Committee Guidelines.* European radiology, 28,

Venkatesan, A.M., et al., 2010. Practice guideline for adult antibiotic prophylaxis during vascular and interventional radiology procedures. *Journal of Vascular and Interventional Radiology*, 21(11), pp. 1611–1630.

Section five

# Person-centred care in breast imaging

# 12 Person-centred care in breast imaging

*Joleen K. Eden, Johanna E. Mercer,*
*Catherine A. Hill and Claire E. Mercer*

## Supporting a person centred approach in breast services

Person-centred care supports the priority of the needs of the individual within the breast imaging setting. It is multi-faceted and is considered to: support empathy within communication, assist shared decision making, provide comfort within environments and offer respect (Itri, 2015; Paluch et al., 2022). Measurement of success is often undertaken though evaluations of services by practitioners themselves (Paluch et al., 2022) and it is becoming more common for individuals to use such reviews to start selecting service providers (Prang et al., 2021).

Practitioners, and those who work within breast imaging settings, are often in a distinctive position whereby they encounter those in their care continually throughout their journey, from initial screening mammography, to recall and delivery of results. This continuity of care is unique to the breast screening programme (NHSBSP), whereby individuals become familiar with the role of the practitioners through repeated interaction. It has been reported that mammographers, like angiographers, score highly for emotionality and well-being in global emotional intelligence (Mackay et al., 2012) which could be seen to support effective person-centred approaches within the mammography setting.

Those entering breast services, whether it be through the breast screening or the breast symptomatic route, can often have multiple visits to services and the importance of person-centred approaches to care are imperative to ensure both continuity of care and re-attendance to breast services in the future. The two streams of breast services are unique and staff work across both services within their working day; this can pose challenges in terms of being able to support a seamless patient-centred approach and managing staff within these services related to their anxiety.

The following section will discuss the two services.

DOI: 10.1201/9781003310143-17

## Breast screening services

The NHS Breast Screening Programme (NHSBSP), founded in 1988 in the United Kingdom (UK), invites for a three-yearly mammogram those aged 50 up to their 71st birthday (NHS Digital, 2022). Regular mammographic screening enables the detection of small abnormalities to be identified in asymptomatic individuals, in the absence of a palpable clinical finding. For some individuals, particularly those with a strong familial history, screening can offer reassurance. Some may even feel protected that they have taken the opportunity to look after themselves. However, regardless of how reassuring a normal outcome can be, the examination itself may cause undue worry and distress for some individuals. Some can often feel vulnerable, challenged by preconceived ideas, stories from others, body dysmorphia and prior medical experiences that can all influence the examination. A person-centred approach within a screening setting is imperative to ensure a comfortable experience and support re-attendance at screening.

Since the introduction of the NHSBSP, smaller impalpable abnormalities are found routinely through mammography. The increase in early detection is reported to reduce mortality; however, controversially, there is a debate on the overdiagnosis of malignancy which may prove insignificant if left untreated (Marmot et al., 2013) and this continues (Marmot, 2020). Regardless, once an abnormality is found, treatments are offered with the individual's consent, rather than leaving progression to chance. Some experts argue that the treatment of non-invasive ductal carcinoma in situ (DCIS) may cause undue anxiety with unnecessary surgical procedures; however, it is widely viewed that if left untreated DCIS is a precursor to invasive disease. The national Sloane Study aims to provide evidence on the progression of atypical cells to ascertain those at increased risk (Jenkinson et al., 2022). Therefore, within breast screening services it is important as mammography healthcare professionals we are honest and transparent with individuals to enable engagement, informed choice, and shared decision-making within a person-centred approach to care.

Following breast screening the mammography images within the UK are blind double reported, by two interpreters, with a consensus or arbitration approach to any discrepancies. Once a potential abnormality is identified, a triple assessment process begins, including clinical breast examination (CBE), imaging (mammography and ultrasound) and intervention if necessary (PHE, 2016). Technological advancements with digital mammography systems have enabled smaller screen-detected invasive carcinomas of less than 15mm. With the advancement of technology, digital breast tomosynthesis allows for better evaluation of perceived mammographic abnormalities, reducing the requirement for some needle intervention in normal cases, and increasing confidence in the assessor. Some NHSBSP centres can offer further diagnostic techniques such as contrast-enhanced spectral mammography (CESM) and magnetic resonance imaging (MRI) biopsy. Up to 9% of those who attend routine mammography will require further assessment, with such interventions. However, the majority will have a normal/benign outcome (McCann, Stockton and Godward, 2002). These are known as false positives and may cause psychosocial consequences of increased distress (van der Steeg et al., 2011). Those with a false positive result may experience a similar anxiety level to those receiving a diagnosis (Gøtzsche and Jørgensen, 2013). In Ireland, their breast screening programme found a correlation in reduced re-attendance associated with greater invasive tests (Fitzpatrick et al., 2011).

For some individuals who require a biopsy, it may be a traumatic experience, and may reduce the likelihood of future attendance. However, for others it may highlight the importance of screening (Fitzpatrick et al., 2011). The disparity in re-attendance following assessment illustrates the importance of effective communication and a strong advocate for patient-centred approaches. It can be identified that whilst an audit of a single breast unit found 15% of those who received a false positive diagnosis failed to reattend (Nightingale, Murphy and Borgen, 2015), contrastingly, a European meta-analysis reviewing 340,000 attendees for breast-screening in Denmark, Norway, UK and France and found no significance in re-attendance than those who received a "normal" result (Brewer, 2007). When undergoing the biopsy itself it has been demonstrated that having some preprocedural information prior to the day can improve overall patient satisfaction (Lutjeboer et al., 2015).

Within screening, attendance and compliance are acknowledged as consent, however with the caveat of maintaining this throughout. Individuals may withdraw at any time, rendering an incomplete examination. Understanding the examination is helpful to the individual and by communicating effectively, the practitioner can adapt the technical aspects of mammography to acquire the best diagnostic images possible. The adapted technique is often required for those with mobility and difficulties in maintaining positioning. Limited understanding through cultural or language barriers may impair the examination, but the adoption of translation services within the NHS and transcription of leaflets adapted for specific needs has helped improve understanding. Implications of recall should be conveyed to enable all to understand what may follow and alleviate any unwanted anxieties and support a person-centred approach to care.

## Breast symptomatic services

The symptomatic breast service enables patients to be referred from the general practitioner (GP) within two weeks if they have any concerns (NCIN, 2022). The two week wait (2ww) standard is used for any pathway of suspected cancer. The referral to a one-stop breast clinic allows for triple assessment with discharge on the same day if no intervention is required, thus reducing anxiety. For those that do require intervention, an appointment will be made for their results. Similarly to the NHSBSP, all individuals undergoing interventional biopsies will be discussed at the multidisciplinary team (MDT) meeting to ascertain an agreed patient outcome and pathway.

Within the symptomatic service, patients may encounter a Consultant Breast Surgeon, Advanced Nurse Practitioner or Physician Associate to ascertain the clinical history, provide a clinical examination and justify the request for imaging. Those patients under 40 years of age will have an ultrasound, without the need for a mammogram unless indicated as suspicious, either clinically or on ultrasound. During the imaging, patients will usually be seen by either a Consultant Breast Radiologist, Consultant Radiographer or Advanced Practitioner Radiographer. Often, during imaging, normal and benign results will be discussed, however indeterminate or suspicious findings warranting biopsy may delay the outcome until confirmed on histopathology. In cases concerning malignancy, the differential diagnosis should be discussed with the patient at the time of the biopsy, also known as a "warning shot". For some patients, for example, those with atypia but no definitive diagnosis, a follow-up biopsy may be required for conclusivity. In patients where atypia is found during a breast biopsy, a further biopsy with extensive sampling may upgrade the outcome of the disease by approximately 30% (Pinder et al., 2018). For some, atypia may not necessitate surgery, however mammography surveillance is required to monitor any progression of disease. Following MDT, those patients awaiting their biopsy result requiring further management will often be seen by the Consultant Breast Surgeon to discuss the appropriate treatment pathway, without further delay.

For such patients, anxiety is often presented and can be identified by the practitioner early. Patient anxiety for those individuals with symptomatic abnormalities may present in many ways – fear, anger, grief – so it is important to communicate effectively to recognise these signs. Patients who appear silent and withdrawn may welcome the opportunity to discuss their thoughts and feelings, and within busy breast clinics the practitioner may fail to identify this. Likewise, patients and family members that direct their anger towards waiting times and lack of definitive information may be fearful of a cancerous outcome for their loved ones, and unable to express this. Patients may have already encountered loss through cancer which may resurface feelings of grief and turmoil, with some having an additional personal familial history risk of breast disease.

Most patients want the complete process to be over quickly, and yet they realise the importance of the diagnostic tests and associated interventions. A hospital environment can be worrisome, with some highlighting anxieties regarding their experience of the mammogram itself; fear of the unknown or the potential that something may be found. A complete range of emotions may co-exist, with concerns of the ionising radiation, or the notion of possible damage occurring to their implant, even potential risk to their unborn foetus. Regardless, all patients should receive the opportunity to discuss their worries with practitioners in a comfortable environment. Practitioners should seek informed consent and appreciate patient decisions may not concur with their own views and opinions.

## Breaking bad news

Bad news in breast-imaging can relate to surgery, cancer or recurrence. However, a negative outcome can be perceived in many ways not appreciated by others. BBN is defined as: "any information which adversely and seriously affects an individual's view of his or her future" (Buckman, 1984). For example, the idea of potential chemotherapy side effects may be worse than the surgery itself. A considerable proportion of individuals interpret breast cancer to be life-threatening (Graydon et al., 1997). "Curable" is a term best avoided within the oncology setting with "treatable" preferred. It is important to provide hope without causing unrealistic expectations.

The terminology staff use can be as harmful as how it is communicated. When healthcare professionals speak of cancer, phrases such as "fought" and "lost the battle" are often used, like it is a war, however these can be unhelpful and insensitive to patients. Patients may not feel "brave" or wish to be labelled as a "survivors" and so it is important not to assume any individuals' feelings. It is suggested that healthcare professionals working in a specialist oncology setting have the skills, experience and confidence to effectively communicate (Kalber, 2009).

Radiology has evolved with advancements in technology and the traditionalisms of minimal patient contact are something of the past (Smith and Gunderman, 2010). Within a person-centred approach, we now advocate for individuals to be directly involved in their own care and management which requires complex and challenging discussions.

However, how much information should be disclosed at any one time is uncertain, particularly when patients are informed of unwelcome news which may prevent information recall due to altered perception, thereby increasing anxiety (Jefford and Tattersall, 2002). Many experienced clinicians may struggle to discuss unwelcome news and deflect emotions by focusing on the facts (Maguire, 1999). Clinicians discussing uncomfortable feelings and difficult conversations may experience anticipated anxiety of their own, and deflect difficult conversations (Gilligan et al., 2017; Adler, Riba and Eggly, 2009). Effective communication within such times is essential considering ineffective communication accounts for two thirds of clinical errors (Cooper and Frain, 2017).

## Communicating within a person-centred approach

The way staff communicate with screening and symptomatic clients is key to an effective person-centred approach. A large multisite study representing cancer patients in the UK focused on the communication of information, and revealed 87% wanted all information, both positive and negative

(Jenkins, Fallowfield and Saul, 2001). Patients are increasingly educated due to the accessibility of information; however questionable accuracy from unreviewed sources can lead to misunderstandings. A person-centred approach is considered the most effective with shared decision-making promoting patient engagement (Elwyn et al., 2010; Harvey et al., 2007; Kitson et al., 2013).

There are many courses and models which can effectively support staff training in the workplace, such as NHS Connected (2008). Advanced Communication Skills Training (ACST) can incorporate different communication models; however, the SPIKE model (Baile, 2000) is most often implemented for breaking bad news. Unlike medical training which identifies the Calgary-Cambridge model for clinical interviews, ACST is intended for all NHS cancer care professionals, and although initiated in the NHS Cancer Plan (Department of Health, 2000), reiterated in the Cancer Reform Strategy (Department of Health, 2007).

There is no standardised guidance on communication skills within the mammography educational programmes, despite being considered a core component of practice. The only stipulation is for a diagnosis of breast cancer to be given in the presence of a clinician or nurse specialist (NICE, 2002) and individuals should be communicated of the outcome of their imaging in less than three appointments from the initial screen (inclusive of all required biopsies) to minimise anxiety and undue delay (Public Health England, 2017). At an advanced clinical level, accredited advanced communication skills training is required for those working in a NHSBSP setting (NICE, 2002). As core members of the multi-disciplinary team, breast practitioners should routinely attend the MDT meetings to discuss diagnostic patient management decisions (Public Health England, 2017).

## Practical techniques to assist a person-centred approach: Impact on health of practitioner

There are many factors which can arise that can impact the health of the practitioner; having limited support in the workplace, being placed into a challenging situation such as an individual reacting negatively during a procedure, finding out difficult news about your patient, or having multiple challenging cases during your working day. In these instances, practitioners may find that they start to experience stress, anxiety or even burnout (Mainiero & Parikh, 2019; Mercer, 2022), all of which can lead to negative effects on both the mental and physical health of practitioners. It is important to be aware of the warning signs of those factors coming into effect, and to understand how to both tolerate and ease those feelings to reduce the likelihood of practitioner illness and/or errors, both of which can occur when practitioners experience stress, anxiety or burnout (Hall et al., 2016; Patel et al., 2019; Mercer, 2022).

*Managing your own health and getting the appropriate support as a practitioner is important to be able to support an effective person-centred approach.*

There are three main ways to achieve this:

1. Be aware of warning signs
2. Get support from the workplace
3. Practice self-help techniques

**Be aware of your warning signs**

The first step in reducing feelings of anxiety and stress, or helping manage burnout, is to be aware of when these factors are happening or are about to happen. To do this, practitioners should take some time to identify what those warning signs are, so the next time they occur they can identify this as a sign to slow down or seek support. Common symptoms of anxiety, as noted by the DSM V (American Psychiatric Association, 2013), include, but are not limited to, feeling: fatigued, restless, irritable, having tense muscles, disturbed sleep, difficulty concentrating, nervousness, rapid breathing, sweating, trembling, and an increased heart rate. These are also similarly presenting with stress and burnout.

Whilst being aware of what the common signs of anxiety, stress and burnout are, it is important to take the time to identify what the individual's warning signs are. This can be achieved using the five areas model (Williams, 2012). Mostly used in cognitive behavioural therapy (CBT), this model helps break down problems into smaller, more easily identifiable, sections, consisting of thoughts, feelings, behaviours and physical sensations. However, if you are already aware of one of these areas, for example if you experience the emotion anxiety, it can also be used to help identify what the other areas consist of and identify a clearer picture of what you may experience when you feel anxious or are beginning to feel anxious. Figure 12.1 is a representative example of a five areas model:

**Get support from the workplace**

Having a supportive work environment can be helpful in preventing staff members from getting overwhelmed or burning out (Aronsson, 2016). Drawing from the literature, there are thirteen suggested factors which can help ensure a workplace is a supportive environment for staff members (Collins, 2016), with eight key factors when put into practice (Aronsson et al., 2017; Dyrbye & Shanafelt, 2016; Riley et al., 2021). These factors are detailed in Table 12.1.

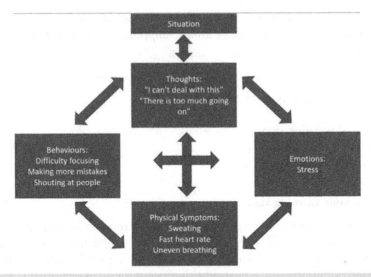

This example focuses on the emotion 'stress'; if for example you notice you are raising your voice or are getting thoughts that there is 'too much going on', this could be one of your warning signs which indicate that you are feeling stressed.

If you get stuck with completing your model, it can be helpful to ask yourself, 'what do I normally do/think/feel, when I get stressed/anxious?'. You may wish to look back at a previous time when you were stressed and consider what behaviours you did, or did not do during that time, or what thoughts were occurring in your mind. It may also be beneficial to consider what was occurring around you at the time when you experienced this emotion, to see if you can identify a specific situation that causes these aspects to take route. For example, you may identify that whenever you are at work on a Tuesday you notice these emotions and behaviours occur more frequently, indicating that it is maybe something occurring on that specific day that is causing the stress to intensify.

This formulation will likely not be the same for each of you, and it may be that your individual models for stress, anxiety and burnout differ, or in some cases are exactly the same. As long as you start identifying those thoughts, behaviours and physical sensations you experience with these emotions, and if possible, consider situations that you know contribute to that emotion, this can help you to be more aware of when you are getting to a stage where you need to seek out some support.

*Figure 12.1* A representative example of a five areas model around the emotion "stress".

*Table 12.1* Factors to promote a supportive workplace environment

| Organisational Culture | Psychological and Social Support | Clear Leadership and Expectations | Civility and Respect |
|---|---|---|---|
| Providing a fair, honest and trustworthy environment | Co-workers and Supervisors provide a supportive environment with a good understanding of employee mental health | Effective leadership and support which keeps staff informed of changes, provides feedback and clear communication | Respectful and considerate environment with individuals from all backgrounds fairly treated |
| **Psychological Job Demands** | **Growth and Development** | **Recognition and Reward** | **Workload Management** |
| Environment which provides a good fit between job requirements and employee emotional and interpersonal competencies | Providing support and encouragement for employee skill development | Acknowledging and appreciating employee success and efforts | Providing enough time and resources for successful accomplishment of tasks and responsibilities |
| **Engagement** | **Work/Life Balance** | **Psychological Protection from Violence, Bullying and Harassment** | **Protection of Physical Safety** |
| Providing an environment which encourages employee motivation and engagement | Recognition that balance is important | Providing a supportive environment which handles incidents in a supportive and appropriate way | Taking health and safety concerns seriously, considering how the work environment can impact mental health, assessing and altering the demands of the job and its environment for employee benefit |
| **Other Chronic Stressors as Identified by Workers** Acknowledging staff stressors and adapting as appropriate | | | |

It is the workplace's responsibility to ensure that each of these conditions are met. Time should be taken to focus on each factor and consider ways to aid in their improvement or ensure that there are no areas which are lacking. This could be doing something as simple as providing regular one-to-one meetings for staff (Fischer et al., 2019; Oates, 2018), ensuring any changes in the workplace are openly and frequently communicated throughout the team (Schulz-Knappe et al., 2019), providing/supporting adaptive needs on an individual case-by-case basis (Graham, 2019), ensuring training and development needs are met (Rodriguez & Walters, 2017), or even considering the development of serenity rooms where staff can go to relax/unwind for a few moments during a difficult time within a working day (Salmela et al., 2020). Ensuring these aspects are met can vastly aid in the reduction of stress and anxiety in practitioners (Mercer, 2022). Serenity rooms can often be supported in combination with a clinical room; such combination of a relaxing working environment may be seen to support practitioners' health and wellbeing, and thus aid a person-centred approach to care.

### Practise self-help techniques

What the organisation does is often outside of staff members' control, however individual staff members can put in place several techniques to additionally aid the reduction of stress and anxiety and improve overall wellbeing; these can help take attention away from what is causing the stress/anxiety, refocusing it on aspects such as breathing, which can slow the heart rate and reduce tension. There are many different types of relaxation techniques such as autogenic training (Holland et al., 2017), sense imagery (Smith, 2002), progressive muscle relaxation (Torales et al., 2020), and mindfulness (McConville et al., 2017). An example of a common relaxation technique used can be seen in Figure 12.2 (National Health Service, 2022).

There are also techniques such as problem solving (Smith, 2002) and cognitive restructuring (Clark, 2013), which tackles the issue causing stress or anxiety more head on. For more information on how to practise these techniques refer to "Managing Anxiety in Mammography: The Client and the Practitioner", which discusses these in more depth (Mercer, 2022).

### Practical techniques to assist a person-centred approach: Impact on health of individual receiving care

Once staff have supported their own health, they can maintain, focus and be able to support a person-centred approach. It is important to ensure a person-centred approach is provided to individuals; understanding the signs and symptoms of their anxiety or distress can help with this. It can enable the practitioner to adjust the situation the individual is in, so it has less of a negative impact on them. Those undertaking breast imaging or a breast procedure may have anxiety and distress and present in similar ways to practitioners;

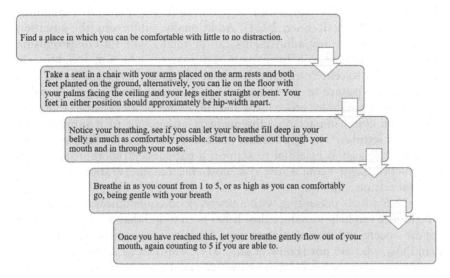

*Figure 12.2* Relaxation technique to help reduce anxiety and stress.

the difference being they may not express that they are anxious, and the practitioner will have to pay close attention for any warning signs. This could present as trembling, sweating, rapidly breathing, being irritable, avoiding the situation, or expressing they wish to leave or are in pain. There are also other factors such as being tired and having an increased heart rate which are tell-tale signs of distress/anxiety, however these may be harder to identify (APA, 2013; Mercer, 2022).

If an individual is presenting with anxiety, there are two main ways practitioners can help to support this, which comes in the form of reducing and preventing. Reducing considers what techniques can be used in the situation an individual is presenting with anxiety or distress when the practitioner is there. Preventing, on the other hand, will occur when the individual is not present, and the practitioner or workplace considers what techniques or actions can be done or taken to prevent the anxiety/distress from occurring when the individual does show up, or even before or after they attend.

## Reducing

Pain and discomfort have been found to be linked to anxiety (Maimone et al., 2020; Mercer et al., 2022), therefore, by targeting ways to reduce pain in individuals, this will naturally help to reduce their anxiety. If an individual presents with, or expresses, pain during the session, one of the best ways to help reduce their distress around this is to acknowledge that their pain is real and normalise that some pain can occur during the procedure at times. By simply acknowledging this and not dismissing it, anxiety around the pain can reduce, which can help minimise any pain felt (Maimone et al., 2020; Pai & Rebner, 2021). Additionally, addressing the situation and determining if any adjustments can be made, without compromising the image, can further help in pain and anxiety reduction (Nightingale et al., 2015). Recent research (Nelson et al., 2022) has highlighted that five-minute exposure to non-binaural and binaural music improves tolerance to mammography procedures and this could be an essential intervention which could impact on subsequent re-attendance to mammography in certain settings. The application prior and during interventional biopsies could be seen as essential to support a patient centred approach, with introduction prior being a low risk, cost-effective intervention within the hospital setting.

For individuals presenting with anxiety, good communication skills as a practitioner can help to reduce this. It has been found that practitioners who appear sincere and empathetic are more likely to ease anxiety in individuals than those who are not (Louw et al., 2014). This can be achieved by using positive communication behaviours, which encompass both verbal and non-verbal behaviours. A practitioner could use empathetic statements, simple language and normalise the situation the individual is in to demonstrate good verbal behaviours. Similarly, they could use a calming tone of voice, smile and maintain eye contact to show positive non-verbal behaviours (Lille & Marshall, 2018; Mercer, 2022; Pai & Rebner, 2021). In addition, providing a good delivery of information around what the procedure entails can further help reduce anxiety in individuals; this can be achieved in the session by the practitioner, or beforehand through the use of informational leaflets (Meguerditchian et al., 2012; Mercer, 2022).

Being aware of some short breathing exercises that you could talk through with an individual who is having a panic or anxiety attack does not take up too much clinical time, and can be useful in reducing any anxiety experienced during the procedure. A good example is the Wehrenberg (2018) in 2, out 2-4-6-8 exercise which takes around a minute to complete.

### Preventing

Research has indicated that aspects such as machine or paddle design can contribute to increased anxiety in those having a mammography procedure; for example, the hard edges of the paddle has been identified as a source of discomfort in screening due to the resultant pinching it can cause (Smith, 2017), therefore having a softer edged paddle may aid in preventing pain from occurring and the resultant anxiety that comes with this. Additionally, mood lighting and colour has been found to influence anxiety in clients (Mercer, 2022); introducing lighting into the mammography room or around the machine, such as blue or green light, can further help prevent anxiety from occurring (Al-Ayash et al., 2016; Pacifici, 2016) and indeed many manufacturers are now introducing this to their machine design.

Another method which can be used to help prevent anxiety from arising in mammography individuals is to provide informative leaflets or psychoeducational sessions to them before arriving at the screening. Keeping individuals informed around their procedure and what to expect has been found to reduce anxiety on multiple occasions (Whelehan et al., 2017; Sharaf & Hafeez, 2019; Setyowibowo et al., 2022). Additionally, simply including explanations in a leaflet of what anxiety is and how this can present, along with a few relaxation techniques to help reduce this anxiety, can further aid in preventing anxiety from occurring during the procedure, and can allow the individual to subdue any anxiety that does arise more effectively (Mercer, 2022).

## Alleviating professional stress

Emotional stress is known to be a major cause of communication failures in healthcare, as highlighted in the Francis Report (Hughes, 2013). Task performance will decrease in a stressful environment, which may be detrimental to an individual or whole service (Orfus, 2008). As highlighted, burnout and compassion fatigue can manifest as emotional or physical exhaustion (Bragard et al., 2010). This is often cited within radiology by sonographers communicating unwelcome and life-changing information to patients and their families (Thomas, O'Loughlin and Clarke, 2017). Regular debriefings will help to alleviate burnout and should be considered within the service. It should not be trivialised that increasing workload pressures may severely impact stress levels, reducing overall satisfaction; however, a supportive management structure and workplace environment is essential.

## Conclusion

The provision of a person-centred approach is essential within the breast imaging setting to ensure the supportive advancement in services. It is not until staff are supported effectively that a person-centred approach can be successfully delivered. Patient-centred care can therefore only be fulfilled with an integrated, holistic approach to professional well-being.

## References

Adler, D.D., Riba, M.B. and Eggly, S., 2009. Breaking bad news in the breast imaging setting. *Academic Radiology*, 16(2), pp.130–135.

Al-Ayash, A., Kane, R.T., Smith, D. and Green-Armytage, P., 2016. The influence of color on student emotion, heart rate, and performance in learning environments. *Color Research & Application*, 41(2), pp.196–205.

American Psychiatric Association, 2013. 5th ed. American Psychiatric Association: Diagnostic and Statistical Manual of Mental Disorders, Arlington.

Aronsson, G., Theorell, T., Grape, T., Hammarström, A., Hogstedt, C., Marteinsdottir, I., Skoog, I., Träskman-Bendz, L. and Hall, C., 2017. A systematic review including meta-analysis of work environment and burnout symptoms. *BMC Public Health*, 17(1), pp.1–13.

Baile, W.F., Buckman, R., Lenzi, R., Glober, G., Beale, E.A. and Kudelka, A.P., 2000. SPIKES – a six-step protocol for delivering bad news: application to the patient with cancer. *The Oncologist*, 5(4), pp.302–311.

Bragard, I., Etienne, A.M., Merckaert, I., Libert, Y. and Razavi, D., 2010. Efficacy of a communication and stress management training on medical residents' self-efficacy, stress to communicate and burnout: a randomized controlled study. *Journal of Health Psychology*, 15(7), pp.1075–1081.

Brewer, N.T., Salz, T. and Lillie, S.E., 2007. Systematic review: the long-term effects of false-positive mammograms. *Annals of Internal Medicine*, 146(7), pp.502–510.

Buckman, R., 1984. Breaking bad news: why is it still so difficult? *British Medical Journal* (Clinical research ed.), 288(6430), p.1597.

Clark, D.A., 2013. Cognitive restructuring. *The Wiley handbook of cognitive behavioral therapy*, pp.1–22. Wiley.

Collins, J., 2016. Assembling the pieces: an implementation guide to the national standard for psychological health and safety in the workplace. CSA Group; 2014.

Cooper, N. and Frain, J. eds., 2017. *ABC of Clinical Communication*. John Wiley & Sons.

Department of Health, 2000. The NHS Cancer Plan. Department of Health (September), pp. 1–98. NHSCancerPlan.pdf (thh.nhs.uk).

Department of Health, 2007. Cancer reform strategy. Department of Health.

Dyrbye, L. and Shanafelt, T., 2016. A narrative review on burnout experienced by medical students and residents. *Medical Education*, 50(1), pp.132–149.

Elwyn, G., Laitner, S., Coulter, A., Walker, E., Watson, P. and Thomson, R., 2010. Implementing shared decision making in the NHS. *Bmj*, 341.

Fischer, J., Alpert, A. and Rao, P., 2019. Promoting intern resilience: individual chief wellness check-ins. *MedEdPortal*, 15, p.10848.

Fitzpatrick, P., Fleming, P., O'neill, S., Kiernan, D. and Mooney, T., 2011. False-positive mammographic screening: factors influencing re-attendance over a decade of screening. *Journal of Medical Screening*, 18(1), pp.30–33.

Gilligan, T., Coyle, N., Frankel, R.M., Berry, D.L., Bohlke, K., Epstein, R.M., Finlay, E., Jackson, V.A., Lathan, C.S., Loprinzi, C.L. and Nguyen, L.H., 2018. Patient-clinician communication: American Society of Clinical Oncology consensus guideline. *Obstetrical & Gynecological Survey*, 73(2), pp.96–97.

Gøtzsche, P.C. and Jørgensen, K.J., 2013. Screening for breast cancer with mammography. *Cochrane Database of Systematic Reviews* (6), 1–73.

Graham, K.M., McMahon, B.T., Kim, J.H., Simpson, P. and McMahon, M.C., 2019. Patterns of workplace discrimination across broad categories of disability. *Rehabilitation Psychology*, 64(2), p.194.

Graydon, J., Galloway, S., Palmer-Wickham, S., Harrison, D., Bij, L.R.V.D., West, P., Burlein-Hall, S. and Evans-Boy den, B., 1997. Information needs of women during early treatment for breast cancer. *Journal of Advanced Nursing*, 26(1), pp.59–64.

Hall, L.H., Johnson, J., Watt, I., Tsipa, A. and O'Connor, D.B., 2016. Healthcare staff wellbeing, burnout, and patient safety: a systematic review. *PloS one*, 11(7), p.e0159015.

Harvey, J.A., Cohen, M.A., Brenin, D.R., Nicholson, B.T. and Adams, R.B., 2007. Breaking bad news: a primer for radiologists in breast imaging. *Journal of the American College of Radiology*, 4(11), pp.800–808.

Holland, B., Gosselin, K. and Mulcahy, A., 2017. The effect of autogenic training on self-efficacy, anxiety, and performance on nursing student simulation. *Nursing Education Perspectives*, 38(2), pp.87–89.

Hughes, G., 2013. Mid-Staffordshire – the Francis Report. *Emergency Medicine Journal*, 30(6), pp.432–432.

Itri, J.N., 2015. Patient-centered radiology. *Radiographics*, 35(6), pp.1835–1846.

Jefford, M. and Tattersall, M.H., 2002. Informing and involving cancer patients in their own care. *The Lancet Oncology*, 3(10), pp.629–637.

Jenkins, V., Fallowfield, L. and Saul, J., 2001. Information needs of patients with cancer: results from a large study in UK cancer centres. *British Journal of Cancer*, 84(1), pp.48–51.

Jenkinson, D., Freeman, K., Clements, K., Hilton, B., Dulson-Cox, J., Kearins, O., Stallard, N., Wallis, M.G., Sharma, N., Kirwan, C. and Pinder, S., 2022. Breast screening atypia and subsequent development of cancer: protocol for an observational analysis of the Sloane database in England (Sloane atypia cohort study). *BMJ open*, 12(1), p.e058050.

Kalber, B., 2009. Breaking bad news–whose responsibility is it? *European Journal of Cancer Care*, 18(4), pp.330–330.

Kitson, A., Marshall, A., Bassett, K. and Zeitz, K., 2013. What are the core elements of patient-centred care? A narrative review and synthesis of the literature from health policy, medicine and nursing. *Journal of Advanced Nursing*, 69(1), pp.4–15.

Lille, S. and Marshall, W., 2018. *Mammographic Imaging*. Lippincott Williams & Wilkins.

Louw, A., Lawrence, H. and Motto, J., 2014. Mammographer personality traits–elements of the optimal mammogram experience. *Health SA Gesondheid*, 19(1).

Lutjeboer, J., Burgmans, M.C., Chung, K. and van Erkel, A.R., 2015. Impact on Patient Safety and Satisfaction of Implementation of an Outpatient Clinic in Interventional Radiology (IPSIPOLI-Study): a quasi-experimental prospective study. *Cardiovascular and Interventional Radiology*, 38(3), pp.543–551.

Mackay, S.J., Hogg, P., Cooke, G., Baker, R.D. and Dawkes, T., 2012. A UK-wide analysis of trait emotional intelligence within the radiography profession. *Radiography*, 18(3), pp.166–171.

Maguire, P., 1999. Improving communication with cancer patients. *European Journal of Cancer*, 35(14), pp.2058–2065.

Maimone, S., Morozov, A.P., Wilhelm, A., Robrahn, I., Whitcomb, T.D., Lin, K.Y. and Maxwell, R.W., 2020. Understanding patient anxiety and pain during initial image-guided breast biopsy. *Journal of Breast Imaging*, 2(6), pp.583–589.

Mainiero, M.B. and Parikh, J.R., 2019. Recognizing and overcoming burnout in breast imaging. *Journal of Breast Imaging*, 1(1), pp.60–63.

Marmot, M., 2020. Health equity in England: the Marmot review 10 years on. *Bmj*, 368.

Marmot, M.G., Altman, D.G., Cameron, D.A., Dewar, J.A., Thompson, S.G. and Wilcox, M., 2013. The benefits and harms of breast cancer screening: an independent review. *British Journal of Cancer*, 108(11), pp.2205–2240.

McCann, J., Stockton, D. and Godward, S., 2002. Impact of false-positive mammography on subsequent screening attendance and risk of cancer. *Breast Cancer Research*, 4(5), pp.1–9.

McConville, J., McAleer, R. and Hahne, A., 2017. Mindfulness training for health profession students – the effect of mindfulness training on psychological well-being, learning and clinical performance of health professional students: a systematic review of randomized and non-randomized controlled trials. *Explore*, 13(1), pp.26–45.

Meguerditchian, A.N., Dauphinee, D., Girard, N., Eguale, T., Riedel, K., Jacques, A., Meterissian, S., Buckeridge, D.L., Abrahamowicz, M. and Tamblyn, R., 2012. Do physician communication skills influence screening mammography utilization? *BMC Health Services Research*, 12(1), pp.1–8.

Mercer, C., Hogg, P. and Kelly, J. eds., 2022. *Digital mammography: A holistic approach*. Springer Nature.

Mercer, J.E., 2022. Managing Anxiety in Mammography: The Client and the Practitioner. In *Digital Mammography* (pp. 137–153). Springer, Cham.

National Health Service. (2022). Breathing Exercises for Stress. Available from: www.nhs.uk/mental-health/self-help/guides-tools-and-activities/breathing-exercises-for-stress/ [Accessed 5 October 2022].

Nelson, D., Berry, R., Szczepura, K. and Mercer, C.E., 2023. Assessing the impact of binaural and non-binaural auditory beat intervention to pain and compression in mammography. *Radiography*, 29(1), pp.101–108.

Nightingale, J.M., Borgen, R., Porter-Bennett, L. and Szczepura, K., 2015. An audit to investigate the impact of false positive breast screening results and diagnostic work-up on re-engagement with subsequent routine screening. *Radiography*, 21(1), pp.7–10.

Nightingale, J.M., Murphy, F.J., Robinson, L., Newton-Hughes, A. and Hogg, P., 2015. Breast compression–an exploration of problem solving and decision-making in mammography. *Radiography*, 21(4), pp.364–369.

Oates, J., 2018. What keeps nurses happy? Implications for workforce well-being strategies. *Nursing Management*, 25(1).

Orfus, S., 2008. The effect test anxiety and time pressure on performance. *The Huron University College Journal of Learning and Motivation*, 46(1).

Pacifici, S., 2016. Decreasing anxieties in women undergoing mammography. *J Women's Health Care*, 5(314), pp.2167–0420.

Pai, V.R. and Rebner, M., 2021. How to minimize patient anxiety from screening mammography. *Journal of Breast Imaging*, 3(5), pp.603–606.

Paluch, J., Kohr, J., Squires, A. and Loving, V., 2022. Patient-centered care and integrated practice units: embracing the breast care continuum. *Journal of Breast Imaging*, 4(4), pp.413–422.

Patel, R.S., Sekhri, S., Bhimanadham, N.N., Imran, S. and Hossain, S., 2019. A review on strategies to manage physician burnout. *Cureus*, 11(6).

Pinder, S.E., Shaaban, A., Deb, R., Desai, A., Gandhi, A., Lee, A.H.S., Pain, S., Wilkinson, L. and Sharma, N., 2018. NHS Breast Screening multidisciplinary working group guidelines for the diagnosis and management of breast lesions of uncertain malignant potential on core biopsy (B3 lesions). *Clinical Radiology*, 73(8), pp.682–692.

Prang, K.H., Maritz, R., Sabanovic, H., Dunt, D. and Kelaher, M., 2021. Mechanisms and impact of public reporting on physicians and hospitals' performance: A systematic review (2000–2020). *PloS one*, 16(2), p.e0247297.

Public Health England (PHE), 2016. Breast screening: clinical guidelines for screening assessment.

Public Health England, 2017. NHS Breast Screening Programme: Guidance for breast screening mammographers. [Online] NHSBSP PHE gateway number: 2017607, 3rd edition. Available from: www.gov.uk/government/publications/breast-screening-quality-assurance-for-mammography-and-radiography.

Riley, R., Kokab, F., Buszewicz, M., Gopfert, A., Van Hove, M., Taylor, A.K., Teoh, K., Martin, J., Appleby, L. and Chew-Graham, C., 2021. Protective factors and sources of support in the workplace as experienced by UK foundation and junior doctors: a qualitative study. *BMJ open*, 11(6), p.e045588.

Rodriguez, J. and Walters, K., 2017. The importance of training and development in employee performance and evaluation. *World Wide Journal of Multidisciplinary Research and Development*, 3(10), pp.206–212.

Salmela, L., Woehrle, T., Marleau, E. and Kitch, L., 2020. Implementation of a "Serenity Room": Promoting resiliency in the ED. *Nursing 2020*, 50(10), pp.58–63.

Schulz-Knappe, C., Koch, T. and Beckert, J., 2019. The importance of communicating change: Identifying predictors for support and resistance toward organizational change processes. *Corporate Communications: An International Journal*, 670–686.

Setyowibowo, H., Yudiana, W., Hunfeld, J.A., Iskandarsyah, A., Passchier, J., Arzomand, H., Sadarjoen, S.S., de Vries, R. and Sijbrandij, M., 2022. Psychoeducation for breast cancer: A systematic review and meta-analysis. *The Breast*, 62, 36–51. doi: 10.1016/j.breast.2022.01.005. Epub 2022 Jan 12. PMID: 35121502; PMCID: PMC8819101.

Sharaf, A.Y. and Hafeez, N.A., 2019. Effect of nursing interventions on pain and anxiety among women undergoing screening mammography. *International Journal of Novel Research in Healthcare and Nursing*, 6(3), pp.454–69.

Smith, A., 2017. Improving patient comfort in mammography. *Hologic*: WP-00119 Rev, 1.

Smith, J.C., 2002. Stress management: A comprehensive handbook of techniques and strategies. Springer Publishing Company.

Smith, J.N. and Gunderman, R.B., 2010. Should we inform patients of radiology results? *Radiology*, 255(2), pp.317–321.

Thomas, S., O'Loughlin, K. and Clarke, J., 2017. The 21st century sonographer: role ambiguity in communicating an adverse outcome in obstetric ultrasound. *Cogent Medicine*, 4(1), p.1373903.

Torales, J., O'Higgins, M., Barrios, I., González, I. and Almirón, M., 2020. An overview of jacobson's progressive muscle relaxation in managing anxiety. *Revista Argentina de Clinica Psicologica*, 29(3), pp.17–23.

Van der Steg, A.F.W., Keyzer-Dekker, C.M.G., De Vries, J. and Roukema, J.A., 2011. Effect of abnormal screening mammogram on quality of life. *Journal of British Surgery*, 98(4), pp.537–542.

Wehrenberg, M., 2018. The 10 best-ever anxiety management techniques: understanding how your brain makes you anxious and what you can do to change it (second). WW Norton & Company.

Whelehan, P., Evans, A. and Ozakinci, G., 2017. Client and practitioner perspectives on the screening mammography experience. *European Journal of Cancer Care*, 26(3), p.e12580.

Williams, C., 2012. *Overcoming anxiety, stress and panic: A five areas approach.* CRC Press.

# 13 Patient care in breast imaging

## A UK context

*Judith Kelly and Peter Hogg*

## Introduction

Individuals attending the NHS for breast imaging enter through one of two distinctly different pathways, namely NHS Breast Screening Programme (NHSBSP, www.gov.uk/topic/population-screening-programmes/breast) and as 'individuals being referred from the family doctor (commonly referred to as General Practitioner (GP) within the UK). NHSBSP invites women aged 50–70 for triennial mammograms. Such women are usually asymptomatic, though not always. Women over 70 are not routinely invited but are encouraged to self-refer if they wish. Individuals (of any gender) with a perceived breast symptom which may represent pathological change are referred through their family doctor to a specialist breast clinic for assessment by a range of different professionals (including breast surgeons, breast care nurses, radiologists, radiographers, and pathologists) so that the appropriate diagnostic work up can take place in a timely manner. This is important not only for patients' physical well-being but understandably crucial for psychological health too.

Given the two entry pathways outlined above, breast imaging using mammography or ultrasound (or both) for initial diagnostic work up consequently presents practitioners with a complex case mix, comprising patients and clients. Patients include those who already have an existing breast problem which is being investigated further, or they have a definitive diagnosis and are placed onto an appropriate treatment pathway. Alternatively, they may be new into the healthcare system having been referred as symptomatic via their family doctor due to a suspicious or equivocal [breast] finding on physical examination. Patients presenting for breast imaging will have varying levels of understanding and anxiety regarding any breast pathology they may have. By contrast, clients are usually asymptomatic, 'well women' and because of this they are not classified as 'patients'. Generally speaking, clients attend for routine breast screening mammography anticipating that nothing abnormal will be found and that they can get on with life normally afterwards. In the context of breast imaging, clients and patients have similar and different

DOI: 10.1201/9781003310143-18

care requirements and individual concerns. Both groups will be considered in this chapter. Prior to imaging, practitioners will be adequately aware of whether they are to image a patient or client since attendees are separated into different clinics, dependant upon their entry pathway.

This chapter outlines a UK perspective for patient care in breast imaging. The chapter commences by setting the political scene and relevant historical factors which influence contemporary practice, for example. Major changes in healthcare delivery, and more significantly regarding the personnel providing the care were instigated within the UK before the turn of the twenty-first century. The [UK] NHS and Community Care Act (NHSE 1990) initially introduced significant changes to health care provision by moving to a system of internal markets. Traditional methods of health care delivery were challenged, causing blurring of professional boundaries, especially between radiographers and radiologists. To meet growing demand, many healthcare organisations created new roles for nurses and professions allied to medicine, such as radiography. Multiple [UK] Department of Health (DoH; 1992; 1995; 1997; 2000a and 2000b) publications outlined major plans to modernise the NHS to meet demand into the twenty-first century. Innovation was seen as integral to modernisation and the development of a workforce capable of delivering high-quality patient-centred care. A key objective was to make cancer and health services more responsive to patients needs by enabling staff to renegotiate their roles, responsibilities, and practise across traditional professional boundaries, ensuring seamless delivery of care. Against this background an innovative staffing model was established within the radiography profession and piloted in the National Health Service Breast Screening Programme (NHSBSP) (NHSBSP 2002). This model is now fully implemented across the UK and based on team working, called the 'Four-tiered structure' and comprised of:

- assistant practitioners
- state-registered practitioners (diagnostic radiographer)
- advanced practitioners
- consultant practitioners

Assistant practitioners are trained to undertake mammography imaging under the direction of qualified diagnostic radiographers who have specialised in mammography. Typically, their training lasts for two years and is at sub degree level. [Diagnostic] radiographers have a broad base training at degree level which lasts three years. On graduation they can specialise in mammography imaging by completing a postgraduate qualification lasting a year. After this, the mammography qualified radiographer can engage in further postgraduate study, often leading to a master's degree, and become an advanced practitioner in breast imaging. The advanced practitioner can perform an extended range of duties, including image reading (reporting), performing and interpreting ultrasound examinations and undertaking image-guided breast interventional procedures under the supervision of a Consultant Radiologist/ Radiographer. Beyond this, the advanced practitioner may study further post-graduate courses and become a consultant radiographer practitioner. Such consultant radiographers carry the same diagnostic and patient care respon-sibility as a radiologist within breast imaging. The consultant radiographer in breast imaging generally carries all the responsibilities of an advanced prac-titioner plus many more, working without direct supervision and in many ways their role can be indistinguishable from that of a consultant radiolo-gist. The consultant radiographer plays a full role in multidisciplinary team meetings (MDTMs) where patient diagnostic work up and treatment plans are decided, they are often engaged in research, teaching within clinical and academic settings – which frequently involves teaching a multidisciplinary audience including junior radiologists; administering medicines and demon-strating leadership in innovation and service developments. Prescribing of medicines is currently in a transition phase and we anticipate that within one to two years consultant radiographers will be able to do this on the same footing as a consultant radiologist. Occasionally difficult conversations with patients involving Duty of Candour or Disclosure of Audit when a patient may have had a delayed diagnosis are required and carried out by con-sultant radiographers. Interestingly, breast imaging has the highest number of consultant radiographers compared with any other imaging or therapy modality within the UK. It is worth noting that an early influencing factor in the development of advanced and consultant radiography practice was that the number of UK radiologists failed to keep pace with the greatly expanded workload within imaging departments (RCR 2002.) This is the result of many new and increasingly sophisticated imaging methods, demands for interven-tional procedures and overall rising numbers of referrals for all aspects of imaging due to an increasing dependency on imaging by clinicians within the healthcare system in decision making and patient pathways. Patient expectations inevitably also play a part in contemporary healthcare which is driven by accessibility to multiple online resources, not least social media sites too.

With the high level of responsibility that consultant radiographers hold in breast examination, from imaging the patient to discussing treatment plans with surgeons, the range of care they must provide to clients is extensive and must be comprehensive. Not only are patients/clients able to receive more timely care (which must be of at least the same standard as that provided by a consultant radiologist) but there are significant NHS cost savings in terms of staff remuneration since even the highest paid [consultant] radiographers do not receive the same salary as a medical consultant.

Building on the introductory context, the chapter will now illustrate how breast imaging services have progressed in the development of patient-centred care thus far and challenges will be identified regarding current perceived unmet and future needs.

## Patient pathways

Having gained an understanding of the career structure and responsibilities of radiographers within UK breast imaging let us now examine the current position of patient pathways. As already indicated, individuals enter the UK breast imaging pathway initially through one of two main routes: NHSBSP routine invitation; and Symptomatic Services.

## NHSBSP routine invitation

From the outset the NHSBSP was configured to be as convenient as possible to encourage maximum uptake by women. For example, this included the siting of mobile breast screening vans in the community such as supermarket car parks and shopping centres, away from hospital sites where possible to avoid the association with illness/health problems and the stresses of parking on hospital sites. That said, some women in remote geographical areas may still have some distance to travel to access screening. NHSBSP literature was also translated into multiple different languages to ensure that non-English speakers were not disadvantaged due to a lack of understanding of the process in a multicultural population. All women who are registered with a family doctor and have a postal address receive an invitation every three years. Family doctor practices are targeted on a rotational basis such that women from a particular geographical area tend to receive their invitations over a particular, short time period and once all eligible women have received an invitation, the next family doctor practice is targeted. This invitation cycle continually repeats itself. Each designated breast screening unit in the UK has its own screening population and is responsible for delivering a screening service to the point of diagnosis if a woman is found to have a mammographic abnormality and recalled for diagnostic work up.

Following some criticism, a number of years ago, that the screening programme overall was too coercive and didn't explain the risks as well as benefits of screening the NHSBSP has sought to encourage women to make an informed choice regarding whether to attend for screening or not. The current invitation includes information which seeks to provide a balanced approach and outline in lay terms that whilst screening does save lives from breast cancer it does also carry risks (www.gov.uk/phe/breast-screening-leaflet). The main risk is that cancers may be found that would never have caused harm in a woman's lifetime, i.e. they would have died of some other condition. This is termed 'Overdiagnosis' and defined as 'detection of cancers on screening that would not have become apparent were it not for the screening test' (IARC 2002). Obviously if that were the case the woman (and her family) would be spared the painful, stressful experience of being given a cancer diagnosis, undergoing unpleasant treatment and then carrying on their lives as a 'cancer' patient with the worry that the cancer may return. An additional important point explained is that screening does not detect all cancers that may be present either because it is not visible or occasionally cancers are missed by the 'image readers'. Image readers is a collective term for those who interpret images to make diagnoses and comprises radiographers, radiologists and others (e.g. surgeons).

Women over seventy are not automatically invited for screening but are provided with information if they attend following their last routine invitation. This explains that they can request screening for as long as they wish along with the risks/benefits of doing so (www.gov.uk/phe/breast-screening-71-or-over). An HTML version of the leaflets is available to be viewed and downloaded in large print, a screen reader can provide an audio version and a braille version is available too, as are translations into multiple different languages.

In recent years it has been increasingly acknowledged that several client groups are disadvantaged in accessing breast screening services and this leads to reduced uptake of invitations and then more likely several late, symptomatic cancer presentations. This can result in poorer outcomes for such women – since screening operates on the premise that early detection confers a better prognosis and cosmetic outcome, with less aggressive treatment required. The reasons certain groups may be or feel disadvantaged can be variable; for example, physical difficulty for wheelchair users accessing the mobile screening van which has steps for access, language and understanding issues (already mentioned), socio-economically deprived communities where healthcare is often not prioritised by the local population. Perceived problems can be overcome; for example, wheelchair users may arrange to have their screening at a static site with easy wheelchair access (lifts and wider doors) and healthcare organisations have Language Line facilities and translators available by telephone to assist with communications.

To address problems, additional funding has been made available to recruit personnel called Cancer Screening Improvement Leads (CSILs) or Breast Screening Coordinators (BSCs) whose role is to identify specific needs in their local screening populations, implement strategies appropriately and monitor the outcomes. Such personnel do not come from any specific healthcare professional or administrative group but need to demonstrate suitable education and/or experience/knowledge of health promotion/screening services to be deemed competent to undertake the role. The overall aim is to improve uptake and reduce inequalities in relation to populations ability to access screening services and encourage attendance. Different geographical, cultural, socioeconomic populations and those with physical limitations or learning difficulties have different needs and thus inequalities in accessing screening services.

Women with physical limitations/wheelchair users and the designated carers of those with learning difficulties are informed in the screening invitation that a longer appointment time of approximately 15 minutes (or longer) can be allocated if the screening office is notified of this need in advance of their attendance. (Standard screening appointment slot is usually 6–8 minutes.)

More recently, many screening centres are working very flexibly to offer a range of weekend and evening appointments to maximise convenience for the working population who may find difficulty being granted time away from work to attend during conventional working hours. The screening backlog caused by the Covid19 pandemic (2020/2021) expedited this practice as screening units sought to catch up on lost capacity for their populations. A simple but important advantage of evening/weekend appointments is the availability of onsite car parking which is much more problematic during the core hours of weekdays if screening is delivered by static hospital sites rather than a mobile facility.

Another benefit of evening/weekend appointments on hospital sites is that other routine outpatient clinics are not being held so breast imaging departments are quieter and there is more waiting room and client changing space.

### Screening service client survey

The NHSBSP undertakes local surveys once every three years to obtain and act upon the feedback of service users. Since every screening service has local variations, the surveys are conducted and analysed for the local populations. Most recently (2022) this included:

1. Questions related to the Covid19 pandemic, the information provided in the invitation literature and whether clients had concerns for their safety when attending or were satisfied with the measures in place within the screening facility.
2. Further questions relate to ascertaining women's awareness of the Screening Unit's website and women are encouraged to feedback on how they rate it.
3. Another section requests feedback relating to the suitability/convenience of timing and venue of appointments and whether women found it easy to change these if necessary.
4. The mammogram experience – closed questions to ascertain whether an adequate explanation was given prior to the examination being performed; whether the need for compression was described; whether the client found the examination painful; whether they felt the aftercare was appropriate and if they felt treated with respect.
5. Post mammogram examination – was the client told when she would receive the result of the screening test?
6. Finally, clients are invited to include any further comments they wish to make in free text.

Analysis of the 2022 survey at one screening unit revealed little negative feedback and was as follows:

- Respondents felt well cared for and were given clear explanations about the mammogram procedure.
- Mammograms were, overall, perceived to hurt 'a little'.
- Overall respondents felt appointments suited their requirements in terms of time and venue.
- Covid19 Safety Literature was received by nearly all respondents.
- Covid19 safety measures taken by the Programme were deemed adequate by most respondents.

### Screening attendance

Upon arrival at a breast screening facility (mobile van or static site) a client is greeted by one of four possible staff members namely a clerical officer, Assistant Practitioner (AP), Radiographer Practitioner (RP) or Advanced Practitioner (Adv P.) The invitation they've received gives an overview of breast screening including the test itself, what happens on the day of the test, the result and guidance on breast self-examination so they have had the opportunity to know in advance exactly what to expect. The appointment slot allocated is of necessity short (standard 6–8 minutes) and a practitioner must explain the test, obtain and document information about any current breast symptoms/concerns, previous breast surgery/cancer and perform the imaging satisfactorily during this time before discharging the client. A consultant radiography practitioner will not usually have any involvement in the screening test aspect of the screening pathway.

### Screening results

All screening mammograms are double read (two appropriately trained professionals review the images and input a binary decision, either normal or abnormal, with the latter requiring recall for assessment of the perceived abnormality). Readers will include a combination of: radiologists; breast physicians; consultant radiographers or advanced practitioners. Different screening centres will operate various protocols in terms of reaching consensus regarding which clients are finally recalled where there is discordance, and some will seek further opinions on cases where both readers have input a recall decision. The aim is to keep recall rates low to avoid creating significant anxiety unnecessarily (false positives) and overburden available assessment capacity.

### NHSBSP recalls for assessment/repeat imaging

Women are recalled for further imaging either because the original screening images are deemed undiagnostic in some way (Technical recall, TR) or because the image readers have indicated the presence of a possible cancer (Recall,

RC). All recalls (TR/RCs) are tightly monitored closely by the NHSBSP with percentage thresholds indicated above which screening programmes should not exceed. This is an important aspect of quality assurance. Every effort must be made to keep both TRs and RCs within the acceptable range as any recall causes significant anxiety for women and impacts upon capacity resources for breast imaging units.

Women recalled for assessment of a perceived abnormality are offered support to help alleviate anxiety by being given contact details of Clinical Nurse Specialists (CNS) to discuss their concerns both before and after their attendance appointment. The role of the CNS in breast screening is comprehensive and detailed guidance is provided by the NHSBSP and Royal College of Nursing (RCN) which states it should include:

1. An important contact for women throughout the assessment process
2. A professional resource for colleagues and the wider primary care team
3. An important member of the multidisciplinary team (MDT) in the screening service

The CNS provides specialist support for women recalled to assessment and to manage and alleviate any anxiety and distress experienced by such women. To achieve this the CNS will provide appropriate, high-quality information to women and help them understand it so they can make a more informed choice about treatment options. They must deliver this care and information in a sensitive and caring manner which preserves the dignity of women. Discussions should be held at an appropriate time with due regard to women's emotional, physical, social and educational needs.

Most women who are recalled to assessment experience some psychological consequences whether the ultimate outcome of assessment is normal, benign or malignant. Evidence shows that women can experience significant distress at every stage of the screening process, including the time between receiving the recall letter and attending the recall appointment. Undergoing a biopsy deepens the level of anxiety women experience. They may remain concerned about their health for some time after such a procedure. Consequently, a recall to assessment has been shown to adversely affect a woman's future attendance at screening appointments (Bond, Pavey, Welch et al. 2013.)

The CNS should attend any pre-assessment briefing alongside the imaging team (includes radiologists; consultant radiographers; Adv P; RPs and APs) and have access to the patient notes so they can deal with woman's queries about the reason for recall. The mammograms for cases to be assessed are reviewed and discussed by the team and a plan agreed for the imaging work up that will be carried out.

It is important that a CNS sees women after any biopsy procedure. This is to make sure that the woman fully understands what has been discussed with the health professional carrying out the biopsy and can ask questions.

The training of Adv Ps and consultant radiography practitioners over the last twenty years has vastly increased the capacity of screening units to meet service demands and directly benefit the patient by enabling a full assessment diagnostic work up in a single attendance instead of several visits being necessary. This obviously speeds up the pathway which is beneficial for the physical and mental well-being of the woman. This training usually involves post-graduate masters' level courses delivered by higher education institutions (HEIs) with the practical training delivered at the students hospital base. Radiographers can extend their role if they wish and the opportunities are there to perform mammographic image interpretation, breast ultrasound, image guided interventional procedures, all of which were previously performed by consultant radiologists. As imaging is becoming more complex and sophisticated this now means consultant radiographers being able to interpret Tomosynthesis, Contrast Enhanced mammography and some report Breast Magnetic Resonance Images following additional training. The creation of the AP role to perform mammograms has provided more scope for such role advancement.

### Breaking bad news

In addition to the imaging and interventional aspects of the patient pathway, the nature of breast imaging inevitably includes giving patients bad news on a regular basis and is often undertaken by a consultant radiographer during a clinical session. It is very important to give patients hope in such dark moments and reasons to believe they can be optimistic that a good outcome will be achieved even if the next several months are difficult.

Women found to have breast cancer following breast screening are often more shocked than those diagnosed symptomatically as they usually have no clinical symptoms and feel 'well'. In these cases, the cancers are mostly smaller (15mm or less on imaging) and it is helpful to indicate an approximate (small) size as this gives a sense of perspective, a realistic hope that it can be treated successfully and they will be able to resume a normal life again, albeit a 'new' normal. Another useful approach when breaking bad news is to reassure the woman that she made the right decision to attend screening (whilst asymptomatic) because (having been found to have breast cancer) she's given herself the best chance of a good outcome.

Women diagnosed symptomatically often have larger cancers and sometimes it has already been demonstrated on imaging to have spread to axillary nodes but again it is important to couch bad news giving with hope, a vital issue in mental/emotional health which influences how women handle the stresses of diagnosis and treatment. Again, reassurance to the woman that she is now in the treatment care pathway and that she will be well cared for usually helps to allay some of the fears at an incredibly anxious time (which includes the woman's family too.)

## Symptomatic services

Patients with a new breast symptom are referred from their family doctor for a specialist breast assessment to a hospital outpatient clinic. The aim is for all such patients to receive an appointment within two weeks of the family doctor consultation though in practice this is not always possible. They immediately enter a (potential) cancer pathway where waiting times are monitored strictly, and every attempt is made by the hospital services to avoid patient breaches. The development of 'Fast track, One-stop or Rapid Access' clinics (the title varies depending on the organisation, but the concept is the same) has evolved over the last twenty-five years in the UK. Such clinics aim to perform all the necessary breast diagnostic workup in one attendance, which speeds up the waiting time for patient diagnoses and enables treatment to commence promptly.

Patients arrive in the breast imaging department for initial imaging (either mammograms/ultrasound or both), depending on their specific symptom and age. Mammograms (which may include Tomosynthesis or Contrast Enhanced Mammography) are performed by a RP (or AP for conventional mammograms) who will explain the procedure and obtain verbal consent prior to commencing the imaging. This is important to ensure maximum patient understanding, compliance and therefore high-quality mammograms in order to facilitate accurate interpretation.

Most symptomatic patients also undergo an ultrasound examination so will subsequently be transferred onto the care of an Adv P, consultant radiographer/radiologist/breast physician who will complete the initial imaging workup and any interventional procedures which may be indicated. A detailed radiological report is dictated immediately onto the hospital administrative system which the referring clinician can access and be appraised of the initial imaging result in each case. A CNS is always present during these clinics to provide their specialised care as necessary. Many patients imaging care is completed at this point, but some may require to return on another day for additional imaging such as magnetic resonance imaging (MRI) to obtain more information on disease extent in order to inform the most appropriate treatment pathway.

## Breast imaging challenges within the UK

This section of the chapter will reflect on the current service provision with a person-centred approach in mind, with proposals being suggested on how to address some perceived problems. It should be noted that, at the time of writing, the world has experienced the Covid19 pandemic (2020–2022) which has created enormous demands on healthcare resources, pressures to deal with the backlog of work which was postponed during the pandemic and the inevitable staff fatigue that has resulted. Consequently, the challenges to improve and expand services are even greater than one would normally expect.

Some example deficiencies identified in a more person-centred approach in current service delivery include:

1. Insufficient engagement with carers (family member/paid helper who regularly looks after a sick/elderly/disabled person), patient advocates (a person who helps guide a patient through the healthcare system, including screening, diagnosis treatment so they receive the information required to make decisions about their healthcare) and communities from which patients are drawn. For example, some carers/advocates bring women for screening who are extremely infirm, unwell or with learning difficulties such that they're unable to comply with the requirements of producing a diagnostic mammogram, believing they are acting in the woman's best interests. This can cause distress which could be avoided if carers/advocates were appraised in advance of such requirements which should probably be included in more detail in the invitation letter and liaison with family doctor practices.

2. Reduced engagement in breast screening services by some ethnic populations and socio-economically deprived groups. For example, breast cancer screening uptake in London varies by specific ethnic group for first and subsequent invitations, with White British women being more likely to attend. The variation in the uptake for women from the same ethnic groups in different geographical areas suggests that collaboration about the successful engagement of services with different communities could improve uptake for all women (Jack, Møller, Robson & Davies 2014).

   Another example is that overall uptake of breast screening is lower in women living in more deprived areas of London, but this association does not apply in the same way to all screening areas. Further investigation and regular audit of local practice is needed to understand why women are not attending, both to inform service development and decrease inequalities in early diagnosis (Jack, Robson & Davies 2016).

3. Lack of engagement of key stakeholders in the design of services and equipment. It is probably fair to say that, traditionally, breast imaging facilities and indeed equipment design was carried out by professionals who were remote from breast imaging staff and the service users (clients/patients) without sufficient collaboration. Consequently, insufficient attention to certain detail regarding how things would actually work in practice was paid. For example, imaging room layout, space utilisation, lighting and sinks, waiting and changing room siting are often suboptimal for workflow which is crucial for efficiency and throughput.

   Some equipment manufacturers have acknowledged the importance of engaging with imaging staff/service users and responding to feedback in recent years but overall, imaging professionals need to find a way of making their voices (and the service users') heard and opinions sought in the design of the services in which they operate.

4. The problem of challenging entrenched professional behaviours that might be resistant to change.

5. Difficulties surrounding the recruitment and retention of sufficiently trained staff from all the professions involved in breast imaging (including Radiographic Support Workers) to develop and sustain patient-centred service delivery. There are no quick fixes to these problems and the current skill and overall staffing shortages within the NHS are well publicised, with agency staff being frequently required to keep services going. Staff shortages impact on the ability to train existing staff in new skills or even staff wishing to embark on a career in breast imaging. Various solutions have been looked at, including recruitment from overseas and allowing male practitioners to train and perform mammograms but the latter has been ruled out, fearing this may deter some women from attending screening.

6. The desirability to identify audits that encompass elements of patient-centred care with the aim of continuously improving services.

7. The need to ensure all four tiers in the staffing structure remain well informed and educated so they are equipped with the skills to provide the person-centred care appropriate to their roles and contribute to the continuous improvement of care in their local services. This should be both through initial training and Continuing Professional Development (CPD.) As always, this requires resources which, in the current climate, are scarce.

## Duty of candour

The intention of the duty of candour legislation is to ensure that providers are open and transparent with people who use services. It sets out some specific requirements providers must follow when things go wrong with care and treatment, including informing people about the incident, providing reasonable support, providing truthful information and an apology when things go wrong (PHE 2016).

There are circumstances when a person who has been screened may experience severe or moderate harm. In breast screening this will be because the condition screened for has not been detected and it is not treated early enough to improve the outcome for the patient. A review (audit) should be carried out to understand why this has occurred. If the audit reveals something has gone wrong in the screening process, then this should be treated as a notifiable safety incident and duty of candour regulations will apply.

The timing of exactly when to apply duty of candour once it has been identified that something has gone wrong will vary. Sometimes a patient will request a review at the point of diagnosis as they may have been screened recently (within the last 12 months or so) and want to know if the cancer might have been visible then. The best time for information disclosure is likely to become clear during or after treatment, by which time a rapport may have built up between the patient and the clinician responsible for the treatment or intervention. It is obviously important that clinicians are flexible, and any information disclosure occurs at a time suited to the patient's needs.

The decision regarding who has this conversation with the patient will be made within the imaging department, but it may fall to a consultant radiographer who will ideally have undergone training in breaking bad news and advanced communication skills since this is likely to be a difficult task. There is no right or wrong way of conducting this, but it is important to collaborate with surgeons/pathologists and oncologists and be fully appraised of the patient's medical situation in terms of prognosis so that an honest conversation takes place. A CNS should be available for support if required too and the patient should be given the opportunity for a friend, relative, carer or advocate to be present at the discussion and the patient should always be given an opportunity to voice their comments and concerns to reassure them their feelings and views are being listened to and taken seriously.

## Reflection on the role of the radiographer

As the earlier sections have illustrated, the role of the radiographer within breast imaging has undergone a revolution in the last three decades, driven by service need, patient expectations and the shortage of radiologists. Consultant radiographers no longer perform mammography but operate in the same way as radiologists in breast imaging, carrying their own caseload, making complex decisions which inform patient management in the pathway and playing a key role in MDTs. The consultant and advanced practitioner roles are now embedded in breast imaging services, most of which now operate the four-tier structure model of delivery. Role developments do not happen quickly. They require years of hard work and investment in training/mentoring to ensure such practitioners are well prepared for the rigours of additional responsibility and accountability which require a resilience and mental/psychological strength beyond that which was previously needed in the RP role.

## References

Bond M, Pavey T, Welch K, Cooper C, Garside R, Dean S & Hyde C (2013) Psychological consequences of false-positive screening mammograms in the UK. *Systematic Review. Evidence-Based Medicine* April; volume 18: number 2.

DoH (1992) – Department of Health; The Patient's Charter: Department of Health, London.

DoH (1995) – Department of Health; The Patient's Charter and you. HMSO, London.

DoH (1997) – Department of Health. The new NHS: Modern: Dependable. Department of Health, London.

DoH (2000a) – Department of Health. The NHS Cancer Plan: a plan for investment: a plan for reform. Department of Health, London.

DoH (2000b) – Department of Health. Meeting the challenge: a strategy for the allied health professions. Department of Health, London.

IARC (2002) IARC handbooks of cancer prevention, breast cancer screening. International Agency for Research on Cancer, World Health Organisation. Lyon: IARC Press; p. 144.

Jack R, Møller H, Robson T & Davies E (2014) Breast cancer screening uptake among women from different ethnic groups in London: a population-based cohort study. *BMJ Open* http://dx.doi.org/10.1136/bmjopen-2014-005586. Vol 4:1.

Jack R, Robson T & Davies E (2016) The varying influence of socioeconomic deprivation on breast cancer screening uptake in London. *Journal of Public Health*, Volume 38, Issue 2, Pages 330–334, https://doi.org/10.1093/pubmed/fdv038.

NHSBSP (2002) – National Health Service Breast Screening Programme. New ways of working in the Breast Screening Programme. First Report on Implementation. NHS Cancer Screening Programmes, Department of Health, London.

NHSE (1990) – National Health Service Executive: NHS & Community Care Act. HMSO, London.

Public Health England (PHE). NHS Screening Programmes Guidance on applying Duty of Candour and disclosing audit results. Incorporating disclosure of audit guidance and adapted from 'Disclosure of audit results in cancer screening: advice on best practice 2006.' PHE Version 1.0/ September 2016. PHE publications gateway number: 2016343.

RCR (2002) – Royal College of Radiologists. Clinical radiology: a workforce in crisis. The Royal College of Radiologists, London. www.gov.uk/topic/population-screening-programmes/breast (Accessed 19/05/23)

www.gov.uk/phe/breast-screening-leaflet. Pub 06/2013. Current version 05/2021.

www.gov.uk/phe/breast-screening-71-or-over. Pub 01/2007. Current version 01/2019.

# Section six

# Person-centred care in Magnetic Resonance Imaging

# 14 MRI

## Understanding and enhancing the patient journey

*Darren Hudson and Christine Heales*

**What is MRI?**

Magnetic Resonance Imaging (MRI) has, over the last 30 years, established itself as an important diagnostic tool for many clinical applications within modern healthcare practice (Bornert and Norris; 2020, van Beek et al., 2019). Benefits of MRI compared with other imaging techniques include the fact that it does not use ionising radiation, it provides excellent differentiation between tissues and is able to create images in any body plane (McRobbie et al., 2017). Hence, increasing numbers of patients and service users are referred for MRI scanning (NHSE, 2020; RCR, 2017).

MRI utilises a strong magnetic field which is typically created by a superconducting magnet. This magnetic field, combined with timed exposures to radiofrequency pulses, is how images are created. In order to produce a uniform magnetic field, the superconducting magnet is constructed as a tunnel, known as the 'bore' of the magnet, within which the patient lays for their scan. In order to receive the signal created by exposure to the radiofrequency pulses, ancillary equipment called 'coils' are placed within close proximity to the patient (McRobbie et al., 2017). MRI examinations consist of a series of image acquisitions and so patients are required to lay still within the scanner bore for periods of time that can range from 15 minutes to an hour or more (McRobbie et al., 2017). Since the early days of MRI, magnet design has become more patient friendly with bores now being shorter in length and with bore diameters of 70 cm (Brunnquell et al., 2020) rather than 60 cm being common place. Advances in sequence and coil technology have also helped reduce acquisition times so that the time spent within the scanner is less (Bornert and Norris, 2020; Brunnquell et al., 2020). Nevertheless, having an MRI scan can still pose significant challenges for some people.

Whilst there is a ready recognition that people with significant degrees of claustrophobia may find it difficult to undergo MRI (with consequences ranging from non-attendance to incomplete scans, through to potentially reduced quality images) other groups of patients merit consideration. Scan

DOI: 10.1201/9781003310143-20

length can be challenging for patients experiencing physical pain, both due to the need to adopt specific positions to be able to 'fit' within the coils and the bore of the scanner, and because of the need to then maintain a fixed position for the duration of the scan. Body habitus can sometimes result in a patient not being able to have a scan at all, or being scanned without the coils, impairing image quality. Perhaps unsurprisingly, children may find it difficult to collaborate for a range of reasons ranging from fear of the procedure through to inability to keep still for the periods of time required. Similar barriers to access to MR imaging may also apply to adults with learning disabilities or dementia, as well as neurodiverse individuals. Hence a complete understanding of the challenges facing all ages and demographics of people referred for MRI so that reasonable adjustment can be made is key to improving the MRI experience for everyone.

### Understanding the patient journey

Patient experience is an integral part of high-quality care and should be considered to have equal value with clinical effectiveness and clinical safety (Department of Health, 2008). Patient experience is also known to impact upon health outcomes (Doyle et al., 2013) and consequently is often a consideration within regulatory frameworks such as the UK's Care Quality Commission (CQC) (CQC, 2022) as well as independent accreditation schemes such as the Quality Standard in Imaging (QSI) (UKAS, 2022). But what is patient experience? The Beryl Institute is a global community of healthcare practitioners and lay members focusing on enhancing patient experience (Wolf, 2022) which it defines as 'the sum of all interactions, shaped by an organisation's culture, that influence patient perceptions across the continuum of care' (pg8) (Wolf et al., 2014). Applying this definition to an imaging modality, such as MRI, could mean that care is deemed to commence the moment the patient receives their appointment and continues throughout until the patient receives their results. To understand what healthcare providers can put in place to maximise patient experience, it therefore follows that there first needs to be an understanding of this complete patient journey. Each stage of the journey plays a role in how a patient may experience anxiety and/or perceive their overall experience.

### The patient journey model

The benefit of thinking about the entire patient journey is that it enables staff and service providers to consider patient experience in MRI holistically through consideration of each discrete element. This should then translate into enhanced satisfaction and reduced anxiety (Munn et al., 2015) which is important as having a poor experience can potentially create apprehension around future appointments or healthcare provision.

The patient journey model breaks the patient journey into three components, each of which can be considered as discrete entities, but each being of equal value in relation to the whole patient experience (Hudson et al, 2023):

1. The warm welcome
2. The positive procedure
3. The fond farewell

So, what are these three stages? The warm welcome commences with the receipt of any patient information and appointment details through to arrival and initial waiting within the MRI department. The positive procedure is the process of undergoing the MR imaging itself, and the fond farewell is the aftercare including the 'send-off' from the department. Each element will now be discussed in further detail below, along with suggested strategies to maximise patient experience.

**The warm welcome**

The warm welcome encompasses all early aspects of the patient's experience from the receipt of information about their appointment and the procedure through to arriving in the MRI department to await their scan. This stage in the patient journey can be associated with low emotions or a heightened state of anxiety. Something as straight-forward as well-constructed and accessible patient information in a range of appropriate formats can reduce anxiety (Munn et al., 2015; Powell et al., 2015, Tazegul et al., 2015; Tugwell et al., 2018). The conventional format of written information may be effective to convey key information (what, when, where) but may not have much impact upon how prepared a patient feels (Tugwell et al., 2018). Audio-visual media can more accurately represent what is involved when having a scan and can therefore have more of a role in helping patients prepare to cope with the procedure (Ahlander et al., 2020). Likewise, video resources can both reduce anxiety and enhance experience (Powell et al., 2015, Tugwell et al., 2018). Virtual reality (VR) is also emerging as a tool that has potential benefits in relation to patient preparation for MRI (Hudson et al., 2022c).

As technology evolves and it becomes more straightforward to have patient information available in a range of formats, there may be the opportunity for a greater focus on patient choice, as it is likely that individuals have their own preference as to the most suitable format for them (Kemp, May et al, 2021). Ultimately it is the provision of suitable information that is relevant, easy to understand and meets an individual's needs that is important (Carlsson and Carlsson, 2013; Munn and Jordan, 2011; Tugwell-Allsup and Pritchard, 2018).

Upon arrival in the MRI department, in-person explanation by members of the MRI team reduces any anxiety (Tazegul et al., 2015; Tugwell et al., 2018) especially when tailored to the individual to ensure sufficient understanding (Carlsson and Carlsson, 2013). Radiographers consider themselves to be well attuned at recognising the social signs of possible anxiety in patients attending for MRI (Hudson et al., 2022b) meaning they have the opportunity to adapt their initial communications in a way that reassures and supports the patient.

Waiting in the department to be safety screened and/or prepared for their scan can induce or enhance anxiety; this can be exacerbated if there are also time delays which can also result in patients having a more negative perception of their care overall (Kranzbühler et al., 2018). It is therefore important to keep waiting patients, and the people accompanying them, notified about any delays, and likely wait times.

The physical environment of the waiting area is also important (Hyde and Hardy, 2021b) and there is increasingly better understanding of how environments can help or hinder individuals with particular needs (SCIE, 2022). For example, people with dementia benefit from contrasting colours between the seating and the floor and wall, or around door frames. Whilst hospital toilet facilities often have handrails and emergency pull cords, using contrasting colours for the toilet seat, as well as dementia-friendly signage (such as for toilets and the exits) can all help with independence. Creating a waiting area that is dementia friendly and that can be navigated with a degree of autonomy can help with general mood thereby enhancing an individual's overall experience when attending for any imaging procedure, as can considering the levels of ambient noise (SCoR, 2020).

Environmental adaptations may also be beneficial for neurodiverse patients such as those with autism (Stogiannos et al., 2022). For example, autistic people may experience discomfort when noise levels are high, and simple adjustments such as turning off or lowering the volume of any background music can be helpful (Matusiak, 2022). Whereas people with dementia require bright lighting, this may be unhelpful for some autistic people. Taking the time to find out the needs of individuals is often the solution and having lights that can be dimmed is helpful. Paediatric patients tend to like fun, creative environments with toys or other activities available to occupy them during any waiting periods. It is unlikely that all requirements for all patients can be met within a single, small waiting area and compromise is likely to be needed. For example, it may be that it is possible to use a colour scheme that is helpful for people with dementia, keep the walls clear and free of distraction, but also have ample toys available for children that are put away when not in use. Being aware of how the environment can help or hinder patient experience can at least mean that the numerous factors are considered.

### The positive procedure

The 'Positive Procedure' refers to the scan itself and is often the point where there is most interaction with clinical staff (Hudson et al, 2023). This part of the journey commences upon entry to the scan room and is when the patient first sees the scanner and the imaging environment which, in and of itself, can be alien and daunting for them (Törnqvist et al., 2006). This first viewing of the MRI scanner may also be a trigger for claustrophobia and being positioned and placed within the bore of the scanner can also contribute to heightened anxiety (Enders et al., 2011). These negative experiences are then compounded by the noise and duration of the scan, further exacerbated for people who are not physically comfortable (either due to pain or body habitus). There is a need for radiographers to establish sensitive ways of interacting with patients who are unable to fit within MRI scanners. There is currently a lack of research about what such communication may look like but anecdotally this is an area that many radiographers find challenging. A related unanswered question is whether some patients may not attend their appointments if they have concerns about fitting into the scanner and there is also a need for engagement and research with such patients/service users about how best to meet their needs in this area.

Anxiety about undergoing any medical imaging procedure is common but is generally found to be greater for MRI than, say, ultrasound (Forshaw et al., 2018). This is likely due to claustrophobia, and MRI radiographers observe anxiety in their patients on a weekly, if not daily, basis (Hudson et al., 2022b). Whilst circa 2% of the general population are thought to experience claustrophobia this sensation may account for up to 1% of MRI scans being abandoned part way through (Hudson et al., 2022a; Wardenaar et al., 2017). Whilst some patients with claustrophobia may be able to tolerate their MRI scan, others may struggle resulting in scans degraded by patient movement, partially complete scans, and some patients may not be able to undergo MRI at all (either on the day or by avoidance, i.e. not attending their appointment). Any of these scenarios can impact adversely on the patient's health outcomes; there may be delays whilst arranging second attempts or alternative imaging or diagnostic tests. In some cases, alternative diagnostic imaging or tests may not be as informative. Everyone undertaking MR imaging will, therefore, have a range of skills to best support people with claustrophobia through their scan.

Taking time to show patients the scanner, to let them familiarise themselves with the environment, and showing them the coils prior to placing them over the patient can all be helpful, particularly for people who are claustrophobic (Hudson et al, 2023). This is where patient-centred care is key, with patients needing to feel that they are being listened to and having their needs responded to (including adaptation of position, and the use of aids for both immobilisation and comfort where needed). This means that radiographers should be able to use alternate patient positioning and/or coils when necessary and make any necessary adjustments to the scan protocols accordingly. Radiographers can tend to focus on the functional aspects of the procedure (Hyde and Hardy, 2021a; Munn et al., 2016; Reeves and Decker, 2012) such as patient flow and obtaining diagnostic images efficiently but it is important that patients themselves are not made to feel rushed or unimportant (Hudson et al, 2023). Something as simple as appropriate eye contact and calling the patient by their preferred name can make all the difference (Bleiker et al., 2020). It is clear that support from the radiographic staff is important (Carlsson and Carlsson, 2013; Funk et al., 2014; Munn and Jordan, 2011) and makes a significant difference to the patient, both in terms of experience and outcome (Munn et al., 2016).

Ensuring patients can use the buzzer can give a sense of control over the procedure. It is important to agree alternative forms of communication for people who cannot compress the buzzer; in such cases it may be necessary to have a carer/member of staff remain with the patient to communicate on their behalf.

Using the MRI scanner to its best capability (e.g., adjusting air flow, offering music) all, anecdotally, have a positive impact for patients. Likewise, where local protocols permit, enabling a friend or family member to accompany the patient for a scan can also be very helpful (following appropriate safety screening and preparation) (Hudson et al, 2023). Interestingly, the noise of the scanner is, based on self-reporting of patients, a significant factor that affects their well-being, so considering offering double-ear protection may also be helpful (Hudson et al, 2023). Correct fitting ear protection is important, for those who may have existing hearing issues (tinnitus) or suffer from hyperacusis or phonophobia (Stogiannos et al., 2022).

It is also important not to underestimate how important patients find it to know how long each scan is going to be (Hudson et al, 2023). Regular and consistent use of the intercom is important and the value of this to patients should not be underestimated. Again, where a patient cannot hear over the intercom, other forms of communication should be agreed beforehand, and it may be appropriate for a staff member to remain in the scan room with the patient (albeit in line with local practice and safety considerations). Wherever possible, the MRI service should have advance notice of any particular communication needs for their patients. This includes patients where English may not be their first language, as well as people with sensory impairment such as dementia. Continuing professional development of clinical staff is key in order to utilise our evolving understanding of how best to support our patients, to make use of available expertise (such as Dementia Champions) and to adopt best practice guidance into local practice (Stogiannos et al., 2022).

Many patients employ coping mechanisms to help them through their scan. Hence a patient may be claustrophobic, and find the procedure disturbing, and yet manages to tolerate it, whereas someone else would have to abandon the scan. This means that it is important to offer the same level of consideration to all patients, no matter how seemingly able they are to tolerate the scan. Patients self-report that the support offered by clinical staff and regular contact over the intercom are significant in terms of helping them cope (Hudson et al, 2023). Other strategies that may be useful, but will vary between individuals, include offering MR conditional/MR safe eye masks. Some patients may find breathing exercises helpful, but care must be taken that these don't introduce movement artefact.

Another strategy is, of course, faster scanning. Increasingly scanner interfaces enable operators to select pre-programmed faster versions of routine protocols, but it is helpful for all MRI radiographers to understand how to modify protocols appropriately to make them quicker and yet still diagnostic. This is more than just understanding how to select the shorter versions of scans or making appropriate adjustments to the parameters of the standard scans, but also knowing which sequences to prioritise. If a patient is unlikely to tolerate the entire scan – which sequences and planes would yield the most useful diagnostic information?

Finally, a pharmacological approach may be required. This can range from the use of anxiolytics to sedation, right through to general anaesthesia (GA). Whilst these approaches are considered effective from the perspective of enabling a diagnostic scan to be obtained there is of course a risk associated with them hence increasing caution about their use (Sozio et al., 2021). Whilst sedation or GA may be essential to enable an MRI scan to proceed, it is not without risk. As with any medication, anaphylaxis is a possibility (Ma et al., 2020), as is emergence delirium from GA (Hoch, 2019). Hence, MRI services would typically only utilise this approach where the MRI scan was felt to be of clear benefit to the patient and using a pharmacological approach the only way of achieving this.

As an aside, for any patient unable to communicate, including those less or non-responsive due to the use of pharmacological agents, must be physiologically monitored. The level of monitoring required will vary depending on the medication used and it goes without saying that (a) the monitoring equipment must be suitable for use in the MRI environment and (b) suitably trained and experienced staff must undertake that monitoring. Care also needs to be taken when positioning and scanning patients under these conditions as they themselves will be unable to raise concerns or highlight any discomfort themselves.

Immediate aftercare after any MRI scan is also important, namely the experience the patient has as they are assisted with getting down from the scanner and leaving the scanner room at the end of the procedure. People may need to sit for a few moments to re-orientate themselves before being steady enough to walk out of the scanner room. This then leads to the fond farewell stage of the patient journey.

## The fond farewell

The fond farewell relates to the patient's MRI experience coming to an end and is more than them just being told the scan is finished. It relates to the information the patient receives (and understands) about any after effects, how they will receive their results and more prosaically, about how to leave the department/hospital.

Typical approaches to enhancing the fond farewell is to acknowledge/ thank the patient for their participation/co-operation with the scan procedure (Murphy, 2009) but it must not be underestimated how much concern and even stress can be associated with anticipation over results (Carlsson and Carlsson, 2013; Engels et al., 2019). Ensuring patients are informed about this process is key. Radiographers themselves can find this part of the process challenging, especially if the scan has shown a significant pathology, and sometimes feel they are brushing the patient off or rushing them out to avoid difficult conversations (Murphy, 2009). Understanding that ensuring patients understand how and when they receive their results does really have a positive impact upon the perception of overall care may help radiographers manage their own feelings of discomfort in this regard (Hyde and Hardy, 2021b). Now that general strategies have been discussed for each stage of the patient journey, particular mention is going to be given to supporting children and adults with learning disabilities through MRI.

## Children

MRI departments that routinely undertake MRI on children will have a range of strategies in place. The 'feed and wrap' technique is generally held to be effective for infants, when a standardised protocol is used (Antonov et al., 2017) and can help avoid the use of sedation and/or anaesthesia in this cohort. The typical age range for children to be scanned under GA or with sedation is from infanthood through to 7 or 8 years old (Carter et al., 2010; Harned and Strain, 2001; Viggiano et al., 2015) with older children with complex needs and some adults also potentially requiring this approach. As discussed previously, there are risks associated with the use of GA or sedation. Play therapy is used to support children through a wide range of healthcare interventions and has a clear role within imaging, and specifically within MRI. Such therapy is typically provided by Play Specialists or specifically trained paediatric radiographers (Bharti et al., 2016; Cavarocchi et al., 2019; Heales and Lloyd, 2022; Morel et al., 2020). Equipment used can range from small-scale models (toy scanners) (Morel et al., 2020) through to real-size play tunnels and Play Simulators ('mock' scanners) (Heales and Lloyd, 2022).

There is a continuum between play therapy and simulation. Play scanners are available which have the advantage of including coaching elements, e.g. the Playful Magnetic Resonance Imaging Simulator (PMRIS) (Domed, Lyon, France) is designed such that whilst the child is inside the toy scanner, they are recorded by motion sensors. They then watch this recording and receive feedback and coaching on keeping still which is an important aspect of an 'awake' MRI (Heales and Lloyd, 2022; DoMed, 2022). Virtual reality is also being explored in this age group; this has the potential to include elements of gamification as distraction technique (Rahani et al., 2018; S and M, 2017). This is currently an area of active research.

The aim through any of these approaches is to support children in having an 'awake' MRI, which eliminates the need for either a GA or sedation. The use of such preparation has been shown to be effective with children being less anxious and better able to keep still, thereby eliminating the need for medication (Bharti et al., 2016; Heales and Lloyd, 2022). Such approaches have clear benefit for children and carers but are likely to also be cost effective and time efficient for clinical services also (Heales and Lloyd, 2022). Departments undertaking MRI with children are likely, therefore, to have a range of resources and approaches in place to facilitate child-centred imaging with the most essential component being a core team with paediatric expertise.

## Learning disabilities

People with learning disabilities are more likely to have chronic physical conditions (Care, 2020), some of which, clearly, will benefit from being investigated and/or monitored with MRI. Combining the increased prevalence of such chronic ill-health with difficulties this group of people experience accessing healthcare in the first place, there is, as a result, long-standing data indicating reduced healthcare outcomes, including premature death, when compared with the general population in the UK (Health, 2013). As a result a wide ranging series of recommendations have been made including training for healthcare practitioners around understanding the needs of this group of patients and for the sharing of best practice (Care, 2020). This has also led to the introduction of Learning Disability Adult (LDA) teams throughout the NHS within the UK (Excellence, 2018). There is a clear benefit to any MRI service, therefore, in working with such LDA teams to optimise their provision.

Patient information and documentation should be available in Easy Read formats, ideally being reviewed by service users with the support of the local LDA team. The LDA team can also advise on the level of support that an individual patient may need, ranging from more time on the day, through to pre-visits or simulation. Appointment scheduling should be such that patients are not kept waiting for longer than necessary and a member of the LDA team may be able to attend to support the patient where appropriate. The role of carers should also be valued, they often know the patient the best, and can therefore also provide significant support during the procedure (Stogiannos et al., 2022). It is worthwhile creating a strong working relationship with the LDA team if one is available. This also applies to supporting patients with dementia; many hospitals now have dementia champions and of course professional body guidance is also available (SCoR, 2020).

## MRI safety

Patient care in MRI cannot fully be discussed without considering MRI safety. An in-depth discussion of this topic is beyond the scope of this chapter particularly as there is a wide range of excellent resources already available around safety screening of patients and personnel and maintaining a safe MR environment. Nevertheless, there are aspects of patient care that are worth considering. For example, in the UK the Medicines and Healthcare products Regulatory Agency (MRHA) advice is that the written safety questionnaire is completed and signed by the patient, verified and then countersigned by an MRI radiographer (MHRA, 2021). Anecdotally, most if not all MRI radiographers will have had the experience of a patient recalling more information when the radiographer is verifying the content. This may be because human memory is not a straightforward retrieval system; memories are offloaded when not needed but can be recovered following retrieval clues. In other words, the initial completion of the safety questionnaire by the patient, followed by a period of reflection before the radiographer verifies the information, may enable more information to be recalled. Consideration needs to be given, therefore, if a radiographer is going through the form with a patient, to ensure time is left for that period of reflection. It is important that radiographers understand that safety screening needs to be an active conversation with their patient (Hudson and Jones, 2019).

Secondly, the wording/phrasing and font size of the safety questionnaire all matter. According to UK data, more than 4 in 10 adults may struggle with reading/understanding health content aimed at the general public, with 1 in 6 adults having a reading age lower than that expected of a 9–11 year old (NHS, 2022; NLT, 2022). This means that the safety questionnaire must be well designed to ensure patients understand the questions. MRI radiographers also need to understand that they may need to modify their language when verifying answers given to ensure the patient has fully understood the questions, and therefore been able to give accurate information (Hudson and Jones, 2019).

Another consideration is the general environment within which MRI radiographers practise. We can become acclimatised to the safety considerations and may forget how alarming it may seem to a patient to see large warning signs, equipment with MR UNSAFE labels, and maybe even equipment tethered to the walls. Clinical services are often under time pressure, but the overall patient experience can be significantly improved by just taking time to explain the fundamentals (i.e. the magnet in the scanner is always on) so that the patients understand that we have all these measures in place so that, actually, the MRI department, is a safe place to be. Likewise, taking the time to explain to the patient why we're asking about their medical (implant foreign body) history can also help.

Careful patient positioning is essential, not just to maximise patient comfort and ability to stay still for the required period of time, but there is also a safety component to how people are positioned in the bore, and how coils and any ancillary equipment are placed; again, this is beyond the scope of this chapter and readers are recommended to seek out this information from resources specifically focusing on MRI safety.

## Summary

There is value in understanding the impact of each stage of the patient journey on patient experience of MRI. Every aspect of the process, from the way initial information is presented, through to the way a patient leaves the MRI department, impacts upon their overall experience. Recognising this means that radiographers can consider each element and optimise it. The value of effective communication cannot be overstated; arguably, this is the single most important element in the patient–radiographer interaction and without this, any other supportive measure is unlikely to be as successful. Furthermore, appreciating that our understanding of how to support our diverse patient population is continuously evolving means that we, as radiographers, can embed a culture of enquiry, interprofessional working (to take advantage of others' expertise) and continuous improvement into our practice for the benefit of our patients.

## Acknowledgements

Ruth Evans MBE: Patient Experience Network

## References

Ahlander, B. M., Engvall, J. & Ericsson, E. 2020. Anxiety during magnetic resonance imaging of the spine in relation to scanner design and size. *Radiography (Lond)*, 26, 110–116.

Antonov, N. K., Ruzal-Shapiro, C. B., Morel, K. D., Millar, W. S., Kashyap, S., Lauren, C. T. & Garzon, M. C. 2017. Feed and Wrap MRI Technique in Infants. *Clin Pediatr (Phila)*, 56, 1095–1103.

Beryl_Institute. 2022. *About The Beryl Institute* [Online]. Available: www.theberylin stitute.org/page/About [Accessed August 2022].

Bharti, B., Malhi, P. & Khandelwal, N. 2016. MRI Customized Play Therapy in Children Reduces the Need for Sedation--A Randomized Controlled Trial. *Indian J Pediatr*, 83, 209–13.

Bleiker, J., Knapp, K., Morgan-Trimmer, S. & Hopkins, S. 2020. What Medical Imaging Professionals Talk About When They Talk About Compassion. *J Med Imaging Radiat Sci*, 51, S44–S52.

Bornert, P. & Norris, D. G. 2020. A half-century of innovation in technology-preparing MRI for the 21st century. *Br J Radiol*, 93, 20200113.

Brunnquell, C. L., Hoff, M. N., Balu, N., Nguyen, X. V., Oztek, M. A. & Haynor, D. R. 2020. Making Magnets More Attractive: Physics and Engineering Contributions to Patient Comfort in MRI. *Top Magn Reson Imaging*, 29, 167–174.

Care, I. O. P. 2020. *Best practice on care coordination for people with a learning disability and long term conditions*. Oxford Brookes University.

Carlsson, S. & Carlsson, E. 2013. 'The situation and the uncertainty about the coming result scared me but interaction with the radiographers helped me through': a qualitative study on patients' experiences of magnetic resonance imaging examinations. *J Clin Nurs*, 22, 3225–34.

Carter, A. J., Greer, M. L., Gray, S. E. & Ware, R. S. 2010. Mock MRI: reducing the need for anaesthesia in children. *Pediatr Radiol*, 40, 1368–74.

Cavarocchi, E., Pieroni, I., Serio, A., Velluto, L., Guarnieri, B. & Sorbi, S. 2019. Kitten Scanner reduces the use of sedation in pediatric MRI. *J Child Health Care*, 23, 256–265.

CQC. 2022. *Guidance for Providers* [Online]. Care Quality Commission. Available: www.cqc.org.uk/guidance [Accessed August 2022].

DoMed. 2022. *Playful MRI Simulator* [Online]. Available: www.playful-mri-simulator.com/functionning-mri-child-simulator/ [Accessed].

Doyle, C., Lennox, L. & Bell, D. 2013. A systematic review of evidence on the links between patient experience and clinical safety and effectiveness. *BMJ Open*, 3.

Enders, J., Zimmermann, E., Rief, M., Martus, P., Klingebiel, R., Asbach, P., Klessen, C., Diederichs, G., Wagner, M., Teichgraber, U., Bengner, T., Hamm, B. & Dewey, M. 2011. Reduction of claustrophobia with short-bore versus open magnetic resonance imaging: a randomized controlled trial. *PLoS One*, 6, e23494.

Engels, K., Schiffmann, I., Weierstall, R., Rahn, A. C., Daubmann, A., Pust, G., Chard, D., Lukas, C., Scheiderbauer, J., Stellmann, J. P. & Heesen, C. 2019. Emotions towards magnetic resonance imaging in people with multiple sclerosis. *Acta Neurol Scand*, 139, 497–504.

Forshaw, K. L., Boyes, A. W., Carey, M. L., Hall, A. E., Symonds, M., Brown, S. & Sanson-Fisher, R. W. 2018. Raised Anxiety Levels Among Outpatients Preparing to Undergo a Medical Imaging Procedure: Prevalence and Correlates. *J Am Coll Radiol*, 15, 630–638.

Funk, E., Thunberg, P. & Anderzen-Carlsson, A. 2014. Patients' experiences in magnetic resonance imaging (MRI) and their experiences of breath holding techniques. *J Adv Nurs*, 70, 1880–90.

Harned, R. K., & Strain, J. D. 2001. MRI-compatible audio/visual system: impact on pediatric sedation. *Pediatr Radiol*, 31, 247–50.

Heales, C. J. & Lloyd, E. 2022. Play simulation for children in magnetic resonance imaging. *J Med Imaging Radiat Sci*, 53, 10–16.

Health, D. O. 2008. *High Quality Care For All*. Department of Health. .

Health, D. O. 2013. Six Lives: Progress Report on Healthcare for People with Learning Disabilities.

Hoch, K. 2019. Current Evidence-Based Practice for Pediatric Emergence Agitation. *AANA J*, 87, 495–499.

Hudson, D. M., Evans Mbe, R. & Heales, C. 2023. Journey to the center of the bore: A service evaluation of the patient experience in magnetic resonance imaging. *Journal of Radiology Nursing*.

Hudson, D. & Jones, A. P. 2019. A 3-year review of MRI safety incidents within a UK independent sector provider of diagnostic services. *BJR Open*, 1, 20180006.

Hudson, D. M., Heales, C. & Meertens, R. 2022a. Review of claustrophobia incidence in MRI: A service evaluation of current rates across a multi-centre service. *Radiography (Lond)*, 28, 780–787.

Hudson, D. M., Heales, C. & Vine, S. J. 2022b. Radiographer Perspectives on current occurrence and management of claustrophobia in MRI. *Radiography (Lond)*, 28, 154–161.

Hudson, D. M., Heales, C. & Vine, S. J. 2022c. Scoping review: How is virtual reality being used as a tool to support the experience of undergoing Magnetic resonance imaging? *Radiography (Lond)*, 28, 199–207.

Hyde, E. & Hardy, M. 2021a. Patient centred care in diagnostic radiography (Part 1): Perceptions of service users and service deliverers. *Radiography (Lond)*, 27, 8–13.

Hyde, E. & Hardy, M. 2021b. Patient centred care in diagnostic radiography (Part 2): A qualitative study of the perceptions of service users and service deliverers. *Radiography (Lond)*, 27, 322–331.

Kemp, L., May, K., Smith-Harris, N. & Heales, C. 2021. Patient preparedness for MRI – an evaluation of the perceptions of different resource types.UKIO Congress 2021. 7th-25th Jun 2021. UKIO, 2021 Liverpool.

Kranzbühler, A. M., Kleijnen, M., Morgan, R. & Teerling, M. 2018. The Multilevel Nature of Customer Experience Research: An Integrative Review and Research Agenda. *Int J Manag Rev*, 20, 433–456.

Liszio, S. & Masuch, M. 2017. Virtual reality MRI: Playful reduction of children's anxiety in MRI exams. Proceedings of the 2017 ACM Conference on Interaction Design and Children. 127–136.

Ma, M., Zhu, B., Zhao, J., Li, H., Zhou, L., Wang, M., Zhang, X. & Huang, Y. 2020. Pediatric Patients with Previous Anaphylactic Reactions to General Anesthesia: a Review of Literature, Case Report, and Anesthetic Considerations. *Curr Allergy Asthma Rep*, 20, 15.

Matusiak, M. 2022. *How to create an autism-friendly environment* [Online]. Available: https://livingautism.com/create-autism-friendly-environment/ [Accessed August 2022].

McRobbie, D. W. A., Moore, E. A. A. & Graves, M. J. A. 2017. *MRI from picture to proton*.

MHRA 2021. Safety Guidelines for Magnetic Resonance Imaging Equipment in Clinical Use. In: Agency, M. A. H. P. R. (ed.). Medicines and Healthcare products Regulatory Agency.

Morel, B., Andersson, F., Samalbide, M., Binninger, G., Carpentier, E., Sirinelli, D. & Cottier, J. P. 2020. Impact on child and parent anxiety level of a teddy bear-scale mock magnetic resonance scanner. *Pediatr Radiol*, 50, 116–120.

Munn, Z. & Jordan, Z. 2011. The patient experience of high technology medical imaging: a systematic review of the qualitative evidence. *JBI Libr Syst Rev*, 9, 631–678.

Munn, Z., Pearson, A., Jordan, Z., Murphy, F., Pilkington, D. & Anderson, A. 2015. Patient Anxiety and Satisfaction in a Magnetic Resonance Imaging Department: Initial Results from an Action Research Study. *J Med Imaging Radiat Sci*, 46, 23–29.

Munn, Z., Pearson, A., Jordan, Z., Murphy, F., Pilkington, D. & Anderson, A. 2016. Addressing the Patient Experience in a Magnetic Resonance Imaging Department: Final Results from an Action Research Study. *J Med Imaging Radiat Sci*, 47, 329–336.

Murphy, F. 2009. Act, scene, agency: The drama of medical imaging. *Radiography*, 15, 34–39.

National Institute for Health and Care Excellence. 2018. Learning disabilities and behaviour that challenges: service design and delivery. www.nice.org.uk/guidance/ng93

NHS. 2022. *Health literacy* [Online]. NHS. Available: https://service-manual.nhs.uk/content/health-literacy [Accessed August 2022].

NHSE 2020. Diagnostic Imaging Dataset Annual Statistical Statistical Release 2019/20. NHS England. *In:* ENGLAND, N. (ed.).

NLT. 2022. *What do adult literacy levels mean?* [Online]. National Literacy Trust. Available: https://literacytrust.org.uk/parents-and-families/adult-literacy/what-do-adult-literacy-levels-mean/ [Accessed August 2022].

Powell, R., Ahmad, M., Gilbert, F. J., Brian, D. & Johnston, M. 2015. Improving magnetic resonance imaging (MRI) examinations: Development and evaluation of an intervention to reduce movement in scanners and facilitate scan completion. *Br J Health Psychol*, 20, 449–65.

Rahani, V. K., Vard, A. & Najafi, M. 2018. Claustrophobia game: Design and development of a new virtual reality game for treatment of claustrophobia. *J Med Signals Sens*, 8, 231–237.

RCR 2017. Magnetic resonance imaging (MRI) equipment, operations and planning in the NHS Report from the Clinical Imaging Board. The Royal College of Radiologists.

Reeves, P. J. & Decker, S. 2012. Diagnostic radiography: A study in distancing. *Radiography*, 18, 78–83.

SCIE. 2022. *Dementia Friendly Environments* [Online]. Social Care Institute for Excellence. Available: www.scie.org.uk/dementia/supporting-people-with-dementia/dementia-friendly-environments/ [Accessed 23rd January 2023 2022].

SCoR 2020. Caring for people with Dementia: a clinical practice guideline for the radiography workforce (maging and radiotherapy). Society & College of Radiographers.

Sozio, S. J., Bian, Y., Marshall, S. J., Rivera-Nunez, Z., Bacile, S., Roychowdhury, S. & Youmans, D. C. 2021. Determining the efficacy of low-dose oral benzodiazepine administration and use of wide-bore magnet in assisting claustrophobic patients to undergo MRI brain examination. *Clin Imaging*, 79, 289–295.

Stogiannos, N., Carlier, S., Harvey-Lloyd, J. M., Brammer, A., Nugent, B., Cleaver, K., Mcnulty, J. P., Dos Reis, C. S. & Malamateniou, C. 2022. A systematic review of person-centred adjustments to facilitate magnetic resonance imaging for autistic patients without the use of sedation or anaesthesia. *Autism*, 26, 782–797.

Tazegul, G., Etcioglu, E., Yildiz, F., Yildiz, R. & Tuney, D. 2015. Can MRI related patient anxiety be prevented? *Magn Reson Imaging*, 33, 180–3.

Törnqvist, E., Månsson, Å., Larsson, E. M. & Hallström, I. 2006. t's like being in another world – Patients' lived experience of magnetic resonance imaging. . *Journal of Clinical Nursing*, 954–961.

Tugwell, J. R., Goulden, N. & Mullins, P. 2018. Alleviating anxiety in patients prior to MRI: A pilot single-centre single-blinded randomised controlled trial to compare video demonstration or telephone conversation with a radiographer versus routine intervention. *Radiography (Lond)*, 24, 122–129.

Tugwell-Allsup, J. & Pritchard, A. W. 2018. The experience of patients participating in a small randomised control trial that explored two different interventions to reduce anxiety prior to an MRI scan. *Radiography (Lond)*, 24, 130–136.

UKAS. 2022. *Quality Imaging Service Accreditation* [Online]. United Kingdom Accreditation Services. Available: www.ukas.com/accreditation/standards/quality-standard-imaging/ [Accessed August 2022].

Van Beek, E. J. R., Kuhl, C., Anzai, Y., Desmond, P., Ehman, R. L., Gong, Q., Gold, G., Gulani, V., Hall-Craggs, M., Leiner, T., Lim, C. C. T., Pipe, J. G., Reeder, S., Reinhold, C., Smits, M., Sodickson, D. K., Tempany, C., Vargas, H. A. & WANG, M. 2019. Value of MRI in medicine: More than just another test? *J Magn Reson Imaging*, 49, e14–e25.

Viggiano, M. P., Giganti, F., Rossi, A., Di Feo, D., Vagnoli, L., Calcagno, G. & Defilippi, C. 2015. Impact of psychological interventions on reducing anxiety, fear and the need for sedation in children undergoing magnetic resonance imaging. *Pediatr Rep*, 7, 5682.

Wardenaar, K. J., Lim, C. C. W., Al-Hamzawi, A. O., Alonso, J., Andrade, L. H., Benjet, C., Bunting, B., de Girolamo, G., Demyttenaere, K., Florescu, S. E., Gureje, O., Hisateru, T., Hu, C., Huang, Y., Karam, E., Kiejna, A., Lepine, J. P., Navarro-Mateu, F., Oakley Browne, M., Piazza, M., … de Jonge, P. (2017). The cross-national epidemiology of specific phobia in the World Mental Health Surveys. *Psychol. Med* 47(10), 1744–1760. https://doi.org/10.1017/S0033291717000174

Wolf, J., Niederhauser, V., Marshburn, D. & Lavela, S. L. 2014. Defining Patient Experience. *Patient Exp J*, 1, 7–19.

# 15 Patient care in MRI

*Iain MacDonald*

## Introduction

The role of MRI is continually expanding with new areas of application such as radiotherapy planning, new applications in cardiac imaging and an increasing role in prostate imaging continually being developed. This has developed from the very early days of MRI when the emphasis was on non-traumatic neuroimaging, with the first image of the human brain being made by Clow and Young (1978). Imaging of the body was more problematic as MRI systems had to cope with reduced signal and high sensitivity to motion and therefore only took off in the 1990s (McRobbie, Moore, Graves and Prince, 2017). This section will focus on newer applications of MRI. In radiotherapy (RT), for example, the role of MRI in planning is growing markedly, with a wide range of contrast mechanisms available including the $T_1$ and $T_2$ weighted imaging that have been the primary mechanisms of contrast in tissues since the first development of MRI. Other contrast mechanisms are available in magnetic resonance imaging such as Blood Oxygen Level Dependent (BOLD) imaging, diffusion weighted imaging (DWI) and dynamic contrast enhanced MRI (DCE-MRI) and these can be exploited for RT. This is compared to computed tomography (CT) where the soft tissue contrast is inferior to MRI (Schmidt and Payne, 2015). Cardiac MRI (CMR) has become an effective method of imaging the heart non-invasively. Over the last 20 years, its application has increased as there have been impressively enhanced spatial and temporal resolution capabilities of MRI coupled with increase in imaging speed (Russo et al., 2020). Another key area is the incorporation of diffusion weighted imaging (DWI) particularly to the diagnosis of stroke. Using magnetic field gradients to sensitise the sequence to diffusion, the images can demonstrate the contrast between the normal random movement of water within and between cells in the normal brain and restricted movement of water due to the swelling of ischaemic cells due to stroke.

Imaging of the prostate, the most common cancer affecting men has been improved by the adoption of multiparametric MRI and changed practice of the diagnosis of the disease (National Institute for Health and Care Excellence,

DOI: 10.1201/9781003310143-21

2019). Compared to biopsy alone, the combination of multiparametric MRI and biopsy gives a better identification of prostate cancer that is clinically significant. In a cross-sectional review of literature, Drost et al. (2019) identified that MRI gave 'the most favourable diagnostic accuracy' for clinically significant prostate cancer detection compared to the other methods of detection they reviewed. Another example of the increasing application of MRI in the authors' recent experience is a patient with severe biliary colic who was booked for a scan directly from the emergency department. This patient was scanned within hours of presentation with an urgent Magnetic Resonance cholangiopancreatography examination (MRCP). Until recently, this patient would have been admitted to the ward and had an MRI scan booked for the next day, increasing the time to diagnosis and treatment. This demonstrates the more frequent use of MRI in clinical scenarios leading to improved patient outcomes.

### Acceleration of scan protocols and reduction of motion artefacts

Many patients find it challenging to remain still for the duration of the MRI scan, which proceed for several minutes. As an example, Sartoretti et al. (2019) identify that a brain MRI takes 24 minutes 35 seconds; knee MRI 16 minutes 53 seconds; lumbar spine MRI 18 minutes 11 seconds. From a technical perspective, body tissue can move during the period between the data collection and the next radiofrequency (RF) pulse which (Type 1 artefact) and/or tissue movement through the magnetic field gradients between the RF pulse and data collection (Type 2 artefact). To mitigate this, sequences such as PROPELLER (Periodically Rotated Overlapping ParallEL Lines with Enhanced Reconstruction) (Pipe, 1999) have been developed to use. Propellor MRI is a GE sequence and is also known as BLADE by Siemens, MultiVane by Philips. All these sequences attempt to correct for these Type 1 and Type 2 motion artefacts, essentially removing the blurring that would otherwise occur.

Deep learning is a new technique, utilising machine learning (a subset of artificial intelligence), that has the potential to dramatically reduce acquisition times, while still maintaining the diagnostic quality of images. A good example is by Recht et al. (2020) who found they were able to reduce the acquisition time for a knee examination to five minutes using a 3T MRI scanner and deep learning. Essentially, this technique uses a reduced set of data points and searches for a compressed image in its database that appears consistent with the acquired data. This is known as compressed sensing. Deep learning takes this further by reconstructing images and adding missing information by learning from a much more complex image data set. This allows accelerated protocols, preserving high image quality. The six observers in the study even judged the accelerated images to be of higher overall image quality than the original clinical images when scoring images on artefacts, signal to noise ratio sharpness and overall image quality. Compressed sensing combines parallel imaging and under-sampled data in k-space. Compressed SENSE, a Philips vendor specific term was used by Sartoretti et al. (2019) who looked at this in six body regions including the brain, lumbar spine, shoulder and breast. They found that scan times were accelerated by 20% and that 27% more examinations were performed using Compressed SENSE rather than the conventional protocols. Again, with compressed sensing, the protocol acceleration technique, Bishoff et al. (2022) found that a combination of compressed sensing and PROPELLOR imaging at 3T gave a higher image quality and reduced time of acquisition compared to conventional T2 weighted sequences.

### Safety concerns with MRI – particularly focusing on the potential for injury to patients.

A particularly relevant area when considering patients' experience in MRI is the issue of safety screening. From a patient's point of view this is a particularly unique aspect of the diagnostic process, and many may question why it is necessary. While the idea of safety may seem implicit – an imaging modality based around a powerful magnet, it is useful to be able to clearly identify the risks involved.

It is important at the outset to define exactly what the safety requirements in MRI are. The American College of Radiology Guidance Document on MR Safe Practice (Kanal et al., 2013) identifies the safety zones from I to IV (Figure 15.1). All patients should be safety screened in safety zone II. This acts as the interface between the public area Zone I and the strictly controlled safety zones III and IV.

Zone III is to be strictly controlled as free access by unscreened personnel (be they staff or patients) can result in injury or even death as a result of interactions between individuals and the MR scanner environment, including the scanners static and time varying magnetic fields. Access to Zone III is to be supervised by 'MR personnel', including access to Zone IV which lies within Zone III. This is to ensure the safety/protection of patients, other healthcare staff and the equipment. Typically, MR radiographers are the specifically identified MR personnel ensuring compliance with this aspect. Safety Zone IV, where the MR magnet is situated, is located within the heavily monitored Zone III. There should be direct visual observation of the access routes into Zone IV, which may be by line of sight or video monitoring.

In regards to the safety education of MR personnel, there are two levels – those at Level 1 have passed minimal educational requirements to ensure their own safety as they work in Zone III. This is defined as a lecture or pre-recorded presentation approved by the MR medical director. Others, at Level 2, have deeper knowledge of MR safety by having had more extensive training in broader aspects of MRI safety including, for example, the heating effect of MRI, potential for burns etc. (Kanal et al., 2013). Given this strict demarcation of roles and access to the facilities, it is important to verify

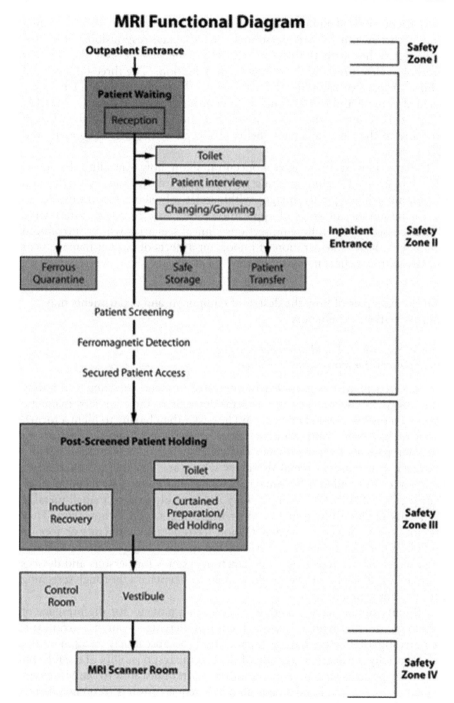

*Figure 15.1* Idealised floor plan for an MR imaging unit.

Source: Kanal et al., 2013.

that patients, indeed all non-MR personnel, must complete a safety screening process carried out by MR personnel. The ultimate responsibility for projectile safety lies with the person who takes the patient over the 5 gauss threshold. This threshold is an important definition: 'The three-dimensional volume of space surrounding the MR magnet that contains both the Faraday shielded volume and the 0.50 mT field contour (5 gauss (G) line)' (MHRA, 2021). This volume is the region in which an item might pose a hazard from exposure to the electromagnetic fields produced by the MR equipment and accessories.

The previous sections have concentrated on how the clinical role of MRI has evolved including some of the latest innovations, with the aim of obtaining high quality images within relatively short timeframes in an increasing number of areas of application. Some key aspects of safety were also discussed. In the following sections clinical scenarios will be introduced which will allow consideration of important aspects of patient management in MRI within a clinical environment.

## Clinical examples of how the design of equipment and departments may improve patient experiences

### Clinical Scenario 1: Could metal detectors be used to screen patients for ferromagnetic objects?

Mr A, a patient who is particularly interested in metal detecting as a hobby asks: 'Why do you not just use a metal detector to look for ferromagnetic objects on my body. Surely that would be easier than having to fill in a form?'

There has been some thought given to the use of metal detectors in screening patients for the presence of ferromagnetic objects on their person. However, conventional metal detectors which are not able to differentiate between ferrous and non-ferromagnetic material are not to be recommended (Kanal et al., 2013). For example, they are not able to detect small metallic fragments in the orbit, spinal cord or heart (e.g. of 2 x 3 mm size). They are not able to detect ferromagnetic/non ferromagnetic implants or foreign bodies. What can be used and are recommended as an *addition* to thorough screening are ferromagnetic detection systems for persons and devices approaching Zone IV. However, these in no way replace a thorough screening process, but supplement this.

All units should have a method of screening patients for the presence of certain factors that would preclude them from having a scan or warrant further investigation before having a scan. This is because it may result in injury, or, potentially the death of an individual. It is the responsibility of the referrer to identify patients with implants or other contraindications to the MR environment before referral and a dedicated MR request form is recommended to be completed (MHRA, 2021).

*Table 15.1* The five primary areas that impact on MR safety

---

*Top five MRI incidents witnessed by 69% of respondents in the previous five years that could have or did impact on patient or staff safety*

---

Patient burns
Projectiles in the MR environment (scan room): phones, hair clips, coins, steel toe
  cap shoes, key, walking stick, prosthesis
MR unsafe device incidents in patients: programmable shunts, aneurysm clips,
  pacemakers and cardiac devices
Heating from external MR unsafe devices
Inappropriate gadolinium administration

---

Source: Nugent, 2019.

In terms of the equipment in the vicinity of the MRI safety zones, it is important to recognise that there are three categories of devices that may be identified in the MRI environment (MHRA, 2021):

**MR Safe:** an item that poses no known hazards resulting from exposure to any MR environment. MR Safe items are composed of materials that are electrically nonconductive, non-metallic, and nonmagnetic.

**MR Conditional:** an item with demonstrated safety in the MR environment within defined conditions. As a minimum, this addresses the conditions of the static magnetic field, the switched gradient magnetic field and the radiofrequency fields. Additional conditions, including specific configurations of the item, may be needed.

**MR Unsafe:** an item which poses unacceptable risks to the patient, medical staff or other persons within the MR environment.

Nugent (2014), a strong advocate for MR safety in her survey of UK MR radiographers found common themes around safety incidents and identified the main reasons for safety incidents (Table 15.1). Nugent's conclusion was that there should be a homogenous education structure where MR staff obtain and maintain their safety standards to improve overall MR safety.

### Clinical Scenario 2: Pacemaker identified on screening the patient

Mrs B, a patient who requires an MRI scan of the knee has a pacemaker and has been referred for a scan. What safety considerations are necessary in this case?

Firstly, cardiac pacemakers have for a long time been considered a contraindication to scanning in MRI. It is important that any pacemaker is considered incompatible (MR unsafe) unless proven otherwise. Some centres are increasingly able to scan patients with these devices, but any centre scanning such patients should have a policy of identification, documentation imaging protocols and aftercare necessary for such patients (MHRA, 2021). Kalin and Stanton (2005) estimate that over a lifetime, 50–75% of patients with an implanted cardiac device (e.g., pacemaker or defibrillator) will need an MRI scan. If these patients are excluded from scans, it denies them this opportunity, affecting the diagnosis and treatment for a disease that could potentially be imaged by MRI, so it is an important area to consider from an ethical perspective. Other imaging modalities may be considered but may not be as effective as MRI. Russo et al. (2017) looked at the risks that are associated with MRI in patients with a pacemaker (1000 patients) or defibrillator (500 patients). This study demonstrated that, if the devices were reprogrammed in accordance with a prespecified protocol, the devices or leads did not fail, there were no deaths and no ventricular arrythmias occurred. Carried out in environments with a 1.5T scanner, the devices were all non-MRI-conditional which means that they were not approved for MRI scanning, though were allowed to be included in this research study. Patients had a variety of indications for MRI – but importantly all were all non-thoracic MR scans. The conditions that the study was undertaken under were that the patient was appropriately screened, and the device reprogrammed with a prespecified protocol. A study by Bireley et al. (2020) identified that in 44 MRI scans of 21 patients, there were no immediate changes in pacemaker function at the time of the MRI scan. Their conclusion was that, with appropriate precautions, this can be carried out safely without affecting pacemaker function, including those with epicardial leads at 1.5T. The precautions include placing the pacemaker in ODO mode (no pacing possible), running an electrophysiology study immediately before the scan and MRI cardiologist and electrophysiology (EP) technical support being in place during the scan to ensure the patient is haemodynamically stable. So, in our clinical example, certainly the patient presenting with a pacemaker in situ, should not be scanned, pending further investigations.

### Clinical scenario 3: Burns occurring in MRI

An elderly patient, Mr P has attended as in in-patient for multiparametric MRI of his prostate. This may be a higher risk examination due to the patient's age, other comorbidities and possibly the presence of a hip replacement in this patient population. What can be done to minimise the risk of a burn in MRI?

Patients can be at risk of receiving a soft tissue burn due to high temperatures occurring because of the effect of the non-ionising electromagnetic fields in the radiofrequency (RF) part of the spectrum that are used to obtain images. A 10-year review in the United States of America (USA) demonstrated that 'thermal events' accounted for the majority of serious injuries to patients (Delfino et al., 2019). In the United Kingdom (UK) 42% of accidents in MRI that happened between 1993 and 2014 were RF burns (Grainger, 2015).

Within the MRI scan room, the patient will experience three forms of energy. The static magnetic field (commonly 1.5T or 3T), gradient fields that are time varying and an RF field. The static magnetic field (Bo) has a 'static' value over time. In terms of the effect of this field, they are mild and sensory – including vertigo and nausea – though many patients do not experience these. The induction of electrical potentials due to moving charges associated with blood flow occurs – the magneto-hydrodynamic effect. However, the major concern with the static field is the torque and translational effects on ferromagnetic objects (McRobbie, 2020).

The time varying magnetic fields induce currents in conducting material by fast switching of the gradient coils on and off. This current can, potentially, generate peripheral nerve stimulation. Conductive loops may also be generated within the human body by crossed arms or legs. This can also occur with other conducting material such as loops of wire (e.g. ECG leads). The loops transfer electrical charge, and burns happen, primarily at points of contact such as clasped hands. Unfortunately, these may not be sensed by the patients at the time of the scan – and therefore the patient may not even be aware of the damage to the tissue being caused (MHRA, 2021). The guidelines state that foam pads, 1–2cm thick should be used to insulate the patient from cables, the bore of the magnet and between the limbs.

There is concern about the heating effect of MRI scans on implanted orthopaedic devices, for example hip and knee replacements. The number of patients who have implanted orthopaedic devices are increasing in frequency in departments, and there are approximately 50 million EU citizens who have a medical implant (Lidgren et al., 2020). Many of these patients will need an MRI at some point in their lifetime. With gradient coil field heating effect, those implants which are (i) large volume and (ii) with the gradient field oriented perpendicular to the plane of the implant produce the greatest heating levels. High frequency gradient switching, and short rise times also contribute to heating effects. With RF heating effects, the antenna effect is predominant, meaning that thin, linear implants have the highest temperature rises (Wooldridge, 2021).

### The impact of Specific Absorption Rate (SAR)

One of the key roles of an MRI radiographer is to monitor and limit the specific absorption rate (SAR) – the deposition of RF energy into the patient's body. The first level is referred to as 'normal' and a higher level is referred to as 'first-level controlled'. The latter requires medical supervision of the patient before scanning can be commenced (Allison and Yanasak, 2015). Some patients present a greater risk for tissue temperature increase and it is interesting to note the wide range of conditions that may predispose patients to this (Table 15.2).

*Table 15.2* Patients at risk of excessive tissue temperature increase

---

*Patients with reduced thermoregulatory capacity*

---

- Cardiac impairment
- Hypertension
- Diabetes
- Old age
- Obesity
- Fever
- Impaired ability to perspire
- Pregnancy (risk for foetal heating)
- Drug regimes that may affect thermoregulatory capabilities (diuretics, tranquillizers, vasodilators)
- Patients in a cast (risk of thermal insulation)
- Patients who are unable to sense or communicate heat sensations (risk is no self-report of comfort during scanning due to patient being unconscious, sedated, locally anesthetized, or confused)
- Patients with extensive tattoos (risk is potential for radiofrequency burns)
- Patients with implanted organs or devices may have much higher risk from exposure to MRI radiofrequency pulses and must be evaluated individually.

---

Source: Adapted from Allison and Yanasak, 2015.

SAR is associated with the electrical current produced in body tissue from the RF pulse and their oscillating electromagnetic fields. When considering sequence choice, Fast Spin Echo has a greater propensity to deliver greater SAR than gradient echo sequences – due to the use of large, frequent radiofrequency pulses. Gradient echo use much smaller RF pulses generally. The SAR displayed on the console is a pure estimate of the value, based on the weight of the patient and the technical characteristics of the pulse sequence selected – with each vendor differing as to how SAR is estimated (Allison and Yanasak, 2015). Therefore, the weight entered for the patient is of paramount importance in estimating the SAR.

To reduce SAR there are two approaches – one is varying the pulse sequence parameters, the other to do with patient comfort in the scanner (Table 15.3).

**Clinical scenario 4: A patient who is nervous about having an MRI scan**

Ms B is a 25-year-old woman who is anxious about having an MRI scan. What can be done in terms of the environment to reduce her anxiety?

Siddiqui et al. (2017) identify that 80,000,000 MRI scans are carried out annually, and 2.5% of these are not completed fully due to anxiety and claustrophobia. This is clearly inconvenient for the affected patients, who may not receive a definitive diagnosis, and may lead the patient to have an MRI under general anaesthetic, increasing costs and potential adverse reactions. Abdominal, pelvic and extremity scans (e.g. knee and ankle) are more likely

*Table 15.3* Methods to limit excessive temperature increase in patients with MRI

---

**Variation in sequence parameters**
- Increase the TR, which can lead to longer scanning times.
- Reduce flip angles (for FSE sequences, use 60–130° refocusing pulses rather than 180° refocusing pulses), which can alter image contrast-to-noise ratio or signal-to-noise ratio.
- Reduce the number of slices in an acquisition, which can lead to longer scanning times.
- Reduce the number of echoes in multi-echo sequences, which can lead to longer scanning times

**Patient comfort approaches**
Control the scanning room temperature and humidity (follow manufacturer specifications), which may affect comfort for lightweight patients.
- Dress the patient in light clothing, which may affect patient modesty.
- Take breaks between high SAR acquisitions or interleave high SAR and low SAR acquisitions to allow patient cooling, which can lead to longer scanning times.
- Be sure the patient ventilation system in the scanner is turned on.

---

Source: Adapted from Allison and Yanasak, 2015.

to be scanned as patients do not go into the bore of the magnet completely and their head is outside (Munn et al., 2015). Feet first scanning techniques reduce the effects of claustrophobia by ten times (Enders, 2011). Perhaps taking a pharmacological drug to relieve the symptoms of anxiety may be sought for the MRI scan – and a General Practitioner (GP) may be able to prescribe a sedative, usually diazepam before the scan (Buckinghamshire Healthcare NHS Trust, 2023). One of the disadvantages of this approach is that it must be taken at exactly the right time as effects are quite short lived as it is designed for short-term relief of severe anxiety (McGavock, 2015). Incomplete sedation can increase the chances of the scan being sub-optimal due to motion.

There have been many attempts at making the MR environment more welcoming for patients as even high-end scanners may not produce high-quality images as patients may not be able to cope with the lengthy acquisition times and restrictive environment of many scanners in use (Nguyen et al., 2020). Although the subject of claustrophobia is particularly pertinent to the enclosed environment in MRI, this has been considered elsewhere in this book. Therefore, this section will concentrate upon the ambience of the magnet room – the audio-visual (AV) experience. A useful study regarding this was carried out in Japan by Shimokawa et al. (2022). This was a relatively small study, with 61 patients, some of whom experienced the Ambient Experience on a Philips MRI scanner (Philips Healthcare, Best, The Netherlands). An example of this environment is given in Figure 15.2.

Patients were able to choose whether to go into the ambient MRI environment or not. If they were ambivalent, they were randomly allocated to either the conventional environment or the ambient experience. Their

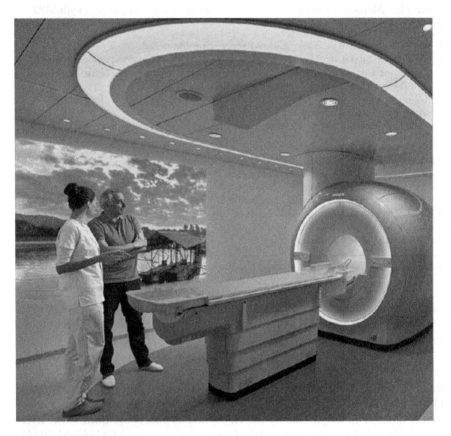

*Figure 15.2* An example of the ambient experience.

Source: Philips.

findings identified that patients with a high anxiety state were more likely to have reduction in anxiety compared to the control group when having a scan in the Ambient Experience (AE) scan situation. From their results of measurements of anxiety state pre and post the MRI scan experience, those who were allocated to the ambient experience group had a reduction in their anxiety noted by 81.8% of patients. Contrasting with this, those in the control group with a high anxiety state who were in the conventional environment had a reduction in anxiety only seen by 42.9%. This would appear to suggest that investing in an environment that is more focused on the anxiety levels of patients would be of benefit to patients in terms of their overall experience of the scan process, and potentially reducing the amount of scans that are either abandoned or sub-optimal in terms of image quality. In the Philips ambient experience system, the patient themselves can select a theme, giving them a semblance of control of their situation and providing a personalisation to their experience.

## The information that is provided to patients before the MRI scan

Information before an MRI scan is very important and it should be concise and clearly outline the procedure to be expected. This is where the use of websites and social media can give much more information prior to the scan. Rosenkratzz et al. (2016) found that in YouTube videos, the top two pages of MRI patient-related content demonstrated that 94% mentioned claustrophobia, 88% noise, 75% the need to remain still, and 68% metal safety. These are all important points to get across to patients, and it is gratifying to see that 100% used a 'patient' in the video and 84% had a radiographer/technologist which makes the video more appropriate to the representation of reality. It is interesting to note that 76% of healthcare settings in the private sector use social media to promote themselves (Rosenkrantz et al., 2016) and, certainly for those who engage strongly in social media, this would be a rich source of information about their MRI scan experience. Information given before and also during a scan procedure can prevent anxiety. There would appear to be many benefits for MRI units to provide a trusted webpage that patients can access. With ease, patients can gather a large amount of information, perhaps be able to ask questions, ease their anxiety and the process of MRI may be made smoother. Potentially this could reduce the need for staff to answer questions and reassure patients when they are in the department.

## Clinical Scenario 5: A patient with dementia needs an MRI scan

An elderly patient from the ward arrived with her sister accompanying her. She had recently been diagnosed with Alzheimer's disease and was, at the time, undergoing an episode of quite acute confusion. The sister asked but was refused entry to the scan room, and unfortunately the patient became quite upset and the scans were undiagnostic. How could this situation be made better so that the patient would have a better experience and more chance of a successful scan?

In general terms the key is communication with patients with dementia. According to Brooker (2012) even though the ability to process language may be impaired in the patient with dementia, being involved in the discussion, and all the non-verbal cues that go with that, rather than being ignored will make the person with dementia feel that they are being included. The radiographers' approach is important, and if the patient feels unimportant or the procedure is rushed, this can cause a resistant attitude in the patient. An MRI unit can induce anxiety in many patients, and patients with dementia can become anxious in an unfamiliar community (Nasisi, 2020). Brooker (2012) suggests that attachment to someone or something that is familiar is often needed when dementia patients become anxious. Disorientation in patients with dementia in situations that are unfamiliar is also seen (O'Malley et al., 2022). From the clinical example, this would be meeting new people and an unfamiliar environment with a noisy MRI scanner. Therefore, it is good practice to allow an appropriately screened relative or familiar carer into the scan room to support the patient during the examination.

### Clinical scenario 6: The critically ill patient

A patient is required to come for an MRI scan with a working diagnosis of a pilocytic astrocytoma. They became comatose and were admitted as an emergency and have undergone an initial CT scan of the brain. The patient is intubated and anaesthetised. The guidance provided by the Association of Anaesthetics, a British- and Irish-based organisation states that MRI requests for those patients who are critically ill must be justified by a radiologist by a review of the request, alternatively, a discussion between the referring consultant clinician and the radiologist (Wilson et al., 2019). There should be a thorough plan for the transfer of patients from the intensive care environment to MRI. This is to minimise the risks posed by the condition of the patient, transfer distance, time in the MRI department (including scan time) and the overall risks that are part of the MRI environment. An agreed appointment time should be set up between the MRI and intensive care staff which includes a sufficient time for staff to prepare and obtain the necessary equipment for transfer.

Wilson et al. (2019) consider the question of consent and critically ill patients should only undertake an MRI procedure for neuro imaging when a diagnosis may reasonably be foreseen which impacts on patient treatment and/or management. By discussion between the referring clinician and radiologist, decisions can be made for the patient if there is a treatment that can support life (Mental Capacity Act, 2005). Safety screening of MRI patients in critical care needs to be very carefully considered. Wilson et al. (2019) identifies that medical history may be lacking in patients who are unconscious – as they are not able to articulate their history directly. Review of previous imaging can be carried out and can be a useful exercise to check any potential implants and perhaps look for a completed safety questionnaire that may have been done before. Wilson et al. (2019) emphasise that the responsibility for screening patients lies with the imaging department. This is, ultimately, the MR responsible person, usually a radiographer with a management role, supported by the MR Safety Expert who provides scientific advice to the MR unit.

Once in the department, the equipment used in dealing with the critically ill patient should be checked for MRI safety. This includes blood pressure monitoring equipment, pulse oximeters and ECG leads. Once the patient is transferred to the MR compatible trolley, a visual inspection of the trolley and the patient should be made before entering safety zone 3 and 4 to ensure that there are no hazardous objects in place. Monitoring and ventilation should be changed over to MRI conditional equipment, ensuring the conditions for safe use are met. This equipment should be maintained by anaesthetic staff on a regular basis, including before an anaesthetised patient attends the department. It has been noted by Enrichi et al. (2020) that MR conditional ventilators do not have the same performance as ventilators that have conventional construction. The anaesthetic team including the anaesthetist and practitioners should be trained appropriately and be aware of the equipment available in the MRI unit together with the emergency protocols (Wilson et al., 2019). Monitoring equipment should have leads and wires that are MRI conditional to avoid the risk of burns. There should also be recognition that the magnetic fields can cause interference in the readings, e.g. in electrocardiography (ECG) (Swart and Rae, 2018). While standard infusion pumps are certainly a projectile safety risk and may malfunction in the MRI environment, there are workarounds that can be made. For example, the use of long infusion lines can allow the pumps to be used outside the scan room by using the waveguide access port. Alternatively, a specifically designed MRI-conditional radiofrequency (RF) shield enclosure or Faraday cage may be used (Weller, 2018).

### Summary of the chapter and implications for day-to-day clinical practice

Within this chapter several scenarios have been used to demonstrate that critical thinking in MRI is imperative to keep patients safe and comfortable and provide a person-centred experience. MRI is a unique environment that can create anxiety in patients, but this can be markedly reduced by improving the environment that the scan is in, such as utilising the ambient experience described. This has benefits to all patients, but particularly those who are anxious, children and the elderly. Safety screening of patients is continually being refined, and the hope is that many more patients with implantable devices will be able to access MRI in the future, improving their diagnosis and treatment.

### References

Allison, J. and Yanasak, N. (2015). What MRI sequences produce the highest specific absorption rate (SAR), and is there something we should be doing to reduce the SAR during standard examinations? *American Journal of Roentgenology*, 205(2), pp.140–140.

Bireley, M., Kovach, J.R., Morton, C., Cava, J.R., Pan, A.Y., Nugent, M. and Samyn, M.M. (2020). Cardiac magnetic resonance imaging (MRI) in children is safe with most pacemaker systems, including those with epicardial leads. *Pediatric Cardiology*, 41, pp.801–808.

Brooker, D. (2012). Understanding dementia and the person behind the diagnostic label. *International Journal of Person Centered Medicine*, 2(1), pp.11–17.

Buckinghamshire Healthcare NHS Trust (2023). *MRI scans for those who are worried about claustrophobia*. [Online]. [Accessed 21 April 2023]. Available from: www. buckshealthcare.nhs.uk.

Clow, H. and Young, I.R. (1978). Britain's brains produce first NMR scans. *New Scientist*, *80*(588), p.b26.

Delfino, J. G., Krainak, D. M., Flesher, S. A., & Miller, D. L. (2019). MRI-related FDA adverse event reports: A 10-yr review. *Medical Physics*, 46(12), 5562–5571.

Drost, F.J.H., Osses, D.F., Nieboer, D., Steyerberg, E.W., Bangma, C.H., Roobol, M.J. and Schoots, I.G. (2019). Prostate MRI, with or without MRI-targeted biopsy, and systematic biopsy for detecting prostate cancer. *Cochrane Database of Systematic Reviews* (4), 1–221.

Enders, J., Zimmermann, E., Rief, M., Martus, P., Klingebiel, R., Asbach, P., ... & Dewey, M. (2011). Reduction of claustrophobia during magnetic resonance imaging: methods and design of the 'CLAUSTRO' randomized controlled trial. *BMC medical imaging*, *11*(1), 1–15.

Enrichi, C., Zanetti, C., Stabile, R., Carollo, C., Ghezzo, L., & Piccione, F. (2020). Use of ventilation bag for the respiratory support during magnetic resonance imaging in Arnold-Chiari ventilated patients, a case report. *The Journal of Spinal Cord Medicine*, *43*(5), 710–713.

Grainger, D. (2015). MHRA MRI safety guidance: Review of key changes and emerging issues. www.sor.org/system/files/article/201504/io_2015_lr.pdf; 2015:42–47 [accessed 29th January 2023].

Kalin, R. and Stanton, M.S. (2005), Current Clinical Issues for MRI Scanning of Pacemaker and Defibrillator Patients. *Pacing and Clinical Electrophysiology*, 28: 326–328.

Kanal, E., Barkovich, A.J., Bell, C., Borgstede, J.P., Bradley, W.G., Jr, Froelich, J.W., Gimbel, J.R., Gosbee, J.W., Kuhni-Kaminski, E., Larson, P.A., Lester, J.W., Jr, Nyenhuis, J., Schaefer, D.J., Sebek, E.A., Weinreb, J., Wilkoff, B.L., Woods, T.O., Lucey, L. and Hernandez, D. (2013), ACR guidance document on MR safe practices: 2013. *Journal of Magnetic Resonance Imaging* 37: 501–530.

Lidgren, L, Raina, D B, Tägil, M and Tanner, K E. (2020). Recycling implants: a sustainable solution for musculoskeletal research. *Acta Orthopaedica*, 91: 125.

McGavock, H. (2015), How Drugs Work: Basic Pharmacology for Health Professionals, Fourth Edition, Taylor & Francis Group, Boca Raton. Available from: ProQuest Ebook Central. [21 April 2023].

McRobbie, D.W. (2020). *Essentials of MRI safety*. John Wiley & Sons.

McRobbie, D.W., Moore, E.A., Graves, M.J. and Prince, M.R. (2017). *MRI from Picture to Proton*. Cambridge University Press.

Mental Capacity Act (2005), [online] Available at:www.legislation.gov.uk/ukpga/2005/9/pdfs/ukpga_20050009_en.pdf. Accessed 16 Feb 2023.

MHRA (2021) *Safety Guidelines for Magnetic Resonance Imaging Equipment in Clinical Us,* viewed 3rd February 2023, https://assets.publishing.service.gov.uk/government/uploads/system/uploads/attachment_data/file/958486/MRI_guidance_2021-4-03c.pdf

Munn Z, Pearson A, Jordan Z, Murphy F, Pilkington D, Anderson A. (2014). Patient Anxiety and Satisfaction in a Magnetic Resonance Imaging Department: Initial Results from an Action Research Study. *Journal of Medical Imaging and Radiation Sciences* 2015 Mar;46(1):23–29.

Munn, Z., Moola, S., Lisy, K., Riitano, D. and Murphy, F. (2015). Claustrophobia in magnetic resonance imaging: a systematic review and meta-analysis. *Radiography*, *21*(2), pp.e59–e63.

Nasisi, C. (2020). Dementia: Psychosocial/Mental Health Risk Factors. *Journal for Nurse Practitioners*, 16(6), pp.425–427.

National institute for Health and Care Excellence (2019). *Prostate cancer: diagnosis and management NICE guideline NG131,* viewed 7 March 2023. www.nice.org.uk/guidance/ng131.

Nguyen, Xuan V., Tahir, Sana, Bresnahan, Brian W., Andre, Jalal B., Lang, Elvira V., Mossa-Basha, Mahmud, Mayr, Nina A., Bourekas, Eric C. (2020). Prevalence and Financial Impact of Claustrophobia, Anxiety, Patient Motion, and Other Patient Events in Magnetic Resonance Imaging. *Topics in Magnetic Resonance Imaging* 29(3):p 125–130, June 2020.

Nugent, B. (2014). The need for MRI safety education in the NHS. *RAD Magazine*, 45, 529, pp22–24.

O'Malley, M., Innes, A., & Wiener, J. M. (2022). (Dis) orientation and design preferences within an unfamiliar care environment: a content analysis of older adults' qualitative reports after route learning. *Environment and Behavior*, 54(1), 116–142.

OECD (2021), *Health at a Glance 2021: OECD Indicators*, OECD Publishing, Paris, https://doi.org/10.1787/ae3016b9-en. Viewed 19 April 2021.

Pipe, J.G. (1999). Motion correction with PROPELLER MRI: application to head motion and free-breathing cardiac imaging. *Magnetic Resonance in Medicine: An Official Journal of the International Society for Magnetic Resonance in Medicine*, 42(5), pp.963–969.

Recht, M.P., Zbontar, J., Sodickson, D.K., Knoll, F., Yakubova, N., Sriram, A., Murrell, T., Defazio, A., Rabbat, M., Rybak, L. and Kline, M. (2020). Using deep learning to accelerate knee MRI at 3 T: results of an interchangeability study. *AJR. American Journal of Roentgenology*, 215(6), p.1421.

Rosenkrantz, A. B., Won, E., & Doshi, A. M. (2016). Assessing the content of YouTube videos in educating patients regarding common imaging examinations. *Journal of the American College of Radiology*, 13(12), 1509–1513.

Russo, R.J., Costa, H.S., Silva, P.D., Anderson, J.L., Arshad, A., Biederman, R.W., Boyle, N.G., Frabizzio, J.V., Birgersdotter-Green, U., Higgins, S.L. and Lampert, R. (2017). Assessing the risks associated with MRI in patients with a pacemaker or defibrillator. *New England Journal of Medicine*, 376(8), pp.755–764.

Russo, V., Lovato, L. and Ligabue, G. (2020). Cardiac MRI: technical basis. *La radiologia medica*, 125(11), pp.1040–1055.

Sartoretti, E., Sartoretti, T., Binkert, C., Najafi, A., Schwenk, Á., Hinnen, M., van Smoorenburg, L., Eichenberger, B. and Sartoretti-Schefer, S. (2019). Reduction of procedure times in routine clinical practice with Compressed SENSE magnetic resonance imaging technique. *PLoS One*, 14(4), p.e0214887.

Schmidt, M.A. and Payne, G.S. (2015). Radiotherapy planning using MRI. *Physics in Medicine & Biology*, 60(22), p.R323.

Shimokawa, K., Matsumoto, K., Yokota, H., Kobayashi, E., Hirano, Y., Masuda, Y. and Uno, T. (2022). Anxiety relaxation during MRI with a patient-friendly audio-visual system. *Radiography*, 28(3), pp.725–731.

Siddiqui, Z., Singh, P., Kushwaha, S., & Srivastava, R. (2017). MRI and fear of confined space: A cause and effect relationship. *International Journal of Contemporary Medicine Surgery and Radiology*, 2(1), 19–24.

Swart, R., & Rae, W. I. D. (2018). Anaesthesia in the MRI suite. *Southern African Journal of Anaesthesia and Analgesia*, 24(4), 90–96.

Weller, G.E. (2018). Anesthesia for MRI and CT. *Anesthesiology: A Practical Approach*, pp.239–254.

Wilson, S. R., Shinde, S., Appleby, I., Boscoe, M., Conway, D., Dryden, C., ... & Wright, E. (2019). Guidelines for the safe provision of anaesthesia in magnetic resonance units 2019: Guidelines from the Association of Anaesthetists and the Neuro Anaesthesia and Critical Care Society of Great Britain and Ireland. *Anaesthesia*, 74(5), 638–650.

Wooldridge, J., Arduino, A., Zilberti, L., Zanovello, U., Chiampi, M., Clementi, V. and Bottauscio, O. (2021). Gradient coil and radiofrequency induced heating of orthopaedic implants in MRI: influencing factors. *Physics in Medicine & Biology*, 66(24), p.245024.

# 16 Person-centred care in Magnetic Resonance Imaging

*Johnathan Hewis and Kevin Strachan*

## Introduction

The extracted quotes shared in the poem in Table 16.1 are de-identified to maintain the privacy and confidentiality of the research participants. This poem is dedicated to the resilient individuals who graciously gave their time to the project by sharing their personal experience.

Magnetic Resonance Imaging (MRI) as a non-ionising radiation modality provides both anatomical and functional imaging capabilities, which have transformed diagnosis, therapeutic management, and surveillance in modern healthcare. In recent decades we have witnessed exponential growth in the number of MRI referrals and scanner installations with an increasing trend towards higher field strength magnets. The clinical role of MRI has expanded beyond being a purely diagnostic modality. PET/MRI as a hybrid technology is now well established within nuclear medicine with growing and diverse clinical applications (Kamvosoulis & Currie, 2021). Similarly, the clinical role for hybrid precision MRI guided radiation therapy is now strongly emerging (Liu et al., 2021), and there is increasing demand for MRI-based interventional procedures, particularly cardiac (Bijvoet et al., 2021). These more bespoke MRI environments bring new additional challenges for the patient and practitioner and broaden the scope of 'who' performs MRI to professions beyond diagnostic radiography. A universal constant across these environments is that MRI is highly hazardous with numerous serious injuries and reported deaths having occurred within this environment (Delfino et al., 2019). This highlights the critical role that MRI practitioners perform in managing MRI safety for their patients, healthcare staff and visitors. It is in this hypervigilant safety-conscious state that MRI practitioners try to balance the delivery of high-quality imaging alongside patient need. It should be recognised that practitioners are navigating an increasingly demanding role under high operational demand (Munn et al., 2014), which is time constrained (Hudson et al., 2022b), often within a reimbursement funding and biomedical service delivery model (Hewis, 2023a), which is efficiency and target driven. Further compounded by global ongoing staff shortages (Poon et al., 2022), long patient waiting lists (McIntyre & Chow, 2020), and

DOI: 10.1201/9781003310143-22

*Table 16.1* Poem dedicated to the resilient individuals who shared their lived experiences of distress in MRI in a qualitative study

My body was sore and vulnerable from surgery the day before.
'Just take this one hour before, that will settle you down'.
There was nothing gentle about it
shoehorned me in
this is hurting
Is it going to be ten more minutes or twenty?
'Here's some earplugs, it will be noisy'.
They close that door and you are on your own
check your dignity at the door.

They changed staff halfway through
when she clicked the microphone,
you could hear background chatter and banter.
Another job
another tick off the list
she was just going through the motions
just another body through the machine.
She said, 'Oh it's not that bad' and she brushed it off
it's like a house of cards, it's going to fall apart.

The fact that I was overweight was an inconvenience to them.
She said, 'You're doing well, you're fine'.
There's no one on the other side
buried alive
came out feeling indignant and discarded
apologising for inconveniencing them
minimal interaction
an absolute failure
they've got the next number to rush in
I hope they don't ask me to do another one.

Source: Hewis & Gallacé 2022. Excerpt reproduced with permission.

an aging population who present with complex, chronic multidimensional healthcare needs (Hewis, 2023a).

An essential component of person-centred care (PCC) is the patient's perception of care and control over their environment (Hyde & Hardy, 2021). Contemporary literature continues to identify clinical MRI as one of the most distressing imaging modalities. Anxiety during medical imaging reportedly ranging from 49–95% with 62% of surveyed MRI practitioners indicating they encounter 'notably anxious' patients daily (Hudson et al., 2022a). Occasional anxiety is a normal emotion and reaction to a threatening event, however if the level of anxiety is disproportional to the event it is identified as problematic (Taylor, 1995). If an individual becomes overwhelmed with anxiety a panic attack may occur (DeMasi, 2004). Research examining patient experience within MRI has often been underpinned by traditional, positivistic scientific methodologies for what is a complex human experience. There is a growing body of qualitative research giving voice to the patient perspective. Distress in MRI has associated cost and time implications that result in decreased efficiency and lower productivity for service providers (Nguyen et al., 2020; Hudson et al., 2022a). Significant technological development has resulted in wider, more open, and more 'patient friendly' scanners, yet neither distress nor claustrophobia have been fully eradicated (Hudson et al., 2022a).

In many ways MRI can be considered as a meeting of two worlds, where human (patient) intersects with high-level technology, mediating by the MRI practitioner during what is often a brief practitioner–patient encounter. Practice within MRI, like other high technology medical imaging specialisations, can quickly become mechanistic and highly task-orientated in nature. Patient care can be easily overlooked unless actively attended to. Staff consider themselves to be 'on the patient's side' (Munn et al., 2014). However, insufficient time is a frequently reported barrier to the provision of patient care (Hudson et al., 2022b), with a perceived requirement to trade-off 'care' against 'getting the patient scanned' (Munn et al., 2014). Therefore, it is critical that MRI practitioners develop a wholistic understanding of the patient experience and their need to be able to provide effective person-centred and individualised support and care.

The chapter will now follow the theoretical patient MRI journey of 'Charlie', drawing upon research by Hewis (2015), Hewis & Gallacé (2022), Homewood & Hewis (2023a), Hewis (2023b) and other relevant contemporary sources.

## The patient journey through MRI

*Before...*

When a Charlie arrives for their MRI scan, their healthcare journey is already well underway. The small snippet of clinical information on their MRI referral does little to capture their unique combination of needs, desires, fears, or vulnerabilities whether physical, emotional, spiritual, cultural, or psychological. If anything, that information might serve to prime the MRI practitioner's inherent assumptions and biases. Patients often arrive with heightened anxiety about the future MRI results (Carlsson & Carllson, 2013; Forshaw et al., 2018; Engels et al., 2019). Fear of results is often an underlying or background anxiety throughout the entire MRI journey. Interestingly, Hudson et al. (2022b) found that practitioners did not perceive anticipation of results as a significant contributor to patient anxiety, demonstrating a mismatch between practitioner understanding and actual patient experience. Charlie might hold hope that the MRI will finally provide an answer for their ongoing symptoms, which have eluded a diagnosis thus far. Or it might confirm their worst fear of cancer recurrence or disease progression. Beyond anxiety of the results, they may experience chronic pain, may have recently lost a loved one, have a phobia of needles, or there may be an older historical vulnerability that unexpectedly re-emerges during this new experience. The MRI examination is just one small step within their bigger healthcare journey.

How informed is the patient before they arrive? The referring doctor used language Charlie did not understand and provided little information about the MRI beyond reinforcing how important it was to get it done. Charlie resorts to searching 'Dr Google' for advice after a worrying chat with their aunt who is highly claustrophobic and had a 'nightmare' experience in MRI years ago (Asante & Acheampong, 2021). Perhaps English is not Charlie's first language, or they are illiterate, so, that wordy information sheet along with the MRI safety screening form was unhelpful (Tazegul et al., 2015). A study by Chesson et al. (2002) found that as little as 14% of patients knew what to expect about their imaging procedure. Being appropriately informed can reduce some pre-procedural anxiety (Tazegul et al., 2015; Tugwell et al., 2018; Ahlander et al., 2018), and is an important tenant of PCC (Hyde & Hardy, 2021). Audio-visual media to include sounds when combined with written information has demonstrated to reduce pre-procedural anxiety and can support coping strategy development (Ahlander et al., 2018; Yaker & Pirincci, 2020). The use of pre-visit to MRI or use of mock scanners are effective to enable familiarisation with the MRI environment, which can be beneficial in a wide array of patient groups to include those with claustrophobia, autism, and children. For patients with chronic and progressive conditions that require regular imaging for surveillance of the condition, their knowledge and understanding of MRI and its role in their condition can be extensive. For example, individuals with Multiple Sclerosis (MS) often prefer to be highly active in the decision-making process and are highly motivated to educate themselves about their condition and MRI (Brand et al., 2014; Freund et al. 2022).

Charlie's journey to their MRI appointment was a significant time and financial undertaking. Charlie lives in rural Australia, which means a few days away from the farm to visit the city, in unfamiliar surroundings and away from their support network. The receptionist was abrupt when Charlie arrived slightly late due to parking challenges. Charlie has also been feeling quite anxious due to waiting several weeks for the scan appointment. Charlie arrives with increased arousal where their state anxiety (sympathetic nervous response to a threat) has been influenced by many factors already. This background level of heightened pre-procedural anxiety raises their vulnerability to experience heightened distress, claustrophobia, and panic during an MRI scan (Thorpe et al., 2008).

Charlie is informed that the list is running behind schedule because 'we had to squeeze a couple of urgent requests in'. They begin to worry if the oral sedative they took one hour ago is going to 'wear off'. The use of oral anxiolytics is prevalent amongst adult patients with varying degrees of success (Hudson et al., 2022b), and associated with potential side effects to include reduced heart rate and blood pressure, slurred speech, confusion, dizziness, and memory loss (Sozio et al., 2021). A recent study determined that oral anxiolytics can be highly effective in specific patient groups undergoing head scans, but that the efficacy of these should be limited to patients with severe claustrophobia only (Sozio et al., 2021).

The pre-MRI ritual then begins with completing MRI safety screening and weight and height measurements. The physical removal of clothing and jewellery or piercings, and other expression of individuality, is sometimes a source of tension between patient and practitioner and can contribute to a loss of identify or deindividualisation (Haque & Waytz, 2012). The gown or scrubs provided might be ill fitting or undignified, while possessions are locked away and perhaps an intravenous cannula is inserted. Needle phobia may coexist with claustrophobia and as a specific anxiety condition can result in significant distress within MRI. It is a common cause of discomfort and distress in children (Staphorst, M. et al., 2015). Changing out of their own clothes can be distressing for certain patient groups. Individuals with autism could be advised to wear cotton only clothing without pockets and zips, which reduce their need to change (Stogiannos et al., 2022).

Charlie sits in their gown in a corridor adjacent to the MRI unit, with the ambient sound of gradients pulsing through the wall. This 'holding area' feels very clinical, cold, sterile, and here Charlie waits alone, deindividualised, contemplating the imminent scan (Haque & Waytz, 2012; Thu et al., 2015). Waiting is a frequently reported contributor to negative MRI scan experience. The phenomenon of waiting transcends the entire MRI journey, waiting for an appointment, waiting at the MRI unit, and waiting for results. For the anxious patient, indeterminate periods of waiting will increase their state of anxiety (Biddiss et al., 2014; Thu et al., 2015; Sadigh et al, 2017), and can reduce their tolerance to complete the scan (Munn & Jordan, 2011).

### During...

Eventually Charlie is called through to the control room where further preparation occurs. Sensory devices (such as hearing aid or glasses) that the individual needs to navigate and interact with the world are removed (if not already locked away) (Haque & Waytz, 2012). Finally, Charlie is escorted into the MRI scan room where a brief explanation is given whilst often concurrently being positioned onto the MRI couch. Seeing the scanner and the MRI environment unsurprisingly has been demonstrated as a common trigger for claustrophobia (Enders et al., 2011; Hudson et al., 2022b). During the journey so far, we have identified many stimuli that have contributed to Charlie's anxiety level, with research suggesting procedural anxiety is highest immediately prior to commencing an MRI (van Minde et al., 2014; Hudson et al., 2022b). Other well-documented high moments of anxiety include the placement of receiver coils around the patient (particularly head and neck coil) and being physically moved into the scanner where they experience the proximity of the bore and its enclosed nature (Van Minde et al., 2014;

Christina et al.,2020). The region of interest, patient position, the use of coils, positioning devices, blankets and the size of the patient additionally contributing to the sense of confinement as the patient reaches the centre of the bore.

The MRI environment and process can be a highly isolating for the patient, physically, emotionally, and psychologically. The patient is physically remote from the practitioner in another room, and the patient is generally separated from any accompanying support person(s) (Asante & Acheampong, 2021).

MRI scans require the individual to keep still for extended periods of time, when they may be in pain or discomfort or feel short of breath (Ellerbrock & May, 2015; Klaming et al., 2014). As an imaging technique, MRI is particularly sensitive to motion during the acquisition stage, which results in artefacts that ultimately impair image quality or can render the examination non-diagnostic (Nguyen et al., 2020). The practitioner will likely provide very clear, but sometimes quite blunt instructions, around the need to keep still, which may also be accompanied using immobilisation devices (Klaming et al., 2015). Most scanners have no visual cues for the progress of time, so the MRI practitioner becomes the 'gatekeeper of time' via the intercom. For Charlie, time becomes an unknown entity that exacerbates their anxiety (Carlsson & Carlsson). Many patients try to develop their own (sometimes maladaptive) coping strategies to monitor the passage of time, such as counting gradient noises. However, the perception of time frequently becomes stretched and distorted and is often perceived to pass very slowly (Funk et al., 2013). For a distressed patient it can feel like there is no immediate end to their current plight.

In this uniquely confined, isolated, noisy, and enforced motionless state, Charlie is left alone with their thoughts, emotions, and physical sensations. The patient can experience heightened awareness of their own body, bodily processes (breathing) or sensations (pain, numbness, itchiness) with an acute awareness of the relationship of their body to the surrounding MRI scanner (Funk et al., 2014). This moment of enforced self-awareness and introspection can be highly confronting.

Being in distress during an MRI is therefore not just a manifestation of claustrophobia. Charlie's distress is caused by contemporaneous internal and external contributing and confounding factors resulting in an aggregated level of anxiety arousal with physical, emotional (affective) and cognitive components, which they will either endure or not. Unsurprisingly, there is a correlation between the coping strategies the patient can employ and their levels of distress during MRI (Thorpe et al., 2008). Coping strategies provide the individual with some degree of control and agency (Funk et al., 2014). They are often based on trial and error and might enable the individual to reduce sensory input, facilitate cognitive and emotional regulation and reduce arousal.

The characteristic sounds of the time-varying gradients can be a source of frustration and anxiety (Hudson et al., 2022b). Listening to your preferred music is emotionally arousing, which can result in increased dopamine secretion ('mood enhancing') and lower cortisol excretion ('stress hormone'), which can reduce pain perception and decrease anxiety (Hewis, 2018). Many individuals with autism often have an anxiety-inducing sensitivity to noise, where music choice is particularly critical (Stogiannos, N., 2022). Music choice is a simple intervention that can provide a source of control and agency for the patient and acts as a distraction coping mechanism (Munn & Jordan, 2011). Another in-scan choice might be allowing Charlie to initiate breath hold sequences rather than having to breathe to the rhythm of another (the MRI practitioner) (Funk et al., 2014).

Non-claustrophobic causes of distress are more common; however, claustrophobia is an important consideration for all MRI practitioners and can be challenging to manage (Hudson et al., 2022b). Claustrophobia during MRI is an extensively documented phenomenon (Munn & Jordan, 2011; Dewey et al., 2007; Murphy, 2009; Funk et al., 2014). Since its inception, the MRI environment has become synonymous with the term 'claustrophobia', a fear of confined spaces or more specifically what might happen within that space. The reported incidence of claustrophobia varies widely, ranging from 1.5 to 14% (Mubarak et al., 2015; Nguyen et al., 2020) in part due to varying definitions, measure and retrospective methods employed. Hudson et al. (2022a) conducted a two-year retrospective study across multiple UK sites and scanner types examining 677,988 patient appointments with a reported overall incidence of incomplete examinations due to claustrophobia of 0.76%. This accounted for 32% of all attended but incomplete MRI appointments. Hudson et al. (2022a) found the 'likelihood of claustrophobia increases with females, those between 45–64 years of age… entering the scanner headfirst, and/or having a head scan.' If Charlie is claustrophobic, they may recognise that their fears are irrational, but they are unable to overcome them. Charlie may also have additional phobia(s). Specific phobias often coexist with other mental health-related comorbidities (Mubarak et al., 2015).

Charlie's distress in MRI (whether claustrophobic and non-claustrophobic related) can progress to panic, where there is a catastrophic cognitive misinterpretation of the bodily sensations of fear. In claustrophobia, panic occurs when there is a combination of fearful cognitions and misinterpreted bodily symptoms, e.g., bodily sensations of dizziness, choking and shortness of breath combined with fearful cognition of suffocation (Thorpe et al., 2008). The more time (real or perceived) that Charlie spends experiencing the physical symptoms of panic combined with the cognition of suffocating without control, the more likely they will precipitate to panic. During panic, Charlie experiences overwhelming terror of a perceived existential threat that results in an intense desire to physically escape (flight) the scanner. It is important to recognise that non-claustrophobic individuals have similar associations with space and can feel restricted and can experience panic due to other causes of distress, for example panic is a feature of several other anxiety disorders including post-traumatic stress disorder (PTSD).

Quantifying distress more broadly within MRI, and the extent to which these feelings are experienced, is challenging to measure objectively (Munn et al., 2015). Several validated behavioural and physiological scales exist (Baeyer & Spagrud, 2007), and the Spielberger State-Trait Anxiety Inventory (STAI) (Quirk et al., 1989) is one of the more frequently used anxiety tools within MRI research. Anxiety has also been assessed via visual analogue scales such as the Hospital Anxiety and Depression Scale (Westerman et al., 2004), and physiological measures such as serum cortisol levels, pulse, and blood pressure (Redd et al., 1994; Westerman et al., 2004).

Quantification measures are used as a potential method to predict adverse reactions, as a potential screening tool or to evaluate interventions (Booth & Bell, 2013). A future application for artificial intelligence could be to provide more accurate predictive capabilities for anxiety and claustrophobia for triage and to tailor patient information and scheduling (Hudson et al., 2022a). Charlie's distress can reduce diagnostic quality, caused by their associated movement, or may result in the scan being interrupted or prematurely terminated (Madl et al., 2022). This has significant implications for diagnosis and management (Munn et al., 2015; Nguyen et al., 2020). For the patient with severe claustrophobia, the fear can be so extraordinarily intense that it can lead to avoidance of situations that trigger fear and therefore future avoidance of MRI. Beyond the use of oral anxiolytics, administration of sedation is routinely practised worldwide to assist adults to cope and keep still in MRI (Loh et al., 2016). A subset of severely distressed patients will progress to complete their scan under general anaesthesia (GA) (Madl et al., 2022). MRI can be a particularly challenging for children and compared to adults and are more likely to be restless and can struggle to remain still for extended periods of time. Sedation and GA are currently routinely used in young children for motion-free MRI examinations and distress reduction, particularly from infant to children aged 7 or 8 (Mastro et al., 2019; Heales & Lloyd, 2022). These examinations are predominantly completed without complication; however, both sedation and GA include risks of acute adverse events to include (but not limited to) airway obstruction, bronchospasm, and anaphylaxis (Mastro et al., 2019).

The use of medicated intervention in specialist patient groups is associated with specific complications. For example, in children the use of sedation and GA can result in emergence delirium, which is linked to the development of prolonged maladaptive behaviours such as separation anxiety, night crying and general anxiety (Hilly et al., 2015; Mastro et al., 2019). However, children can complete MRI without anaesthesia when a PCC approach is adopted. Examples of effective strategies include the use of mock MRI practice sessions prior to performing the MRI (Carter et al., 2010) and play simulation (Heales & Lloyd, 2022). Mastro et al. (2019) evaluated a person and family-centred approach for paediatric MRI based on the principles of power sharing, shared responsibility, trust building and active engagement of the child in their own care and management. This non-anaesthesia approach was multifaceted and included an exploratory preparation session with the child and family, use of play therapy techniques and practising coping techniques. The non-anaesthesia intervention had 'significantly lower costs and short procedure times' compared to MRI with anaesthesia (Mastro et al., 2019). A study by Pua et al. (2020) also adopted a more individualised approach to 'awake MRI' for children with autism spectrum disorders that included familiarisation, gamification, and personalised rewards.

In all patients, but particularly children, there can be an ethical dilemma around the use of immobilisation and when does that become restraint. As with any imaging modality, the practitioner must balance respecting the child's voice and wishes alongside acting in their best interest. The use of restraint is not recommended within the context of MRI due to the extended scan time, the resultant significant emotional and psychological distress caused, and that diagnostic images would likely not be acquired.

*After...*

At the end of the procedure Charlie feels immediate relief. However, the focus of their anxiety rapidly shifts to anticipation of the scan results, and what that means for their health and wellbeing (Carlsson & Carlsson, 2013; Engels et al., 2019). Explaining how and when the patient will receive their results is a key consideration within a PCC approach (Hyde & Hardy, 2021).

Despite a shifting paradigm towards person-centred and individualised care, the patient in MRI may experience negative interactions with staff. The patient may experience a lack of care, compassion or empathy and may feel like they are an object, a condition, a body part or as just a number. The patient experience can be described as conveyor belt like, with staff 'just going through the motions' (Munn & Jordan, 2011; Dewey et al., 2007; Tischler et al., 2008). Patients who have experienced severe distress describe a sense of embarrassment, foolishness, shame, or feeling an immense sense of failure, particularly if they were unable to complete their scan partially or fully. Some patients report being rushed and emerge from the MRI unit feeling discarded and abandoned, with limited or no emotional or psychological support and unanswered questions. This void in post-procedure care can manifest as ongoing emotional and psychological distress that can persist for significant periods of time after their scan (days, weeks, months, years). This can leave the individual feeling angry and indignant about their care (or lack of). There is also tentative evidence to suggest MRI can induce claustrophobia in individuals who previously had no fear of spaces or confinement (Lukins et al., 1997; Thorpe et al., 2008). Conversely, successful completion of the examination by anxious individuals can decrease their fear of confined spaces in the future (Harris et al., 2004).

*Dehumanising*

Several aspects of the MRI experience and process could be considered as dehumanising. The phenomenon of dehumanisation lies on a spectrum from blatant to subtle and it can present in multiple forms. Central to all dehumanisation theories is the denial of human mind and agency to another person (Haque & Waytz, 2012), which can be prevalent in medicine and high technology environments to include MRI (Haslam, N. 2006). Haque & Waytz (2012) describe six forms of dehumanisation within a healthcare context that are of relevance to the MRI experience as summarised in Table 16.2.

In a recent survey, MRI practitioners reported concerns that the 'emotional load' of dealing with anxious patients was a barrier to delivering effective and individualised care (Hudson et al., 2022b), which strongly resonates with the concept of moral disengagement (Haque & Waytz, 2012). Challen et al. (2018), found patients with dementia experienced distress and loss of 'personhood' during medical imaging examinations in keeping with the concepts of reduced empathy, impaired agency, and dissimilarity. Each of these examples can be considered as dehumanising for the patient.

*Delivering person-centred care in MRI*

The patient journey in MRI can be one dominated by the phenomenon of distress. However, patient need is much broader than management of anxiety and many of the approaches used to support anxious patients can have broader benefits for other vulnerable or specialist patient groups. A critical tenant of PPC within MRI is effective communication. The critical role of communication is explored through a brief case study inspired by real-world clinical scenarios (see Table 16.3).

It is critical that MRI practitioners recognise that the relationship between the patient and practitioner is really one of the most significant factors contributing to patient experience and scan success after the physical nature of the scan and procedure (Funk et al., 2013; Carlsson & Carlsson, 2013; Hewis, 2015, Hudson et al., 2022b; Hewis & Gallacé, 2022; Homewood & Hewis, 2011).

The challenging nature of the MRI environment and process presents plenty of opportunity for exploring more person-centred and personalised approaches. A range of possible PCC considerations relevant to an MRI are identified in Table 16.4.

The main self-reported barrier to PCC is a lack of time to provide 'the individualised care needed by the patient' (Hudson et al., 2022b). MRI practitioners may also underestimate the impact they can make through PCC (Munn et al., 2014; Al-Shemmari et al., 2022; Hudson et al., 2022b). Other reported barriers include capacity issues, limited staff experience, need for more training and ineffective teamwork (Hudson et al., 2022b). Hudson et al. (2022b) suggest that a reliance on medication such as oral anxiolytics is perhaps used as a 'quick fix' to substitute or perhaps even

*Table 16.2* Forms of dehumanisation in healthcare

| Form of dehumanisation | Definition |
| --- | --- |
| Mechanisation | Treating the individual as an object rather than as a whole person. Where the practitioner perceives the patient as mechanical systems made up of interacting parts resulting in the use of depersonalising acronyms, body parts, or the name of their disease to describe the patient. |
| Impaired agency | A perception that the individual has impaired agency (capacity to plan, intend and act). Loss of agency is an intrinsic by-product of the patient's reason to seek care. They are incapacitated in some form, resulting in reduced agency with a failure to enable (or to include) patients to be active in their clinical decision making. |
| Dissimilarity | People are more likely to dehumanise individuals dissimilar to themselves de-emphasising the patient's humanity. In a healthcare setting 1) illness makes the patient seem dissimilar because it may alter their appearance, behaviours or functioning. 2) The patient is labelled as their illness, which encourages the perception of the individual as an illness rather than as a human. 3) The patient's illness results in an inherent loss of power and agency that facilitates dehumanisation. |
| Empathy reduction | The failure to consider individuals as a full social entity and instead treating them as mechanical systems with component parts, typically sacrificing empathy for cognitive objectivity when solving complex clinical problems. |
| Moral disengagement | Through moral and psychological distancing, the practitioner actively dehumanises the patient as a form of self-perseveration to minimise self-harm. This however ultimately leads to a reduction in empathy. |
| Deindividuating practices | An individual immersed in a group or otherwise anonymised; either deindividuation of the person being perceived (the dehumanised) or of the perceiver (the dehumaniser). Leads to diminished feelings of personal responsibility for actions, which provides licence (reduced culpability) for dehumanisation by denying a person's 'identity' as an individual who is independent and distinguishable from others and capable of making choices. Haque & Waytz describe staff and patient 'uniforms' as an example that index their distinct group membership. |

Source: Haque & Waytz, 2012.

negate the need for investment in more time-consuming and personalised care (Hudson et al., 2022b). Despite more recent research suggesting efficacy in only more extreme claustrophobia (Sozio et al., 2021). This could also reflect a wider lack of engagement in evidence practice and research endemic within the broader profession of diagnostic radiography. This suggest that

*Table 16.3* A person-centred MRI case study

---

Patient Name: Pia

Request: Non-contrast angiography MRI of thoracic aorta.

Clinical indications: 32-Year-old woman with Turner syndrome. Now 16 weeks pregnant with new onset of borderline hypertension. Surveillance scan to check for dissection, reoccurrence of coarctation and to check status of coarctation repair.

Clinical History: Coarctation repair aged 12 and known bicuspid aortic valve. Regular MRI studies throughout childhood and adulthood for surveillance of aorta anatomy and repair status with no further coarctation or dissection. Pia has anxiety and clinical depression since adolescence, which she successfully manages through antidepressant drug therapy. Pia halted her drug therapy 12 months ago when she decided to start a family with her husband. She is currently using a regime of meditation and therapy to help maintain her mental health. Pia discussed the difficulties with conceiving and the risks for an ongoing successful pregnancy with her Obstetrician. Previous MRI (9 months ago) showed that coarctation repair was stable. After an unsuccessful cycle of fertility treatment (IVF) Pia has now successfully conceived with the assistance of Embryo implantation. Her most recent clinic appointment showed signs of gestational hypertension. 1-week home blood pressure monitor indicated sustained mildly hypertensive results. Given the inherent risks of a difficult pregnancy for an individual with Turner syndrome it is considered appropriate to continue with daily monitoring of blood pressure and to arrange frequent MRI imaging of the thoracic aorta throughout the term with ongoing foetal ultrasound and MRI to be considered as pregnancy continues.

Background: Less than 2% of women with Turners syndrome conceive naturally (Abir et al. 2001). Given this statistical disadvantage and the likelihood of a more challenging pregnancy, early evaluation and monitoring is vital. Risk of aortic dissection and severe hypertension is high during pregnancy and complications such as preeclampsia, premature delivery, congenital abnormalities, and small foetus size are all increased (Althagafi et al. 2021). Pia was advised at her most recent clinic appointment that she is suffering from hypertension and an urgent MRI will be requested. Pia has undergone several previous MRI scans without any issues. An aorta surveillance MRI is a non-contrast study, which is typically uncomplicated and quick.

Discussion: Traditionally individuals with Turner's syndrome have been strongly advised to avoid pregnancy due to the high risk of complication (Folsom & Fuqua, 2015). After her scan referral, Pia called the MRI department several times to seek reassurances about the risks involved with the scan. Upon the third time of calling, an experienced member of the booking team took her details and then escalated Pia's concerns to the MRI team. It was the persistence of the patient to an escalation of her concern. A senior MRI practitioner returned her call that lead and using their extended knowledge was able to discuss the potential risks to their unborn baby and what the counter risks might be not attending for the MRI scan. The MRI practitioner was able to reassure the Pia that a family member could accompany then into the scanner if required (MRI safety accepting). Individualised precautions were discussed to include the use of low SAR and low acoustic noise sequences and the use of additional padding to ensure that the RF coil did not lie directly on her stomach or compress the foetus unnecessarily. The one-to-one call also enabled Pia to feel able to raise concerns about her anxiety. Pia was advised that she could bring her own meditative music to listen to during the scan and additional breathing relaxation. Ultimately, Pia attended for her MRI and successfully completed the scan without any additional concerns.

---

*Table 16.3* (Continued)

---

Conclusion: A multi-disciplinary approach to patient centred care does not begin and end with allied health care professionals or physicians. Communication with non-clinical staff can be vital to a positive patient experience within radiology and beyond. Taking the time to research and then discuss concerns with an anxious or concerned patient as well as considering ways and techniques to help alleviate or reduce these anxieties can help to make the journey through MRI less intimidating and stressful.

---

*Table 16.4* Person-centred care approaches within MRI

| | |
|---|---|
| Seeing the individual | Every person is different and complex. Try to understand their unique needs and preferences. Do not assume, listen, promote shared decision making to empower and give agency. |
| Involve carers and support people | Enable them to be present during MRI for support when practicable & safe; for example, decision making normally involves input of other family members for Aboriginal and Torres Strait Islander people. |
| Optimise your communication | Speak their language, use inclusive language, do not infantilise, involve support people where appropriate, enable shared decision making, avoidance of eye contact is a gesture of respect for Aboriginal and Torres Strait Islander people. |
| Provide information and questionaries in diverse formats | Print, digital, online, video, flash cards, braille and other in languages. |
| Offer personalised support pre & post | Face to face, online chat, videoconference, telephone for personalised information and support. |
| Provide opportunity for familiarisation | Pre-MRI site visits, use of mock scanners, walk arounds, practice on off the scanner and going in and out of the scanner, VR simulation. |
| Give choice where practicable | Type of scanner (offering UpRight or open scans for larger patients or those with distress associated with confinement), clothing (gown or scrubs and in a range of sizes and to include provision of hair and head covering MRI safe clothing items), music, adjunct devices (prism glasses), prone or supine, feet or headfirst. |
| Reduce waiting | Waiting associated with appointment, at MRI facilities, for formal report provides an opportunity to reduce anxiety. |
| Offer inclusive waiting areas | Provide a range of seating options, quieter areas for those with a neurodegenerative condition, spaces for children and their support persons. |

*(Continued)*

*Table 16.4* (Continued)

| | |
|---|---|
| Make reasonable personalised adjustments | Adjust care and practices specific to the individual needs where practicable to negate a 'one size fits all' approach. Using quieter sequences for individuals with neurodiversity, or motion resistant and shorter scan times for children. Manipulating lighting (brighter or reducing) or adopting modified techniques. |
| Explore non-pharmacological interventions | Reduce reliance on anxiolytics, sedation and GA and seek alternatives where appropriate. |
| Work collaborative with other professionals | Collaborate with music therapist, play specialists, special education needs practitioners, referrers to provide more holistic greater continuity of care. |
| Educate and train the MRI team | Increase and strengthen their PCC toolkit. Adopt a team approach to include support and administration staff. Examples might include specialised training for vulnerable, under-represented populations, specialised groups to include neuro diverse, LGBTQAI+, individuals with multiple sclerosis, dementia, culturally competent practice, and cultural safety training (first nations and indigenous). |

MRI practitioners need support and greater understanding to successfully implement PCC (Ajam et al., 2017; Santarem et al., 2020).

### Closing remarks

Patient experience within MRI is complex but it can often be a dehumanising and distressing experience. MRI staff practise in a demanding and time-poor work environment, however there should not be a trade-off between care and efficiency. There is significant opportunity to improve patient experience and increase satisfaction, which can also increase clinical outcomes with additional efficiency and financial benefits for service provision. Greater investment in MRI practitioner education and training to increase and strengthen their PCC toolkit is needed. MRI practitioners need to value humanity and recognise the critical importance of the patient interaction within this environment, noting that effective and sensitive communication is an essential but simple PCC approach all practitioners can adopt.

### References

Abir et al. (2001). Turner's syndrome and fertility: current status and possible putative prospects. *Hum Reprod Update*, 7(6):603–10.

Ahlander BM., Engvall J., Maret E., Ericsson E. (2018). Positive effect on patient experience of video information given prior to cardiovascular magnetic resonance imaging: a clinical trial. *J Clin Nursing*, 27(5–6):1250–61.

Ajam AA., Nguyen XV., Kelly RA., Ladapo JA., Lang EV. (2017). Effects of interpersonal skills training on MRI operations in a saturated market: a randomized trial. *J Am Coll Radiol*,14(7):963–70.

Al-Shemmari AF. Herbland A. Akudjedu TN. Lawal O. (2022). Radiographer's confidence in managing patients with claustrophobia during magnetic resonance imaging. *Radiography*, 28(1):148–53.

Althagafi et al. (2021). Pregnancy outcomes among 184 women with turner syndrome. *Am J Obstet Gynecol* 224(2);S313–314.

Asante S. & Acheampong F. (2021). Patients' knowledge, perception, and experience during magnetic resonance imaging in Ghana: A single centre study. *Radiography*, 27:622–626.

Baeyer C. & Spagrud L. (200). Systematic review of observational (behavioral) measures of pain for children and adolescents aged 3 to 18 years. *Pain*. 127(1–2),140–50.

Biddiss E. Knibbe T. McPherson A. (2014). Anxiety in health care waiting spaces: a systematic review of randomized and nonrandomized trials. *Anesth Analg*, 119(2):433–448.

Bijvoet G. et al. (2021). Transforming a pre-existing MRI environment into an interventional cardiac MRI suite. *J Cardiovasc Electrophysiol*, 32:2090–2096.

Booth L. & Bell L. (2013). Screening for Claustrophobia in MRI: A Pilot Study. *Eur Sci J*, 9(18),20–31.

Brand J. Köpke S. Kasper J. Rahn A. Backhus I. Poettgen J. et al. (2014). Magnetic resonance imaging in multiple sclerosis–patients' experiences, information interests and responses to an education programme. *PloS One*.

Carlsson S. & Carlsson E. (2013). 'The situation and the uncertainty about the coming result scared me but interaction with the radiographers helped me through': a qualitative study on patients' experiences of magnetic resonance imaging ex- aminations. *J Clin Nursing*, 22(21–22):3225–34.

Carter AJ. Greer MLC. Gray SE. & Ware R.S. (2010) Mock MRI: Reducing the need for anaesthesia in children. *Pediatric Radiology*. 40 (8),1368–74.

Challen R. Low L-F. McEntee MF. (2018). Dementia patient care in the diagnostic medical imaging department. *Radiography*, 24:533–42.

Chesson RA. Mckenzie GA. Mathers SA. (2002). What do patients know about ultrasound, CT and MRI? *Clin Radiol*, 57(6):477–82.

Christina et al. (2020). Making Magnets More Attractive: Physics and Engineering Contributions to Patient Comfort in MRI. *Top Magn Reson Imaging*, 29(4):167–174.

Delfino J. et al. (2019). MRI-related FDA adverse events reports: A 10-yr review. *Med Phys*, 16(12):5561–5571.

DeMasi F. (2004). The psychodynamic of panic attacks: a useful integration of psychoanalysis and neuroscience. *Int J Psychoanal*, 85, 311–336.

Dewey M. Schink T. & Dewey C.F. (2007). Claustrophobia during magnetic resonance imaging: cohort study in over 55,000 patients. *J Magn Reson Imaging*, 26(1322–1327).

Ellerbrock I. & May A. (2015). MRI scanner environment increases pain perception in a standardized nociceptive paradigm. *Brain Imaging Behav*, 9, 848–853.

Enders J. Zimmermann E. Rief M. Martus P. Klingebiel R. Asbach P. et al. (2011). Reduction of claustrophobia with short-bore versus open magnetic resonance imaging: A randomized controlled trial, 6(8):1–10.

Engels K. Schiffmann I. Weierstall R. Rahn AC. Daubmann A. Pust G. et al. (2019). Emotions towards magnetic resonance imaging in people with multiple sclerosis. *Acta Neurol Scand*, 139(6):497–504.

Folsom L. & Fuqua J. (2015). Reproductive Issues in Women with Turner Syndrome. *Endocrinol Metab Clin North Am*, 44(4): 723–737.

Forshaw KL. Boyes AW. Carey ML. Hall AE. Symonds M. Brown S. et al. (2018). Raised anxiety levels among outpatients preparing to undergo a medical imaging procedure: prevalence and correlates. *J Am Coll Radiol*, 15(4):630–638.

Freund M. et al (2022). Understanding Magnetic Resonance Imaging in Multiple Sclerosis (UMIMS): Development and Piloting of an Online Education Program About Magnetic Resonance Imaging for People With Multiple Sclerosis. *Front. Neurol*, 13:856240.

Funk E. Thunberg P. & Anderzen-Carlsson A. (2013). Patients' experiences in magnetic resonance imaging (MRI) and their experiences of breath holding techniques. *J Adv Nurs, 70*(8),1880–90.

Haque O. Waytz A. (2012). Dehumanization in medicine: causes, solutions, and functions. *Perspect Psychol Sci*,7(2):176e86.

Harris L. Cumming S. & Menzies R. (2004). Predicting anxiety in magnetic resonance imaging scans. *Int J Behav Med, 11*(1):1–7.

Haslam N. (2006). Dehumanization: an integrative review. *Pers Soc Psychol Rev,* 10:252–64.

Heales C. & Lloyd E. (2022). Play simulation for children in magnetic resonance imaging. *J Med Imaging Radiat Sci*, 53:10–16.

Hewis J. & Gallacé N. (2022). When she clicked the microphone, you could hear background chatter and banter. *J Med Imaging Radiat Sci*, 53:194–195

Hewis J. (2015). Do MRI Patients Tweet? Thematic Analysis of Patient Tweets About Their MRI Experience. *J Med Imaging Radiat Sci*, 46:396–402.

Hewis J. (2018). Music and Music Therapy in the Medical Radiation Sciences. *J Med Imaging Radiat Sci*, 49:360–364.

Hewis J. (2023a). A salutogenic approach: Changing the paradigm. *J Med Imaging Radiat Sci. In press.* doi.org/10.1016/j.jmir.2023.02.004.

Hewis J. (2023b). Phenomenology of Distress in Magnetic Resonance Imaging. [Unpublished Doctor of Philosophy Thesis]. Charles Sturt University.

Hilly J. Horlin A.L. Kinderf J. et al. (2015). Preoperative preparation workshop reduces postoperative maladaptive behavior in children. *Paediatr Anaesth*, 25, 990–998.

Homewood H. & Hewis J. (2011). 'Scanxiety': Content analysis of pre-MRI patient experience on Instagram. Radiography, In press: doi.org/10.1016/j.radi.2023.01.017.

Hudson D. Heales C. & Meertens R. (2022a). Review of claustrophobia incidence in MRI: A service evaluation of current rates across a multi-centre service. *Radiography*, 28:780–787.

Hudson D. Heales C. & Vine S. (2022b). Radiographer Perspectives on current occurrence and management of claustrophobia in MRI. *Radiography*, 28:154–161.

Hyde E, Hardy M. (2021). Patient centred care in diagnostic radiography (Part 2): A qualitative study of the perceptions of service users and service deliverers. *Radiography*, 27(2):322–331.

Kamvosoulis P. & Currie, G. (2021). PET/MRI, Part 1: Establishing a PET/MRI Facility. *J Nucl Med Technol*, 49(2):120–125.

Klaming L. van Minde D. Weda H. Nielson T. & Duijm L. (2014). The Relation Between Anticipatory Anxiety and Movement During an MR Examination. *Acad Radiol, 22*(12), 1571–1578.

Liu C. Li, M. Xiao, H. et al. Advanced in MRI-guided precision radiotherapy. *Precis Radiat Oncol*, 6;75–84.https://doi.org/10.1002/pro6.1143

Loh P-S. Ariffin M.A. Rain V. Lai L-L. Chan L. & Ramli N. (2016). Comparing the efficacy and safety between propofol and dexmedetomidine for sedation in claustrophobic adults undergoing magnetic resonance imaging (PADAM trial). *J Clin Anesth 34*(1),216–222.

Lukins R. Davan I.G. & Drummond P.D. (1997). A cognitive behavioural approach to preventing anxiety during magnetic resonance imaging. *J Behav Ther Exp Psychiatry, 28*(2),97–104.

Madl J, Janka R, Bay S, Rohleder. (2022). MRI as a stressor: the psychological and physiological response of patients to MRI, influencing factors, and consequences. *J Am Coll Radiol*, 19(3):423–432.

Mastro K. et al. (2019). Reducing Anesthesia Use for Paediatric Magnetic Resonance Imaging: The Effects of a Patient and Family Centered Intervention on Image Quality, Health-care Costs, and Operational Efficiency. *J Radiol Nurs*, 38:21–27.

McIntyre D. & Chow C. (2020). Waiting Time as an Indicator for Health Services Under Strain: A Narrative Review. *J Healthcare Org, Provis, Financing*, 57:1–15.

Mubarak F. Bain K. & Anwar S.S.M. (2015). Claustrophobia during Magnetic Resonance Imaging (MRI): Cohort of 8 Years. *Int Neuropsychiatr Dis J, 3*(4),106–111.

Munn Z. & Jordan Z. (2011). The patient experience of high technology medical imaging: a systematic review of the qualitative evidence. *Radiography, 17*(4),323–31.

Munn Z. et al. (2014). 'On their side': Focus group findings regarding the role of MRI radiographers and patient care. *Radiography, 20*,246–50.

Munn Z. Moola S. Lisy K. Riitano D. & Murphy F. (2015). Claustrophobia in magnetic resonance imaging: A systematic review and meta-analysis. *Radiography, 21*(1):59–63.

Murphy F. (2009). Act, scene, agency: the drama of medical imaging. *Radiography,*15(1):34–39.

Nguyen X. Tahir S. et al. (2020). Prevalence and Financial Impact of Claustrophobia, Anxiety, Patient Motion, and Other Patient Events in Magnetic Resonance Imaging. *Top Magn Reson Imaging*, 29(3):125–130.

Poon Y-S. et al. (2022). A global overview of healthcare workers' turnover interntion amid COVID-19 pandemic: a systematic review with future directions. *Hum Resour Health*, 20(70):1–18,

Pua K. et al. (2020). Individualised MRI training for paediatric neuroimaging: A child-focused approach. *Dev Cogn Neurosci*,41:100750.

Quirk M. Letendre A.J. Ciottone R.A. & Lingley J.F. (1989). Anxiety in patients undergoing MR imaging. *Radiology, 170*(2),463–6.

Redd W.H. Manne S.L. Peters B. Jacobsen P.B. & Schmidt H. (1994). Fragrance administration to reduce anxiety during MR imaging. *J Magn Reson Imaging*, 4(4):623–6.

Sadigh G. Applegate K. Saindane A. (2017). Prevalence of unanticipated events associated with MRI examinations: a benchmark for MRI quality, safety, and patient experience. *J Am Coll Radiol*, 14:765–72.

Santarem Semedo C. Moreira Diniz A. Heredia V. (2020). Training health professionals in patient-centered communication during magnetic resonance imaging to reduce patients' perceived anxiety. *Patient Educ Counsel*, 103(1):152–8.

Sozio S. et al (2021). Determining the efficacy of low-dose oral benzodiazepine administration and use of wide-bore magnet in assisting claustrophobic patients to undergo MRI brain examination. *Clin Imaging*, 79:289–295.

Staphorst M. Hunfeld J. van de Vathorst S. Passchier J. van Goudoever J. (2015). Children's self reported discomforts as participants in clinical research. *Soc Sci Med*, 142:154–62.

Stogiannos N. et al. (2022). A systematic review of person-centred adjustments to facilitate magnetic resonance imaging for autistic patients without the use of sedation or anaesthesia. *Autism*, 26(4):782–797.

Taylor S. (1995). Anxiety sensitivity: theoretical perspectives and recent findings. *Behav Res Ther*, 33(3), 243–258.

Tazegul G. et al. (2015). Can MRI related patient anxiety be prevented? *Magn Reson Imaging*, 33:180–183.

Thorpe S. Salkovskis P.M. & Dittner A. (2008). Claustrophobia in MRI: the role of cognitions. *Magn Reson Imaging*, 26(8), 1081–1088.

Thu H. Stutzman S. Supnet C. & Olson D. (2015). Factors associated with increased anxiety in the MRI waiting room. *J Radiol Nurs*, 34:170–4.

Tischler V. Calton T. Williams M. Cheetham A. (2008). Patient anxiety in magnetic resonance imaging centres: is further intervention needed? *Radiography*, 14(3):265–6.

Tugwell JR. Goulden N. Mullins P. (2018). Alleviating anxiety in patients prior to MRI: a pilot single-centre single-blinded randomised controlled trial to compare video demonstration or telephone conversation with a radiographer versus routine intervention. *Radiography*, 24(2):122e9.

van Minde D. Klaming L. & Weda H. (2014). Pinpointing Moments of High Anxiety During an MRI Examination. *Int J Behav Med*. 21, 487–495.

Westerman E. Aubrey B. Gauthier D. et al. (2004). Positron emission tomography: a study of PET test-related anxiety. *Can J Cardiovasc Nurs*, 14(2),42–8.

Yaker B. & Pirincci E. (2020). Investigation of the Effect of Written and Visual Information on Anxiety Measured Before Magnetic Resonance Imaging: Which Method is Most Effective. *Medicina*, 56(30):136.

# Section seven

# Person-centred care in ultrasound

# 17 Person-centred care in sonography

*Yaw Amo Wiafe, Benedict Apaw Agyei and Andrew Donkor*

## Introduction

Ultrasound is the second most requested radiologic test after projection radiography, and the most used imaging method outside of the radiology department (NHS, 2016). In the United States, the annual rate of ultrasound examinations increased from 188 per 1000 person in the year 2000, to 386 per 1000 person in the year 2016 (Smith-Bindman et al., 2019). Due to its lack of ionizing radiation, it has also become the imaging procedure most likely to be done on a person before birth. Several studies have found that patients are generally satisfied with ultrasounds, and the main factors that determine satisfaction are communication, face-to-face interaction, waiting time, and courtesy (DiGiacinto et al., 2016; Mulisa et al., 2017). From the humble beginnings of crude composite images, diagnostic ultrasound has established an integral role in medicine (Donald, 1965; Donald et al., 1958). Initially, reporting findings using early equipment was similar to an X-ray, with reporting doctors assessing images after the examination in a separate room. Thankfully, this has been superseded by 'real-time' ultrasound imaging, which provides a significant advantage over other imaging modalities such as CT and MRI due to its dynamic nature. Patients have also benefited from these technological advances with increased resolution, improving diagnostic certainty and the introduction of three-dimensional (3D) obstetric ultrasound, improving maternal bonding (Ji et al., 2005; Pretorius et al., 2001; Roberts, 2012).

The advent of 'real-time' ultrasound imaging has shifted the role of ultrasound practitioners from 'image takers' to imaging specialists. Today, ultrasound practitioners must demonstrate independent clinical reasoning as they autonomously examine anatomical structures and dynamic functionality to look for abnormalities and construct a diagnosis. In addition, ultrasound practitioners must evolve from a diagnostic focus to a person-centred focus, incorporating emotional intelligence (EQ) skills to recognise the needs and preferences of each patient receiving care.

Dissatisfaction can be caused by an inaccurate ultrasound report, which may not be apparent to the patient until later, sometimes months after the

DOI: 10.1201/9781003310143-24

ultrasound. This has led to a number of lawsuits from unhappy patients who felt the ultrasound results either misled or contributed to adverse outcome. Molestation and sexual assault charges by patients have also been made against sonographers who were attending to patients, some of which resulted from poor communication and failure to obtain proper consent. This chapter deals with the best practice approach to person-centred sonography that is aimed at improving patient satisfaction and minimizing lawsuits for negligence. The chapter addresses four relevant topics on person-centred sonography, namely;

1. Patient interaction and communication
2. History taking
3. Implementation of quality patient care protocols
4. Medicolegal and ethical issues

### Patient interaction and communication

In sonography, patient interaction is regarded as the first step to effective patient care (DiGiacinto et al., 2016; Zelna, 2021). Sonographers should not underestimate the importance of being kind and friendly when speaking with the patient. It's been found that the attitude of healthcare providers usually determines patient satisfaction (Yarnold et al., 1998; DiGiacinto et al., 2016).

Moore et al. (2011) found that patient complaints about care providers were mostly related to ineffective communication. Many legal cases involving medical care have also raised this concern. Clever et al. (2008) also found that patients are more satisfied with their care when care providers are friendly and open with them about their medical options and allow them to ask questions. Cousin et al. (2012) also noted that an improved flow of information and communication increased patient satisfaction.

Sonographers should recognize that patients will have different levels of needs and expectations, somewhat like Maslow's hierarchy of needs (Zelna, 2021). While some patients might only need reassurance of safety and security, others would require a sense of love, belonging, and respect depending on their level of self-actualization. The sonographer should attend to each patient's specific needs to ensure optimal satisfaction for all kinds of patients.

A popular patient communication tool recommended for patient interaction in Sonography is AIDET (Zelna, 2021). AIDET is acronym for Acknowledge the patient, Introduction of yourself, Duration of expected time to complete examination(s), Explanation of the procedure, and Thank you to the patient for choosing your facility (Bello, 2017; Burgener, 2017). AIDET is a helpful resource because it promotes patient dignity and respect. Moreover, the fundamental principles of AIDET improves the quality of care provided, and ensures patient compliance and satisfaction.

According to Palombi et al. (2015), the AIDET framework improves consistency in patient-centred care delivery and highlights patient needs and expectations. They found higher patient satisfaction with pharmacy services when AIDET techniques were introduced in a healthcare setting, and patients felt "happy, comfortable, and trusting." Another study, conducted by Edwardson et al. (2016) also found that AIDET had a positive impact on patient safety-oriented change initiatives.

In another study by Elshamy and Ramzy (2011), it showed that the number of patients who thought the information received from the care provider was 'good' or 'very good' went from 11.9% before the AIDET program started to 73.8% after the program started. A study by Zamora et al. (2015) also found that patient satisfaction scores went up by 2.4% and stayed mostly the same after the introduction of AIDET. After seeing that patients were more satisfied because of this, leadership decided to use AIDET across the whole organization. This array of evidence undoubtedly supports the implementation of the AIDET tool in person-centred sonography.

### Acknowledging the patient in accordance with AIDET

Communicating with the patient by name and eye contact is an important component of patient care in sonography (Zelna, 2021). Gillette et al. (1992) found that 45% of patients preferred their first name, while Parsons et al. (2016) found that 99% of patients preferred informal address over formal address. Stapleton (2000) noted that calling patients by their legal name was perceived as a sign of not caring about them on a personal level, which makes people feel disconnected.

The relationship between the care provider and the individual patient can also be strengthened through the use of non-verbal communication. Non-verbal communication includes eye contact, gesture, touch, posture, facial expression, clothing and hairstyle (Montague et al, 2013). Khan et al. (2014) found that 86.1% of patients believe establishing eye contact shows that the care provider is attentive towards them; however, prolonged staring makes them uncomfortable. The importance of gestures like a smile, a warm blanket, and a reassuring touch is emphasized by McMaster and DeGiobbi (2016). van der Westhuizen et al. (2020) noted that, because the sonographer's priority is to concentrate on the diagnosis, it often leaves room for compromised emotional care.

### Introducing yourself in accordance with AIDET

The first good impression to give a patient is to introduce yourself (Granger, 2013). Introducing yourself to the patient puts them at ease and makes them feel more comfortable. It also reduces patient complaints and improves their satisfaction (Granger, 2013). Making sure the patient remembers your name is important for initiating and keeping a good relationship with the patient (Brockopp et al., 1992; Parson et al., 2016). Patel and Cabana (2010) found improved patient satisfaction when healthcare providers introduced themselves to patients. In another study by Gillen et al. (2018) they found that 89.7% of patients felt that an introduction made their healthcare visit better.

Consequently, sonographers should remember to introduce themselves to patients before the start of every exam and explain their role in the healthcare team. Many patients are not clear on what the role of a sonographer involves (Thomas et al., 2017; van der Westhuizen et al., 2020). In a study by Starcevich et al. (2021), they found that when participants were asked the title of the person who performs ultrasound examinations, less than 50% provided the correct response of a sonographer. It is important to explain your role and level of training to the patient because it has an impact on the patient's satisfaction and their willingness to consent for the examination (Santen et al., 2004).

### Duration of examination in accordance with AIDET

Studies have found an association between patient satisfaction and their perception about the duration of the examination (Miller et al., 2013; Soremekun et al., 2011). It is therefore important to inform the patient about the duration of the examination before it starts. Anderson et al. (2007) found that the length of time a patient spends with the care provider and their perception of the quality of the information given to them significantly affected their satisfaction. Lower satisfaction has also been attributed to longer wait times with less time spent interacting with care provider (Camacho et al., 2006). However, an earlier study by Mowen et al. (1993) reported that waiting patients were satisfied when they were told how long they would have to wait. Locke et al. (2011) noted that telling a patient that the examination might delay is the most important thing when it comes to wait time. They found that telling people about delays had a 65% positive predictive value for satisfaction. Care providers should therefore interact with waiting patients and notify them about possible delays. However, the satisfaction of delayed patients also depends on the comfort in the waiting area. Leddy et al (2003). found that patients who rated the comfort in the waiting area as "very poor" versus those who rated it as "very good" showed satisfaction of 58.3% and 96.7% respectively.

### Explanation of the procedure in accordance with AIDET

Prior to undergoing an ultrasound examination, a patient must be informed of any potential risks and benefits, and other options available. They should be encouraged to make an informed decision free from coercion in order to ensure a valid consent process (Starcevich et al., 2021; Jefford and Moore, 2008; Thomson and Moloney, 2017). A study found that more than half of patients knew very little about their imaging procedure (Chesson et al., 2002). Starcevich et al. (2021) noted that participants had incomplete ultrasound examination knowledge, including 30% of patients who could not tell whether ultrasound is radiation-based. Another study in Ghana also found that patients who had little information about ultrasound were more likely to be very nervous about their exams (Antwi et al., 2015). This is against the principles of informed consent.

In order to ensure valid informed consent, the referring physician or the sonographer explains the procedure to the patient and encourages them to ask questions. Previous studies have reported that referring physicians were often unreliable for explaining imaging procedures (Bowden et al., 2017; Chesson et al., 2002), which makes the sonographer the most appropriate source of explanation. Sonographers must ensure this because some patients have described their care during an ultrasound examination as unsatisfactory. They complained that the procedure was not explained to them before it began, leaving them uncertain about what to expect (Thomas, O'Loughlin, and Clarke, 2019).

### *Thank you in accordance with AIDET*

After completing the examination, sonographers should not underestimate the importance of saying thank you to the patient. Studies have shown that over 50% of caregivers experience burnouts that leads to medical errors (Ariely and Lanier, 2015; Hall et al., 2016). An expression of gratitude has been found to lessen burnouts and, by extension, reduces medical errors (Emmons and McCullough, 2003; Hamilton et al., 2018). Since diagnostic errors leads to patient dissatisfaction, saying thank you is important to minimize burnouts that may cause diagnostic errors.

## History taking

History taking is the cornerstone of patient care and an important avenue for building a strong relationship between the patient and care provider. A patient-centred approach can improve patient satisfaction during history taking (Wu, 2013). Rather than strictly following a structured approach, Wu (2013) recommends a flexible approach that promotes patient-centred care. Most of the time, a good interaction with a patient combines empathetic communication with history taking. However, Ohm et al. (2013) found that while history taking and empathetic communication go together, they are separate sides of the same coin. An accurate clinical history can yield up to 80% of clinical diagnosis (Keifenheim et al., 2015). The mains sources of patient clinical history are the referral letter, patient's medical records, interviewing the patient, and clinical observations. Sonographers should maximize the use of all sources of information in order to minimize diagnostic errors.

## Referral letter

The referral letter is the primary source of clinical history in sonography. (Necas, 2018). It is required to ensure the continuity of patient care. The content and quality of information conveyed by the referral letter may influence patient care (Akbari et al., 2005). It has been proven that incomplete referral letters negatively affect patient outcomes (Jiwa et al., 2009).

## Medical records

Another source of history taking is the patient's medical records such as admission and discharge summaries, clinic letters, surgical reports, laboratory tests, and previous imaging investigations which may be available in a handwritten folder or an electronic database. A sonographer may be able to gather useful information about the patient's medical history from this folder (Necas, 2018). Such medical records are a helpful resource that could guide the sonographer in providing a precise differential diagnosis if the referral letter is deficient (Necas, 2010a). When obtaining patient history, the sonographer should review the medical records if they are available, rather than relying solely on the referral letter.

## Patient interview

Patients who are conscious and in the correct mental state to answer questions can also be interviewed directly in the ultrasound room to glean extra pertinent history. Obtaining pertinent information from the patient is important because the presenting complaint may have changed. Moreover, the patient may provide previously unknown clinical information that may be useful to the ultrasound examination and interpretation of the findings (Necas, 2018).

## Clinical observation

The last source of gathering pertinent history is clinical observations made by the sonographer, such as symptoms and perceived areas of pathology (Necas, 2018). The sonographer should conduct a clinical evaluation of the patient whenever necessary prior to performing the ultrasound (Necas, 2010a). Visual evaluation and palpation of specific regions of interest might provide valuable clinical clues for otherwise confusing sonographic appearances that could be misinterpreted (Necas, 2018).

## Implementation of quality patient care protocols

During the examination, the patient should be positioned on a scanning table that provides adequate comfort. Quality patient care is provided through the safe and accurate implementation of a deliberate protocol. A standard guide for sonographers when implementing quality patient care protocol was published by Bendick (2015) with the endorsement of professional bodies in North America, including the American College of Radiology (ACR), American Congress of Obstetricians and Gynecologists (ACOG), American Institute of Ultrasound in Medicine (AIUM), American Registry for Diagnostic Medical Sonography (ARDMS), American Society of Echocardiography (ASE), and Sonography Canada. A section of this guideline is presented in Box 17.1. Similar protocols are also available in other professional jurisdictions, such as protocols of the British Society of Medical Ultrasound (BMUS) in collaboration with the Society and College of Radiographers (SCoR).

---

**Box 17.1**

- Implements a protocol that falls within established procedures.
- Elicits the cooperation of the patient to carry out the protocol.
- Adapts the protocol according to the patient's disease process or condition.
- Adapts the protocol, as required, according to the physical circumstances under which the examination must be performed (e.g., operating room, sonography laboratory, patient's bedside, emergency room, etc.).
- Monitors the patient's physical and mental status.
- Adapts the protocol according to changes in the patient's clinical status during the examination.
- Administers first aid or provides life support in emergency situations.
- Performs basic patient care tasks, as needed.
- Recognizes sonographic characteristics of normal and abnormal tissues, structures, and blood flow; adapts protocol as appropriate to further assess findings; adjusts scanning technique to optimize image quality and diagnostic information.
- Analyses sonographic findings throughout the course of the examination so that a comprehensive examination is completed and sufficient data is provided to the supervising physician to direct patient management and render a final interpretation.
- Performs measurements and calculations according to facility protocol.

(Adapted from Bendick, 2015)

---

*Patient transfer/safety*

The sonographer is responsible for ensuring that transferred patients are safe. They should carefully monitor unconscious or sedated patients, ensure that siderails are up and wheelchairs and stretchers are locked. Proper body mechanics should be used when instructing patients to change positions. If patient restraint is necessary to protect him or her from causing harm, the order should come from the doctor or nurse in charge (Zelna, 2021)

### Patient support

When performing an examination on a patient who is using a support equipment such as intravenous (IV) line, nasogastric tube, or drainage tube, the sonographer should take extra precautions to protect these lines and tubes from being damaged. Patients can be instructed to raise their arms above their heads without pressing down on the affected arm to alleviate any potential strain on the IV line or cannula caused by the transducer or cable (Zelna, 2021).

### Vital signs

Sonographers should be well-trained in monitoring vital signs in order to recognize when a patient's condition is deteriorating. For example, if an adult's heart rate is over 100 bpm or their breathing rate is over 20 bpm, this should alert the sonographer that the patient is in distress and needs immediate help. Knowledge of cardiopulmonary resuscitation (CPR) will be invaluable when a patient on the examination couch loses consciousness and needs to be kept alive until help arrives (Zelna, 2021).

### Infection control

Bacterial contamination of ultrasound probes has been reported to be significantly higher than contamination of public toilet seats and bus poles (Sartoretti et al., 2017).

Pathogens can live on contaminated surfaces of the ultrasound equipment far longer than most people would think (Kramer et al., 2006). Bacteria like Escherichia coli, Pseudomonas aeruginosa, and Staphylococcus aureus can live for several months or longer on dry inactive surfaces. Viruses like hepatitis A, herpes simplex virus (HSV), and rotaviruses can live on these surfaces for a few weeks, while fungi like Candida albicans can actually live for up to 120 days (Nyhsen et al., 2017).

If contamination occurred through a contact with body fluids or other fluid media such as ultrasound gel, the survival of pathogens will be even longer (Nyhsen et al., 2017). The risk of cross-infection is therefore high in the ultrasound room, especially for patients who undergo semi-invasive procedures such as transvaginal examination. Numerous incidents of patients contracting infections following semi-invasive ultrasound procedures have been reported in various countries around the world (Scott et al., 2018). Sonographers have a responsibility to following guidelines for preventing cross-infection (Ridge, 2005; Abramowicz et al., 2017). AIUM recommended guidelines for transducer cleaning is presented in Box 17.2.

**Box 17.2 Ultrasound transducer cleaning and preparation**

The following specific recommendations are made for the cleaning and preparation of all ultrasound transducers. For the protection of the patient and the health care worker, all internal examinations should be performed with the operator properly gloved throughout the procedure. Transducers need to be cleaned after each exam by using all of the following steps proposed by Abramowicz et al:

1. Cleaning of all transducers—Disconnect the transducer from the ultrasound scanner as appropriate. After removal of the transducer cover (when applicable), remove bulk gel or debris from the transducer. Consider the use of a small brush, especially for crevices and areas of angulation, depending on the design of the particular transducer. Use a damp gauze pad or other soft cloth and a small amount of mild nonabrasive liquid soap, eg, household dishwashing liquid or use a wipe to remove any remaining gel (film).
2. Disinfection can be (a) low-level or (b) high-level:
   a. Disinfection of all transducers in external procedures, as well as interventional percutaneous procedures, should undergo low level disinfection (LLD). If contamination of covered transcutaneous transducers with blood or other bodily fluids occurs, it can be eliminated with low-level disinfectants that are effective against mycobacteria and bloodborne pathogens (including hepatitis B virus, hepatitis C virus, and HIV). Human hands are always cleaned with LLD and covered with gloves.
   b. Disinfection of all internal transducers (eg, transvaginal, transrectal, and transesophageal transducers), as well as intraoperative transducers, require high level disinfection (HLD) before they can be used on another patient. High-level chemical disinfectants rely on clean and dry surfaces as wet surfaces dilute the disinfectant; therefore, always dry the transducer before performing HLD.

Note: An obvious disruption in transducer cover integrity does not require modification of this protocol. Because of the potential disruption of the barrier sheath, HLD with chemical agents is necessary. The guidelines herein take into account possible transducer contamination due to a disruption in the barrier sheath.

3. Rinsing—Depending on the employed disinfection agent, the transducer should be thoroughly rinsed and dried after disinfection, following manufacturer guidelines.
4. Storing—Transducers need to be stored in accordance with their disinfection level.
5. Remove gloves, dispose, and wash hands.

(www.aium.org/officialstatements/57)

### *Reporting ultrasound findings*

The ultrasound report is a formal document that forms an important part of patient management plan. Sonographers are responsible for making sure the report is accurate in every way and is ready and available as soon as possible, preferably right after the examination (SCoR and BMUS, 2020). There are medico-legal risks associated with inaccurate, incomplete, or delayed reports (Berlin, 2000; Wallis and McCoubrie, 2011).

The sonographer should make sure that the report answers all of the clinical questions that were raised in the referral letter (Berlin, 2000; Necas, 2010b; Wallis and McCoubrie, 2011; Pool and Siemienowicz, 2019). In addition, the sonographer or interpreting physician must anticipate and respond to clinical questions not specifically expressed on the referral letter (Kranz, 2017). In some instances, reporting the absence of specific findings might reassure the attending physician by emphasizing that the area of concern has received sufficient attention (Necas, 2018).

It is important to document the steps taken in response to unexpected or urgent findings in the report (SCoR and BMUS, 2020a). For example, if a ruptured ectopic pregnancy is found, it can be stated in the report that an arrangement was made for a transfer to the emergency room for further review.

### Medicolegal and ethical considerations

Sonographers are responsible for protecting patients from foreseeable harm in the course of performing their duties. If the patient is harmed because of the sonographer's action or inaction, this constitutes a breach of duty of care. 'Duty of care' refers to a legal and professional obligation owed to the patient. Liability for negligence may result from breaching duty of care (Bryden and Storey, 2011). The legal concept of medical negligence is known as tort, which comes from the Latin word for "to injure" (tortere). If a patient is able to prove that he or she was harmed as a direct result of the defendant's conduct (or lack thereof) the tort lawsuit for medical negligence will be successful against the hospital, the sonographer or the healthcare team. The burden of proof is on the patient to establish that he or she was owed a duty of care that was breached, resulting in an injury that must attract monetary compensation, not just for the cost of harm but also for the pain and loss of comfort (Bryden and Storey, 2011). The primary goal of these compensations is to deter others from committing similar harm on other patients. Consequently, it is essential to draw lessons from the past litigations involving ultrasound examinations.

### Malpractice prevention

Preventing malpractice simply begins with following the laws that regulate the health care system in the country. The majority of ultrasound lawsuits will include the hospital and the key health personnel on the chain of care, such as the sonographer, a supervising physician and referring physician. The hospital is responsible for ensuring that all hired health professionals are following applicable rules and regulations. In the event of ultrasound malpractice, the hospital will be liable for failing to ensure compliance by the healthcare team. The standards that regulate the practice of sonography vary slightly from one country to another. Sonographers should abide by their national regulations. For example, in North American countries, sonographers can only write a preliminary report for an interpreting physician, such as a radiologist, who takes responsibility for communicating the formal report to the referring physician (Edwards and Sidhu, 2017). This lessens the burden of responsibility on the sonographer in the event of malpractice but is also a risk for the interpreting physician, who may only have a second look at cases if the preliminary report suggests an abnormal finding. In countries where sonographers report their findings directly to the referring physician, such as Europe (Edwards and Sidhu, 2017), the sonographer has a greater responsibility to be extra careful in order to minimize incidents of patient dissatisfaction and cases of ligation. Zelna (2021) recommends the seven "C's" sonographers should follow in order to prevent malpractice. The seven "C's" are Competence, Compliance, Charting, Communication, Confidentiality, Carefulness and Courtesy.

### Competence

Competence refers to the ability of the healthcare professional to integrate knowledge, skills, values, and attitudes into clinical encounters (Frank et al., 2010). The sonographer owes the patient a duty of care to be competent. He or she should have completed the appropriate education and training and demonstrated minimum competency through a recognized certification examination (Bendick, 2015). With regards to physicians, *"The law requires a physician to possess the skill and learning which is possessed by the average member of the medical profession… and to apply that skill and learning with ordinary reasonable care. "* (Cannavale et al., 2013). Similarly, sonographers are required by law to possess the skill and learning that should be possessed by a sonographer. A sonographer who lacks the expertise and certification to perform a sonographic specialty (such as being certified in obstetric ultrasound but lacking certification in cardiac ultrasound) should rather direct the patient to another colleague known to be competent in that specialty (Childs et al., 2021). In the event of a lawsuit, the court is likely to invite experts to determine whether the sonographer possessed the appropriate skill and learning for providing a reasonable care. If it is determined that the sonographer was not competent in that area of ultrasound specialty, the patient may have a stronger claim against the healthcare team. Box 17.3 is a lawsuit case of *Mickle v. Salvation Army Grace Hospital Windsor of Ontario, Canada*.

---

**Box 17.3**

In the case of *Mickle v. Salvation Army Grace Hospital Windsor of Ontario, Canada*, a claim was brought against a sonographer for a supposed negligence. Plaintiff mother claimed that the sonographer and radiologist failed to detect and warn her of baby's prenatal abnormality which denied her the right to abort the pregnancy. Regarding the sonographer's duty, the court stated that she was required to possess and use the reasonable degree of learning and skill ordinarily possessed by a sonographer in similar communities and cases. The court stated that, given the rapid advancement of ultrasound technology, the duty must be based on the reasonable level of learning and skill possessed by sonographers in 1991 (the year that the ultrasound was performed). The plaintiff contended that the sonographer performed a substandard prenatal ultrasound examination since she did not examine all four limbs of the pregnancy. As a result, anomalies on the right side of the fetus were not detected. The court felt that expert testimony was necessary in this type of case. As witnesses, both the plaintiff and the defendants presented radiologists specialized in ultrasound, and the defendants also presented an instructor of ultrasound technology. The opinions of the expert witnesses were directly opposite: the plaintiff's witnesses claimed that the ultrasound examination performed in this case fell below the reasonable threshold, whilst the defendant's witnesses claimed that it met the reasonable standard. The court decided to accept the testimony of the defense specialists. The court reasoned that the plaintiff's experts were attempting to impose the standard required for a Level 2 examination to a Level 1 examination. According to the evidence, the examination done was a Level 1 examination, and the AIUM and ACR Guidelines at the time did not mandate that all limbs be assessed when performing this level of ultrasound.

(Health Professions Regulatory Advisory Council: www.hprac.org)

Lessons can be drawn from the case of *Mickle v. Salvation Army Grace Hospital Windsor of Ontario, Canada*. First, it affirms the need for every sonographer to possess a minimum level of competence that can be regarded as reasonable by an expert committee. Second, it also affirms the importance of continuous medical education, as sonographers will be judged by the competence expected of a professional at the time he or she was on duty. Third, it also underscores the importance of following approved guidelines from recognized professional bodies in the jurisdiction.

Competence in clinical ultrasound has been described by Bahner et al. (2012) in four sub-competence stages. They include; 1) understanding the clinical indications for ultrasound examination; 2) having the technical skills to perform and acquire adequate images; 3) having the skills to interpret the images; and 4) having the skills to integrate ultrasound findings into clinical medical decision making. Again, there are good reasons why medical imaging professions like sonography require specific certification for different areas of specialty, such as abdominal sonography, obstetrical/gynecological, cardiac, and vascular technology (Bendick, 2015). This certification strategy enables sonographers to demonstrate their competence in these specific areas in order to minimize the risk of harm to patients that may result in lawsuits.

The issue of competence extends to delegation. When delegating tasks to others, the sonographer should always ensure that the individual is competent to a reasonable level. Delegated duties should not be accepted unless you are capable of carrying them out to a reasonable standard. Gill (1992) discusses a lawsuit case that explains the problems of delegation as presented in Box 17.4.

---

**Box 17.4**

In a lawsuit case involving two sonographers, the sonographer with considerable experience claims that she only did the paperwork for the less experienced one who performed the scan. It was later found that the less experienced sonographer did not demonstrate a pertinent pathology. Although the experienced sonographer claimed she only completed the report sheet with the impression of the other sonographer, she was still implicated in the lawsuit because the form was filled in her handwriting.

(Gill, K., 1992)

---

### Charting

Charting simply refers to medical record keeping. Medical records should be a dialogue between caregivers on the patient's management and progress so that new caregivers can take up where others left off. Thus, each patient encounter must be adequately recorded to document clinical trends, investigations, results, future management, referral, and follow-up (Panting, 2004)

In presenting ultrasound findings, the sonographer or radiologist should ensure that all findings are reported no matter how trivial it may seem. This will grant them immunity in the event of lawsuit. Box 5 presents an example of a lawsuit involving a referring physician and a radiologist who could not be granted immunity because ultrasound findings he regarded as trivial were not reported which led to patient dissatisfaction.

---

**Box 17.5**

In the jury case of a 33-year-old multiparous woman who presented for second trimester ultrasound at 19 weeks and 1 day, the ultrasound report simply stated: "Normal ultrasound with fetus at 19 weeks and 1 day." This was the last ultrasound she had before giving birth. When she finally gave birth at 39 weeks, the baby had a Down syndrome. An expert went back to review the prenatal ultrasound and saw that it showed mild pyelectasis and a calyceal dilation of between 4.3 and 4.4 mm. Another thing that was found was an echogenic intracardiac focus. At trial, the radiologist said that a calyceal dilation of approximately 4 mm is considered normal. Also, an echogenic intracardiac focus is not a useful marker. It is the obstetrician's responsibility to inform the patient that additional tests are required. The obstetrician stated that because the ultrasound report says that everything was normal, he saw no reason to recommend amniocentesis or additional ultrasound studies. The expert for the plaintiff said that calyceal dilatation of more than 4 mm at 19 weeks and 1 day of pregnancy calls for a second ultrasound at 32 weeks to see if the calyceal dilatation is still there. An echogenic focus by itself is not a good sign of Down syndrome on its own. But if there are multiple soft markers for Down syndrome, they should be written down in the report and suggestions should be made to recalculate the patient's risk with amniocentesis, if necessary. So, the patient should have been told to do another ultrasound. The verdict of the jury was that the radiologist had a duty to report the findings to the obstetrician. If he had done so, the duty for further counselling, evaluation, and treatment would have transferred to the obstetrician. The Radiologist was therefore liable for negligence.

(Shwayder, 2019)

*Compliance*

Being compliant is the act of submitting or surrendering your power to another. It is about yielding or giving in to another person's order. On the other hand, being autonomous means having the right to decide or choose without external interference. The patient has the autonomy to decide on accepting or refusing an order from a healthcare provider. The only legally accepted method for obtaining the patient's permission is the informed consent process.

A sonographer risks liability if accused by a patient of attending to them without appropriate consent or for overstepping the permitted boundaries of consent. Assault charge is one of the commonest lawsuits against sonographers (Thomson and Moloney, 2017) . In one of such cases, a male sonographer who performed transabdominal ultrasound on a woman was sued for molestation. Whiles the sonographer was preparing the patient for the examination, the woman pulled her blouse over the area to be scanned. The sonographer reports having explained to the patient that she would lift her blouse higher so that he can scan the appropriate area, but the patient sued for molestation, claiming that she felt dirty after the examination (Gill, 1992)

The risk of liability is much higher with intimate examinations such as transvaginal and transrectal ultrasound examinations. Healthcare professionals using ultrasound for such intimate examinations must ensure that informed consent is obtained and operate within acceptable protocols in order to avoid lawsuits.

*Communication*

In order to minimize liabilities resulting from communication gaps, other channels of communication besides the written report should be encouraged between the sonographer, the patient, and the referring physician. This may include telephoning the referring physician or their nurse to quickly notify them of any suspicious findings that border on further evaluation or decision-making (Berlin, 2010). Box 17.6 is an example of communication gabs that resulted in a lawsuit case.

---

**Box 17.6**

A woman in Virginia had a Doppler sonogram of her right lower extremity, which revealed a deep vein thrombosis, according to the radiologist. The radiologist attempted to call the referring physician but was unable to reach him, so he directed the office secretary to fax the report. The referring physician was not in the office, and the fax was not seen until the following day, when it was reviewed by the doctor's nurse. The nurse called the patient and scheduled an appointment for two days later. However, the patient died the next day as a result of an acute pulmonary embolism. Both physicians were named in a lawsuit for some technical issues they were both liable for.

(Berlin, 2010)

## Confidentiality

While medical records are highly useful for preventing liability, they can also become a source of liability if care is not taken to protect patient confidentiality. Healthcare professionals and their employers owe a duty of care to ensure that the confidentiality, privacy, and security of their patients' medical records are assured (Tariq and Hackert, 2018). In this age of rapidly advancing information technology, this is now true than it has ever been. Modern electronic data resources such as the Digital Imaging and Communications in Medicine (DICOM) and the Picture Archiving and Communication System (PACS) have now become a 'standard of care' in radiology departments. Smartphone-ultrasound integration is now available, and allows ultrasound examinations to be performed by simply connecting a probe to a smart phone. However, these advancing technologies for keeping medical records are becoming a source of liability for breaches of patient confidentiality. One such case is the 2020 lawsuit filed by 298,532 patients against Northeast Radiology and its allies at the New York Southern District Court. The plaintiffs pressed charges of inadequate data security and negligence per se (Davis, 2021). This lawsuit was in connection with an alert that showed more than 130 health systems are actively exposing millions of medical images via PACS and DICOM.

Healthcare professionals using ultrasound should be careful about sharing patients' ultrasound images and sensitive information through smart phones, without obtaining consent from the patient. This can lead to litigation in the future.

## Carefulness

Carefulness refers to "the watchfulness, attention, caution and prudence that a reasonable person in the circumstances would exercise," and failure to meet this standard of care is called negligence (Ashton, 2006). Box 17.7 is the case of Lillywhite, in which the parents of a child with holoprosencephaly filed a lawsuit. It is an apparent lack of attention to details that must be avoided.

**Box 17.7**

Mrs. Lillywhite had an anomaly scan performed for her pregnancy, but the sonographer was unable to locate a cavum septum pellucidum. She consulted with the obstetrician about the situation, and they decided to send her to a fetal medicine professor for further evaluation. Mrs. Lillywhite was so concerned about the results that she paid for a private referral to visit a radiologist who was informed that an earlier scan by a sonographer could not identify appropriate structures of the brain. The radiologist presented a new report which stated that everything in the brain looked normal. At birth, gross abnormalities were found on the baby. The appeals court found that if the radiologist had used reasonable skill and care he would have detected these abnormalities. The verdict went against the defendant.

(Ashton, 2006)

Carefulness also includes being alert when assisting a patient. A sonographer was charged with providing inadequate patient care in one such lawsuit. The patient got injured when he slipped and fell while being assisted to the bathroom (Gill, 1992).

*Courtesy*

The Cambridge dictionary describes courtesy as polite behaviour, or a polite action or remark. Diagnostic errors are the most common reason for ultrasound related litigations (Busardò et al, 2015). But there is also a strong relationship between diagnostic errors and courtesy which could minimize litigations when properly addressed. Bad behaviours such as failing to communicate, disrespecting patients, and ignoring the patient's knowledge have been linked to diagnostic errors (Giardina et al., 2018). In their study on the view of patients who were victims of diagnostic errors, Giardina et al. (2018) found that patients felt healthcare personnel were either 'belittling', 'mocking', or 'behaving rudely' towards them. Some also reported feeling disrespected by healthcare personnel, which resulted in a misdiagnosis of their condition. Some participants also said their knowledge was ignored, using phrases such as "dismissed," "ignored," "would not listen," "did not pay attention," and "did not take seriously" to describe how they felt.

As discussed at the beginning of this chapter, courtesies such as acknowledging the patient with good facial expressions and calling them by their first name, confirming whether their personal view on their clinical symptoms agrees with information provided by referring physicians, are acts of courtesy that can minimize diagnostic errors and litigation.

Sometimes, impolite attitudes also exist within the healthcare team. An arrogant referring physician may dismiss the view of a radiologist, while an arrogant radiologist may also disrespect view of a sonographer which may lead to diagnostic errors and lawsuits (Berlin et al., 2010). The case presented in Box 17.8 is an example of how ignoring the findings of other healthcare team members can result in litigations.

---

**Box 17.8**

…A referring physician told the patient that the ultrasound was normal, even though the sonographer has reported "Structural irregularities that require further evaluation." After birth, the baby had facial and lower limb abnormalities as well as limited cognitive and communication skills. In the pre-trial proceedings, the physician admitted that he had not reviewed the images from the studies or the sonographer's handwritten report about the findings which resulted in a settlement agreement of $1.9 million.

(Shwayder, 2019)

---

On the same issue of ignoring other members of the healthcare team, another study reported that attending physicians ignored 36% of abnormal radiologic results, some of which were lost to cancer treatment (Singh et al., 2007).

### Conclusion

Sonographers should be mindful that, although obtaining diagnostic information is the priority of their practice, poor communication and interaction with the individual patient may not only ruin their main objective but also make the individual patient dissatisfied. Making the individual patient feel valued, acknowledged, and respected by explaining the purpose of the examination and obtaining proper consent paves the way for obtaining relevant clinical history that could minimize diagnostic errors. In order to improve patient satisfaction during and after the ultrasound examination, sonographers should endeavour to implement standard protocols that suit the needs of the individual patient. In doing so, they should also be mindful of medicolegal issues that may arise from dissatisfied patients. This can mostly be prevented if sonographers are abreast with current standards of practice, document all relevant examination events, and make optimal use of all mediums of communication in presenting ultrasound results, without overstepping their boundaries or breaching confidentiality, but carefully ensuring that diagnostic errors are avoided.

# References

Abramowicz, J. S., Evans, D. H., Fowlkes, J. B., Maršal, K., & WFUMB Safety Committee. (2017). Guidelines for cleaning transvaginal ultrasound transducers between patients. *Ultrasound in Medicine & Biology*, 43(5), 1076–1079.

Akbari, A., Mayhew, A., Al-Alawi, M. A., Grimshaw, J., Winkens, R., Glidewell, E., ... & Fraser, C. (2005). Interventions to improve outpatient referrals from primary care to secondary care. *Cochrane Database of Systematic Reviews* (3).

Anderson, R. T., Camacho, F. T., & Balkrishnan, R. (2007). Willing to wait?: the influence of patient wait time on satisfaction with primary care. *BMC Health Services Research*, 7(1), 1–5.

Antwi, W. K., Kyei, K. A., Gawugah, J. N., Opoku, S. Y., & Ogbuokiri, E. I. (2015). Anxiety level among patients undergoing ultrasound examination in Ghana. *International Journal of Medical Imaging*, 3, 6–10.

Ariely, D., & Lanier, W. L. (2015). Disturbing trends in physician burnout and satisfaction with work-life balance: dealing with malady among the nation's healers. In *Mayo Clinic Proceedings* (Vol. 90, No. 12, pp. 1593–1596). Elsevier.

Ashton, V. (2006) 'A Legal Perspective Within the NHS', *Ultrasound*, 14(4), pp. 230–231. Available at: https://doi.org/10.1179/174313406X150069

Bahner, D. P., Hughes, D. and Royall, N. A. (2012) 'I-AIM: a novel model for teaching and performing focused sonography', *Journal of Ultrasound in Medicine*, 31(2), pp. 295–300.

Bello, O. (2017) 'Effective communication in nursing practice: A literature review'. https://core.ac.uk/download/pdf/84798372.pdf

Bendick, P. J. (2015). Scope of practice and clinical standards for the diagnostic medical sonographer. *Journal of Diagnostic Medical Sonography*, 31(4), 197–197.

Berlin, L. (2000) 'Pitfalls of the vague radiology report', *American Journal of Roentgenology*, 174(6), pp. 1511–1518.

Berlin, L.M. (2010) 'Failure of radiologic communication: an increasing cause of malpractice litigation and harm to patients', *Applied Radiology*, 39(1), p. 17.

Bowden, D. J., Yap, L. C., & Sheppard, D. G. (2017). Is the Internet a suitable patient resource for information on common radiological investigations?: radiology-related information on the internet. *Academic Radiology*, 24(7), 826–830.

Brockopp, D. Y., Franey, B. N., Sage-Smith, D., Romond, E. H., & Cannon, C. C. (1992). Patients' Knowledge of Their Caregivers' Names: A Teaching-Hospital Study. *Hospital Topics*, 70(1), 25–28.

Bryden, D. and Storey, I. (2011) 'Duty of care and medical negligence', *Continuing Education in Anaesthesia Critical Care & Pain*, 11(4), pp. 124–127. Available at: https://doi.org/10.1093/bjaceaccp/mkr016

Burgener, A. M. (2017) 'Enhancing Communication to Improve Patient Safety and to Increase Patient Satisfaction', *The Health Care Manager*, 36(3), pp. 238–243. Available at: https://doi.org/10.1097/HCM.0000000000000165

Busardò, F. P., Frati, P., Santurro, A., Zaami, S., & Fineschi, V. (2015). Errors and malpractice lawsuits in radiology: what the radiologist needs to know. *La Radiologia Medica*, 120(9), 779–784.

Camacho F, Anderson R, Safrit A, Jones AS, Hoffmann P. (2006) The relationship between patient's perceived waiting time and office-based practice satisfaction. *North Carolina Medical Journal*, Nov–Dec; 67(6), 409–413. PMID: 17393701.

Cannavale, A. et al. (2013) 'Malpractice in radiology: what should you worry about?', Radiology Research and Practice, 2013.

Chesson, R. A., McKenzie, G. A., & Mathers, S. A. (2002). What do patients know about ultrasound, CT and MRI?. *Clinical Radiology*, 57(6), 477–482.

Childs, J. et al. (2021) 'Professional Competency Framework for Sonographers'.

Clever, S. L., Jin, L., Levinson, W., & Meltzer, D. O. (2008). Does doctor–patient communication affect patient satisfaction with hospital care? Results of an analysis with a novel instrumental variable. *Health Services Research*, 43(5p1), 1505–1519.

Cousin, G., Mast, M. S., Roter, D. L., & Hall, J. A. (2012). Concordance between physician communication style and patient attitudes predicts patient satisfaction. *Patient Education and Counseling*, 87(2), 193–197.

Davis, J. (2021) *PACS vulnerabilities, data breach spur lawsuit against radiology specialists*, SC Media. Available at: www.scmagazine.com/news/compliance/pacs-vulnerabilities-data-breach-spur-lawsuit-against-radiology-specialists (Accessed: 29 November 2022).

DiGiacinto, D., Gildon, B., Keenan, L. A., & Patton, M. (2016). Review of patient satisfaction research to improve patient surveys in medical imaging departments. *Journal of Diagnostic Medical Sonography*, 32(4), 203–206.

Donald, I. (1965). Ultrasonic echo sounding in obstetrical and gynecological diagnosis. *American Journal of Obstetrics and Gynecology*, 93(7), 935–941. doi:10.1016/0002-9378(65)90152-3

Donald, I., Macvicar, J., Brown, T. G. (1958). Investigation of Abdominal masses by Pulsed Ultrasound. *The Lancet*, 271(7032), 1188–1195. doi:https://doi.org/10.1016/S0140-6736(58)91905-6

Edwards, H. M. and Sidhu, P. S. (2017) 'Who's doing your scan? A European perspective on ultrasound services', *Ultraschall in der Medizin-European Journal of Ultrasound*, 38(05), pp. 479–482.

Edwardson, N., Gregory, S., & Gamm, L. (2016). The influence of organization tenure on nurses' perceptions of multiple work process improvement initiatives. *Health Care Management Review*, 41(4), 344–355.

Elshamy, K., & Ramzy, E. (2011). The effect of postoperative pain assessment and management monitoring program on surgical nurses' documentation, knowledge, attitudes, and patients' satisfaction at Mansoura University Hospitals. *Journal of American Science*, 7(10), 500–516.

Emmons, R. A., & McCullough, M. E. (2003). Counting blessings versus burdens: an experimental investigation of gratitude and subjective well-being in daily life. *Journal of Personality and Social Psychology*, 84(2), 377.

Frank, J. R. et al. (2010). 'Competency-based medical education: theory to practice', *Medical Teacher*, 32(8), pp. 638–645.

Giardina, T. D. et al. (2018) 'Learning from patients' experiences related to diagnostic errors is essential for progress in patient safety', *Health Affairs*, 37(11), pp. 1821–1827.

Gill, K. (1992) 'Legal liability and sonography: an update', *Journal of Diagnostic Medical Sonography*, 8(2), pp. 93–96.

Gillen, P., Sharifuddin, S. F., O'Sullivan, M., Gordon, A., & Doherty, E. M. (2018). How good are doctors at introducing themselves?# hellomynameis. *Postgraduate Medical Journal*, 94(1110), 204–206.

Gillette, R. D., Filak, A., & Thorne, C. (1992). First name or last name: which do patients prefer?. *Journal of the American Board of Family Practice*, 5(5), 517–522.

Granger, K. (2013). Healthcare staff must properly introduce themselves to patients. *Bmj, 347*, 1–2.

Hall, L. H., Johnson, J., Watt, I., Tsipa, A., & O'Connor, D. B. (2016). Healthcare staff wellbeing, burnout, and patient safety: a systematic review. *PloS one, 11*(7), e0159015.

Hamilton, C., Osterhold, H., Chao, J., Chu, K., & Roy-Burman, A. (2018). Gratitude and recognition in a hospital setting: Addressing provider well-being and patient outcomes. *American Journal of Medical Quality, 33*(5), 554.

Health Professions Regulatory Advisory Council (2013) 'Diagnostic Sonographers: A jurisprudence Review'. Health Professions Regulatory Advisory Council, Ontario.

Jefford, M., & Moore, R. (2008). Improvement of informed consent and the quality of consent documents. *The Lancet Oncology, 9*(5), 485–493.

Ji, E. K., Pretorius, D. H., Newton, R., Uyan, K., Hull, A. D., Hollenbach, K., Nelson, T. R. (2005). *Effects Of Ultrasound on Maternal-Fetal Bonding: A Comparison of Two- and Three-Dimensional Imaging. Ultrasound in Obstetrics and Gynecology*, 25(5), 473–477. doi:10.1002/uog.1896

Jiwa, M. et al. (2009) 'What is the importance of the referral letter in the patient journey? A pilot survey in Western Australia', *Quality in Primary Care*, 17, pp. 31–36.

Keifenheim, K. E., Teufel, M., Ip, J., Speiser, N., Leehr, E.J., Zipfel, S., Herrmann-Werner, A. (2015) Teaching history taking to medical students: A systematic review. *BMC Medical Education*, 2015 Sep 28;15:159. doi: 10.1186/s12909-015-0443-x. PMID: 26415941; PMCID: PMC4587833.

Khan, F. H., Hanif, R., Tabassum, R., Qidwai, W., & Nanji, K. (2014). Patient attitudes towards physician nonverbal behaviors during consultancy: result from a developing country. *International Scholarly Research Notices, 2014*.

Kramer, A., Schwebke, I., & Kampf, G. (2006). How long do nosocomial pathogens persist on inanimate surfaces? A systematic review. *BMC Infectious Diseases*, 6(1), 1–8.

Kranz, P. G. (2017) 'What's Your Impression? Get to the Point!', *Journal of the American College of Radiology*, 14(4), p. 451.

Leddy, K. M., Kaldenberg, D. O., & Becker, B. W. (2003). Timeliness in ambulatory care treatment: an examination of patient satisfaction and wait times in medical practices and outpatient test and treatment facilities. *The Journal of Ambulatory Care Management*, 26(2), 138–149.

Locke, R., Stefano, M., Koster, A., Taylor, B., & Greenspan, J. (2011). Optimizing patient/caregiver satisfaction through quality of communication in the pediatric emergency department. *Pediatric Emergency Care*, 27(11), 1016–1021.

McMaster, N., & DeGiobbi, J. (2016). It's the little things: small gestures in patient care, big impact. *Journal of Medical Imaging and Radiation Sciences*, 47(1), S25.

Miller, P. et al. (2013) 'Enhancing Patients' Experiences in Radiology: Through Patient–Radiologist Interaction', *Academic Radiology*, 20(6), pp. 778–781. Available at: https://doi.org/10.1016/j.acra.2012.12.015

Montague, E., Chen, P. Y., Xu, J., Chewning, B., & Barrett, B. (2013). Nonverbal interpersonal interactions in clinical encounters and patient perceptions of empathy. *Journal of Participatory Medicine, 5*, e33.

Moore, P., Vargas, A., Nunez, S., & Macchiavello, S. (2011). A study of hospital complaints and the role of the doctor-patient communication. *Revista medica de Chile, 139*(7), 880–885.

Mowen, J. C., Licata, J. W., & McPhail, J. (1993). Waiting in the emergency room: how to improve patient satisfaction. *Marketing Health Services*, 13(2), 26.

Mulisa, T., Tessema, F., & Merga, H. (2017). Patients' satisfaction towards radiological service and associated factors in Hawassa University Teaching and referral hospital, Southern Ethiopia. *BMC Health Services Research*, 17(1), 1–11

National Health Services (NHS) (2016). Diagnostic imaging dataset statistical release. *London: Department of Health*, 421.

Necas, M. (2010a) 'Duplex ultrasound in the assessment of lower extremity venous insufficiency', *Australasian Journal of Ultrasound in Medicine*, 13(4), p. 37.

Necas, M. (2010b) 'Duplex ultrasound in the assessment of lower extremity venous insufficiency', *Australasian Journal of Ultrasound in Medicine*, 13(4), p. 37.

Necas, M. (2018) 'The clinical ultrasound report: Guideline for sonographers', *Australasian Journal of Ultrasound in Medicine*, 21(1), pp. 9–23.

Nyhsen, C. M., Humphreys, H., Koerner, R. J., Grenier, N., Brady, A., Sidhu, P., ... & Claudon, M. (2017). Infection prevention and control in ultrasound-best practice recommendations from the European Society of Radiology Ultrasound Working Group. *Insights into Imaging*, 8(6), 523–535.

Ohm, F., Vogel, D., Sehner, S., Wijnen-Meijer, M., & Harendza, S. (2013). Details acquired from medical history and patients' experience of empathy–two sides of the same coin. *BMC Medical Education*, 13(1), 1–7.

Palombi, L. C., Nelson, L., Fierke, K. K., & Bastianelli, K. (2015). Pilot study of patient perception of pharmacists as care providers based on health screening encounters with student pharmacists. *Journal of the American Pharmacists Association*, 55(6), 626–633.

Panting, G. (2004) 'How to avoid being sued in clinical practice', *Postgraduate Medical Journal*, 80(941), pp. 165–168.

Parsons, S. R., Hughes, A. J., & Friedman, N. D. (2016). 'Please don't call me Mister': patient preferences of how they are addressed and their knowledge of their treating medical team in an Australian hospital. *BMJ open*, 6(1), e008473.

Patel, M. R., & Cabana, M. D. (2010). Parental satisfaction and the ability to recall the physician's name. *Clinical Pediatrics*, 49(6), 525–529.

Pool, F. J. and Siemienowicz, M. L. (2019) 'New RANZCR clinical radiology written report guidelines', *Journal of Medical Imaging and Radiation Oncology*, 63(1), pp. 7–14.

Pretorius, D. H., Uyan, K. M., Newton, R., Hull, A., James, G., Nelson, T. (2001). Effects of US on Maternal–Fetal Bonding: 2d vs. 3d. *Ultrasound in Obstetrics and Gynecology*, 18, F35–F35. doi:10.1046/j.1469-0705.2001.abs20-2.x

Ridge, C. (2005). Sonographers and the fight against nosocomial infections: how are we doing? *Journal of Diagnostic Medical Sonography*, 21(1), 7–11. s://doi.org/10.1111/1754-9485.12756

Roberts, J. (2012). 'Wakey Wakey Baby': Narrating Four-Dimensional (4d) Bonding Scans. *Sociology of Health & Illness*, 34(2), 299–314.

Santen, S. A., Hemphill, R. R., Prough, E. E., & Perlowski, A. A. (2004). Do patients understand their physician's level of training? A survey of emergency department patients. *Academic Medicine*, 79(2), 139–143.

Sartoretti, T., Sartoretti, E., Bucher, C., Doert, A., Binkert, C., Hergan, K., ... & Gutzeit, A. (2017). Bacterial contamination of ultrasound probes in different radiological institutions before and after specific hygiene training: do we have a general hygienical problem?. *European Radiology*, 27(10), 4181–4187.

SCoR and BMUS (2020) 'Guidlines for professional ultrasound practice'.

Scott, D., Fletcher, E., Kane, H., Malcolm, W., Kavanagh, K., Banks, A. L., & Rankin, A. (2018). Risk of infection following semi-invasive ultrasound procedures in Scotland, 2010 to 2016: A retrospective cohort study using linked national datasets. *Ultrasound*, 26(3), 168–177.

Shwayder, J. (2019) 'Ultrasound errors to avoid: How important is the report?' www.contemporaryobgyn.net/view/ultrasound-errors-avoid-how-important-report

Singh, H. et al. (2007) 'Communication outcomes of critical imaging results in a computerized notification system', *Journal of the American Medical Informatics Association*, 14(4), pp. 459–466.

Smith-Bindman, R., Kwan, M. L., Marlow, E. C., Theis, M. K., Bolch, W., Cheng, S. Y., ... & Miglioretti, D. L. (2019). Trends in use of medical imaging in US health care systems and in Ontario, Canada, 2000–2016. *Jama*, 322(9), 843–856.

Soremekun, O. A., Takayesu, J. K., & Bohan, S. J. (2011). Framework for analyzing wait times and other factors that impact patient satisfaction in the emergency department. *Journal of Emergency Medicine*, 41(6), 686–692.

Stapleton, F. B. (2000). MY name is jack. *JAMA*, 284(16), 2027–2027.

Starcevich, A., Lombardo, P., & Schneider, M. (2021). Patient understanding of diagnostic ultrasound examinations in an Australian private radiology clinic. *Australasian Journal of Ultrasound in Medicine*, 24(2), 82–88.

Tariq, R.A. and Hackert, P.B. (2018) 'Patient confidentiality'. StatPearls. www.ncbi.nlm.nih.gov/books/NBK519540/

Thomas, S., O'Loughlin, K. and Clarke, J. (2019) 'Sonographers' communication in obstetrics: Challenges to their professional role and practice in Australia', *Australasian Journal of Ultrasound in Medicine*, 23(2), pp. 129–139. Available at: https://doi.org/10.1002/ajum.12184

Thomas, S., O'Loughlin, K., & Clarke, J. (2017). The 21st century sonographer: role ambiguity in communicating an adverse outcome in obstetric ultrasound. *Cogent Medicine*, 4(1), 1373903.

Thomson, N. and Moloney, P. (2017) 'Protection against allegations of sexual assault when undertaking ultrasound examinations', *Ultrasound*, 25(1), pp. 58–61.

van der Westhuizen, L., Naidoo, K., Casmod, Y., & Mdlethse, S. (2020). Sonographers' experiences of being a caring professional within private practice in the province of Gauteng. *Health SA Gesondheid*, 25(1).

Wallis, A. and McCoubrie, P. (2011) 'The radiology report—are we getting the message across?', *Clinical radiology*, 66(11), pp. 1015–1022.

Wu, B. J. (2013). History taking in reverse: beginning with social history. *Consultant*, 53(1), 34–6.

Zamora, R., Patel, M., Doherty, B., Alperstein, A., & Devito, P. (2015). Influence of AIDET in the improving quality metrics in a small community hospital-before and after analysis. *Journal of Hospital Administration*, 4(3), 35–38.

Zelna, L. (2021) 'Registry Review . Sonography Patient Care Addendum.' Society of Diagnostic Medical Sonography.

# 18 Person-centred ultrasound experiences

## Shelley Thomson, Lyndal Macpherson, Samantha Thomas and Anja Christoffersen

### Recognising the unique relationship between ultrasound practitioners and patients

As clinicians, ultrasound practitioners have a unique and essential role in diagnosis. The often lengthy period spent in the ultrasound room during a dynamic examination usually far exceeds the time patients spend with their referring doctor – consequently, many patients bond with their ultrasound practitioner. Intrinsically, most ultrasound practitioners recognise this unique situation leads to developing a trusting relationship. Being part of someone's healthcare journey, whether for joyful reasons such as a pregnancy or more challenging times such as an unexpected diagnosis or treatment, these interactions should be valued and can leave a sustained positive or negative impression. Recognising the potential impact of the role, ultrasound practitioners should make every effort to deliver a positive patient experience, putting the patient's needs and physical and mental comfort first – before, during and after the ultrasound examination.

### The ultrasound practitioner

The terms describing ultrasound professionals vary by country. Specialist ultrasound doctors are known as *sonologists*. Health professionals with specialised education and training in diagnostic ultrasound are *ultrasonographers, sonographers,* or *sonographer practitioners*. The specialty training includes the physics and principles of using ultrasound equipment and the medical and pathological knowledge of assessing human anatomy and disease. In Australia and the UK, the occupational term for a qualified and accredited health professional who carries out ultrasound examinations is a *sonographer.* Sonographers and sonologists have an interdependent relationship. Sonologists are responsible for the final patient report (RANZCR, 2020).

In many countries, the sonographer's role is non-existent, with specialist ultrasound doctors performing and reporting all ultrasound examinations. However, in countries such as the United Kingdom, many sonographers

DOI: 10.1201/9781003310143-25

have a more autonomous professional role by providing the official report independent of the sonologist. While acknowledging a variation in terms, there is a consensus that the role of an ultrasound practitioner has evolved to become a specialist frontline practitioner with direct, one-on-one patient interaction and diagnostic responsibility. Throughout this chapter, we will refer to all health professionals trained formerly in ultrasound as **ultrasound practitioners.**

### Expectations of a skilled ultrasound practitioner

The outcome of an ultrasound examination varies with the operator's skill. Dedicated training and education in the performance of ultrasound and the interpretation of the findings are essential (ASUM, 2018; Salomon et al., 2011). The role of the ultrasound practitioner incorporates considerable pressure to diagnose whilst viewing the screen alongside the patient concurrently, requiring the ultrasound practitioner to manage communication with the patient whilst visualising, diagnosing and capturing representative images. Over time the role has evolved from an automated one to incorporating the complex psychomotor skills (Nicholls et al., 2014) required to manipulate a real-time transducer to optimally find, obtain and record 'normal' anatomy or 'abnormal' pathologies (Edwards and Liu, 2018).

### Patient expectations

Research and patient insights reveal the need for ultrasound practitioners to consider patients' expectations of the ultrasound experience, as the outcome depends on their behavioural attributes. Patient expectations include consideration of patient comfort and dignity, trauma-informed practice, and knowledge of lifting techniques to assist immobile patients safely to avoid adverse outcomes. In addition, patients want clear explanations and the opportunity to ask questions and discuss procedures before providing consent. Cultural and religious considerations must be acknowledged, and the examination must be adapted to meet the patient's needs. Key behavioural expectations include listening and open and inclusive communication.

The importance of good communication during an ultrasound examination is a critical component of the patient experience. Ultrasound professionals play a vital role in developing a partnership with their patients. Their communication skills determine the level of trust, compassion, understanding and connection developed when they are together. Essential skills include non-verbal and verbal communication, eye contact, tone of voice, active listening, two-way communication, body language, jargon and pausing to give patients time to respond. Communication skills are critical (refer to Chapter 19). The core focus of this chapter is to showcase all other aspects of delivering a person-centred experience aligned with patient expectations.

## Adopting a person-centred approach

In many countries, a person-centred approach is the minimum standard requirement and is a mandate, not an aspiration. The definition of person-centred is delivering respectful and responsive healthcare to each patient's preferences, needs, choices and values and includes listening to, informing and involving patients in their care (ACSQHC, 2017).

Globally, respected institutions have developed principles of a person-centred approach, including recognised frameworks such as:

- The Institute of Medicine's (IOM) six dimensions of person-centred care (IOM, 2001);
- The Picker Institute's eight principles of person-centred care (The Picker Institute, 1987);
- The Health Foundation; UK Person-Centred Principles (The Health Foundation, 2016);
- The Australian Commission on Safety and Quality in Health Care's Consumer Charter of Healthcare Rights (ACSQHC, 2019);
- The Australian Commission on Safety and Quality in Health Care's National Standard 2: Partnering with Consumers (ACSQHC, 2017).

## What does a person-centred ultrasound approach look like in practice?

For ultrasound practitioners, the person-centred approach may vary depending on the needs, circumstances and preferences of the individual patient receiving care. To provide clarity of expectations and in consultation with patients, the authors have applied the Picker Institute's person-centred care framework to help ultrasound practitioners understand what matters most to patients. These Eight Principles embody Picker Institute's conviction that all patients deserve high-quality healthcare and that patients' views and experiences are integral to improvement efforts (The Picker Institute, 1987). The eight principles clarify what constitutes high-quality, person-centred care during ultrasound examinations.

---

**The Picker Institute's eight principles to deliver person-centred ultrasound care**

1. Respect for the patient's values, preferences and expressed needs.
2. Coordination and integration of care.
3. Information, communication, and education.
4. Physical comfort.
5. Emotional support and alleviation of fear and anxiety

6. Involvement of family and friends.
7. Continuity and transition.
8. Access to care.

Applying the Picker Institute's Eight Principles For Delivering Person-Centred Ultrasound Care includes:

1. Respect for patient's values, preferences and expressed needs – treat each patient with dignity, respect and sensitivity to their needs, intersectionality, cultural values, and autonomy. Involve patients in decision-making, recognising they have their unique values and preferences.
2. Coordination and integration of care – patients may feel vulnerable and powerless in the face of illness and identify three areas in which care coordination can reduce these feelings. Wherever possible, assist patients with coordinating care services across (1) clinical care; (2) ancillary and support services and (3) front-line patient care.
3. Information, communication, and education – patients express concern that they are not entirely informed about their condition, prognosis, or the process of diagnosis and care. To counter this fear, ultrasound professionals can focus on three kinds of communication: (1) information on clinical status, progress and prognosis; (2) information on processes of care and (3) information to facilitate autonomy, self-care and health promotion.
4. Physical comfort – the environment and level of physical comfort patients report significantly impacts their ultrasound experience. Three areas are significant for ultrasound professionals to consider: pain management during the examination, assistance with activities throughout the examination, and navigating surroundings and environment where the ultrasound is taking place.
5. Emotional support and alleviation of fear and anxiety – the fear and anxiety associated with the diagnostic process, the subsequent diagnosis and living with the illness can be as debilitating as the physical effects. Ultrasound professionals should pay particular attention to stress and anxiety over (1) physical status, treatment and prognosis; (2) the impact of the illness on the patient and their family, and (3) the financial impact of their condition.
6. Involvement of family and friends – ultrasound professionals should recognise the important role of family and friends in the ultrasound experience. Family dimensions of person-centred care include (1) involving family and close friends in decision-making to the extent the patient chooses; (2) supporting family members as caregivers; (3) providing cultural, language or cognitive interpretation; and (4) recognising the needs of family and friends as mental health and social support.

7. Continuity and transition – patients express concern about their ability to care for themselves after leaving a healthcare service. Meeting patient needs in this area may require ultrasound professionals to (1) understand detailed information regarding medications, reactions, physical limitations, pain management and emotional needs; (2) coordinate and plan treatments and services beyond the ultrasound examination and (3) provide information to access to support services related to the ultrasound examination. Services may include clinical, social, physical and financial support.

8. Access to care – patients may need to know how to access care to reduce associated stress and anxiety. For patients, access to care extends beyond the examination itself to include before and after the ultrasound examination. Patients indicate the following are essential to them:

- Access and clear directions to the location of the ultrasound examination
- Availability of transportation and parking
- Ease of finding the location via clear signage
- Ease of scheduling appointments
- Availability of appointments when required
- Availability of specialists or specialty services when a referral is made

## Patients' ultrasound experiences are inconsistent

The authors want to acknowledge patient advocates' vital contribution and knowledge sharing. These lived experiences are necessary to be informed by evidence. The lived experiences have provided a range of patient perspectives and highlighted the wide variation and inconsistency of ultrasound experiences.

Positive patient experiences during ultrasound examinations often describe the rapport an ultrasound professional has built with the patient. Patients describe being comfortable, having clear expectations and explanations and having the ability for two-way communication. Tanya, a patient advocate who has had a range of different ultrasounds, describes her positive ultrasound experience:

> *What makes the experience better for all the ultrasounds I have had is the person greeting me when I arrive, taking time to get a brief history before they start, and talking to me and keeping me informed throughout the ultrasound examination.* Tanya, patient advocate

Conversely, poor experiences during ultrasound examinations often relate to *not* investing time before commencing the ultrasound examination, including a greeting, setting expectations, and taking a history. This pre-examination component of the experience should include asking the patient what they would like to contribute and communicate, and from their perspective, how the ultrasound practitioner can create a more comfortable environment. Anja, a patient advocate, describes her ultrasound experience below:

> As someone with complex and abnormal/unusual anatomy, I have had many ultrasound examinations where the ultrasound practitioner has become quiet and continued the ultrasound examination with a concerned look on their face. This has made me concerned, too, that they have found something sinister. They become very focused on scanning, almost as if I was just a body and not a person – forgetting to engage with me or be aware of their body language.
>
> They leave the room and ask other people questions about what they see, not telling me why or asking me a question. They have seen my unusual anatomy, which has not been adequately described on the referral, and they have been too concerned to ask me.
>
> This has led to some negative experiences, leaving ultrasound examinations without answers to my questions as they have not been able to visualise something and haven't asked. It also breaks my rapport with the ultrasound practitioner when they seem disengaged, as ultrasounds are often tense times trying to understand what has been going wrong. Anja, patient advocate

## The consequences of poor experiences

Working in a diagnostic ultrasound can often be a stressful, high-pressure environment. Reducing communication with the patient is a way to improve efficiency, mainly when other patients are waiting. Delivering a poor patient experience has negative consequences for person-centred care, including deterioration of the relationship, trust and empathy for both patients and ultrasound practitioners.

Most ultrasound practitioners have had to deal with the pressures of competing priorities, responsibilities and tasks, as well as efficiency and time constraints. The fast pace of healthcare activities can make person-centred care implementation difficult (Moore et al., 2017). Within the ultrasound environment, published evidence is needed to identify the factors affecting people's reactions and responses to stress and time constraints. However, anecdotal evidence from patients and ultrasound practitioners suggests they experience the following negative impacts within a high-stress, efficiency-focused workplace:

For ultrasound practitioners:

- A high-pressure environment contributes to a high turnover of staff and burnout

- A focus on bookings, throughput and efficiency contributes to a feeling of salespeople working in a high-production factory
- There is a misalignment to purpose due to the focus on volume and efficiency rather than care and outcomes
- A loss of care connection devalues professionalism
- There is no time to gather a clinical history or communicate with patients, which contributes to a lack of clinical and workplace satisfaction

For patients:

- There is no time to build trust and respect with ultrasound professional
- There is no time to ask questions related to symptoms and conditions
- The overall ultrasound experience is negative, which can be detrimental to the patient's health

The online review below illustrates the patients' dissatisfaction with their rushed ultrasound experience.

> ★ ☆ ☆ ☆ ☆  5 years ago
>
> Reception staff were friendly and professional. The radiologist was terrible however, went for our 8 week pregnancy scan they gave the measurements so quickly as if it was for an ankle injury! If you are going for your first child go somewhere else, you deserve the experience to be special and not rushed. Will definitely go elsewhere for subsequent scans.
>
> 👍 2

## The critical components of a positive ultrasound experience

Patients consistently indicate a positive experience when engaging with the ultrasound practitioner and establishing a bond or relationship throughout the examination. Patients acknowledge that this relationship forms via verbal communication and non-verbal body language.

In addition, patients express a desire for ultrasound practitioners to have empathy beyond the clinical ultrasound examination itself, considering that the preparation for an examination can be consequential and tense, and anxiety and stress are ever present while waiting for the outcome.

Patients have a very reasonable expectation of receiving 'care' when visiting 'healthcare' environments. Patients expect to be treated with gentleness and respect, informed and communicated with, and given as much choice and autonomy as possible to ensure a positive care experience.

Tanya, a patient advocate, describes two contrasting experiences below and how she was supported to feel more comfortable and empowered during transvaginal ultrasounds:

*The ultrasound that made me most nervous was my internal pelvic ultrasound, which I have had twice. I did not know what to expect. The first time it was awkward and very uncomfortable; the second time was great. During the second ultrasound, the ultrasound practitioner did several things differently. She talked to me about the procedure before she started. Then, as she scanned, she explained what I saw on the screen as she went. That gave me a sense of control, even though, in reality, I had no control. Knowing what she was doing, especially for an invasive and private procedure, helped.* Tanya, patient advocate

## Why embrace a person-centred approach?

In addition to the clear ethical rationale and the requirement to comply with quality standards to deliver safe, high-quality health care, there are also some practical reasons for ultrasound professionals to adopt a person-centred care approach.

**Involving patients as active partners improves care and outcomes for everyone.**

There is growing evidence that approaches to person-centred care, such as shared decision-making and self-management support, can improve a range of factors, including patient experience, care quality and health outcomes (The Health Foundation, 2016).

Benefits extend to ultrasound practitioners, too; as patient engagement increases, ultrasound practitioners' performance and morale significantly increase. Adopting a person-centred approach connects ultrasound practitioners to their purpose as carers, supports their professionalism, and can be a powerful mechanism to counter burnout (Teisberg et al., 2020). By focusing on the outcomes that matter most to the patient, ultrasound practitioners are intrinsically motivated to positively impact their patient's health instead of defining their success based on traditional volume and efficiency metrics.

**The benefits of ultrasound practitioners adopting a person-centred approach include:**

- A greater sense of purpose as a healthcare practitioner
- Increasing motivation, morale and professionalism
- Contributing to patient outcomes is a powerful mechanism to counter burnout

Examples of initiatives that seek to partner with patients to improve their care and ultrasound experience include:

- Encouraging staff to introduce themselves by name when meeting a new patient is the first step towards providing compassionate care and building trust and rapport while raising the profession's profile (refer to Chapter 19).
- Asking the patient to share their goals and what matters to them. As a result, both parties can consider actions to help progress and achieve these goals. The resulting conversation combines the ultrasound professional's clinical knowledge with the areas the patient knows best: their preferences, personal circumstances, social circumstances, goals, values and beliefs (The Health Foundation, 2016). In addition, the conversation sets expectations about how the examination aligns with and progresses outcomes that matter to the patient.
- Focusing on continuous improvement of the patient experience helps develop a method to capture and collate the experiences of patients and ultrasound professionals and then bring them together to co-design service improvements. Initiatives may consider two aspects of care: (1) how care is organised (care processes) and (2) human interactions (staff interactions with patients and their families). This evidence-based approach helps ultrasound practitioners re-frame what they do and see things from the patient's perspective. Working with patients and families helps to identify small changes that often make a big difference to each patient's ultrasound experience.

### Considerations when adapting person-centred care to an individual

In modern-day practice, ultrasound practitioners must adapt their approach to meet the needs of individual patients whilst considering ethical, social, moral and consent requirements and remaining compliant with professional, quality, safety and sterilisation standards. Below are some key considerations when adapting to a person-centred approach.

### Meeting the needs of individual patients

Picker Principle One of Person-Centred Care is *Respect for the Patient's Values, Preferences and Expressed Needs*. For ultrasound practitioners, this requires a conscious effort to understand each person's requirements and eliminate discrimination or personal bias.

Consider that each patient is an individual and will have different needs and expectations for their ultrasound examination. Ultrasound practitioners can gain much information from the referral before meeting the patient. For example, age considerations may impact consent, or more time may be needed to support an elderly patient or someone with a disability that impacts their mobility (though do not assume this is the case).

Intellectual, psychosocial and physical disabilities may not be immediately evident, and cultural and religious needs may not always be forthcoming. Before an introduction or establishing rapport with the ultrasound practitioner, the patient may wait to share their needs and may seem anxious about the examination or concerned about the implications of the results. Take time to explain the examination and allow for questions or concerns to be raised. For example, a female patient may not be comfortable having an internal examination with a male ultrasound professional for deeply religious or cultural reasons. This response is not personal or related to professionalism but focuses on the patient's social, moral or religious needs. Consider options and collaborate with the patient to find a way forward. Identifying an appropriately trained chaperone, an alternate ultrasound professional, or rearranging the booking time can accommodate the patient's needs.

Patients recommended for transvaginal ultrasound examinations should be offered without discrimination to all females (assigned at birth). This approach extends to transgender men, transgender women and those who identify as asexual, intersex or non-binary. Clearly explain the examination and at all times explicitly obtain consent.

The needs of patients with intellectual disabilities will sometimes force the ultrasound practitioner to adapt their usual protocols and think outside the box. The ability to adapt is demonstrated below during a renal ultrasound examination of an adult patient with an intellectual disability. The ultrasound practitioner adapted to the patient's needs at that time and place by abandoning the need for the patient to lie on the examination table. The carer provided the following feedback:

*Thank you for understanding the needs of the patient and completing the examination while the patient stood up. She was determined to remain standing. Normally we would have had to reschedule her examination with no guarantee that she would be compliant next time. Your ability to adapt to the patient allowed us to ensure her health needs were met, while she remained compliant in completing the examination.*
*Sarah, patient carer.*

In addition, consider the maturity of a child with a growth hormone disorder; they may appear older than their intellectual age suggests. The child may be scared and not understand our instructions as we had initially expected. Therefore it is easy to get frustrated by their behaviour. Rather than continuing, consult a person familiar to the patient – a parent, guardian, or carer – to advise on the approach to achieve the best outcome for the patient. This person-centred approach could avert a traumatic experience, adversely impacting the current situation and leaving a negative lasting impression for possible future examinations.

## Mental health illness or complex trauma

Consider the needs of patients who may be managing challenges that are not visible, such as mental health illness or complex trauma. These patients are more prevalent than ultrasound practitioners may immediately recognise. International studies estimate that 62–68% of young people are exposed to at least one traumatic event by age 17 (Copeland et al., 2007; McLaughlin et al., 2013).

Many people who have experienced trauma show remarkable resilience; however, medical situations and healthcare professionals with whom they come in contact may be triggers. They may feel unsafe or left without choice and autonomy. This situation inadvertently re-traumatises the patient due to the health professionals' lack of awareness, knowledge and training around the particular sensitivities, vulnerabilities and triggers of those who have experienced trauma (Kezelman C, 2014). To take a person-centred approach, consider adopting 'trauma informed practice' which commits to doing no harm and adopting five core principles – safety, trustworthiness, choice, collaboration and empowerment, and respect for diversity (Kezelman C, 2014).

Recognise that each patient has different mechanisms to manage triggers and sensitivities, and ultrasound practitioners should ask if there is anything they can do to support them rather than make assumptions. In one example, when asked, a patient indicated that playing their favourite music genre in the background and dimming the lights helps them to relax.

In another example, Jess, a patient advocate, shares their experience of receiving an ultrasound where the ultrasound practitioners did not take their needs, as someone with a disability and autism, into account:

*During a severe gallstone and gallbladder infection, I had to have an ultrasound. At the time, the pain was still unable to be treated as we had not obtained IV access, and my body was in sensory overload.*

*The ultrasound tech entered the isolation room (bonus of COVID+ patient in a nearby ED bed). They immediately started turning on lights and moving my carefully placed tube feeding supplies without stopping to ask despite the fact they were not in her way. She then banged her way about the room, jostling and slamming my bed back and forth without care, warning or even acknowledgment of my groans of discomfort. Eventually, once she had finished and my pain was significantly worse, which, combined with sensory overload, had sent me back, and I could only take short breaths, groan or completely shut down in silence.*

*She continued to behave sternly, and her actions showed she was asserting her authority about who was in charge of the room. She squirted the gel on and immediately shoved that ultrasound wand into my abdomen. It felt like a sledgehammer had been slammed into me. She continued this rough approach, and on palpating (or ramming), they went into the area where my gallbladder was tender. I involuntarily jumped, yelped, guarded my abdomen and started to whimper with each breath. All she said was 'sore!' but not as a question, more as a statement as if she was yelling 'four' on a golf course.*

*She then continued the investigation of my upper right abdominal quadrant, ignoring the bodily cues indicating I needed a moment to recover. On top of this, I had to clamp my eyes shut due to light sensitivity and the lights being on, causing migraine and sensory overwhelm. Despite what I was experiencing, she turned on all the lights. The room was well-lit by natural light, but she did not seem to care.*

*There were no pauses, warnings, breath reminders, or words or conversation. Furthermore, there were no warnings or informing me what the sonographer would do next, let alone ask permission to move to the next area or to continue. The only words I remember the sonographer saying to me the whole time were 'sore' and her stern statement about having the lights on.*

*On completion, she marched out and left the lights on, knowing I could not move from my bed and my buzzer did not control them. I managed to whimper my way through, pulling the blankets back over my head and curling myself in the foetal position, which I am now aware I do during my autistic breakdowns.*

Jess, patient advocate

Likewise, a disability may not always be obvious or necessary to share or explain. Suppose patients are encouraged to partner in their care conversations and feel comfortable as they build rapport with the ultrasound professional. In that case, they often disclose challenges or limitations that may impact both parties.

*Coordination And Integration Of Care* for ultrasound practitioners requires utilising our knowledge of the healthcare system, hierarchies and medical specialists to assist patients in navigating their journey. Ultrasound practitioners must recognise that patients encounter various resistance, roadblocks and challenges navigating a healthcare system unfamiliar to them. Sharing knowledge can help people progress rather than fall through the cracks.

Anja's ultrasound experience below illustrates the positive impact an ultrasound practitioner had by adopting a person-centred approach. This small gesture of assisting in coordinating care helped Anja achieve the outcome that mattered to her, immeasurably enhancing her experience and a positive lasting impression of the ultrasound practitioner and practice.

> *I had a pelvic ultrasound to assess pain that had been getting more intense over the last few weeks. The sonographer scanned quickly and then stopped, letting me know the ultrasound examination was complete. I was desperate for answers, and there had not been much communication during the ultrasound examination. I asked the sonographer if they had seen anything; otherwise, they could keep looking as the pain was terrible, and I needed a solution. They could tell I was concerned.*
>
> *The sonographer told me they had found a reason for my pain and asked if I could wait in the waiting room while they called my doctor. Usually, I would be told to go and wait a few days for the results before getting medical care. Instead, on this occasion, while I was sitting in the waiting room, my specialist called me to let me know they had found a large cyst. Then the sonographer came out to assist me in going downstairs to the Emergency Department for a streamlined admission. Within 12 hours, I was having emergency surgery. I believe this would not have happened without the initiative taken by the sonographer and their effort to partner with the referring doctor and me to get the outcome that was so important to me.* Anja, patient advocate

### Patient consent

*Information, Communication and Education* require ultrasound practitioners to clearly explain the medical ultrasound examination and the possible benefits and risks so the patient can give informed consent. The ultrasound practitioner's privileged role will be upheld by working in partnership with the patient and demonstrating effective two-way communication, respect and trust. An informed decision-making approach recognises the patient's right to accept or refuse healthcare based on being given adequate information and is a good and essential part of that process.

The guiding principles of informed consent are (ASUM Consent, 2018):

- Consent is given in a voluntary manner
- Informed
- Patient's capacity to understand the information presented, have questions answered and any potential consequences

- Currency of consent has not been diminished due to a lapse in time or change in circumstance
- Consent is specific and covers the ultrasound examination/procedure to be performed

Where patient capacity or age does not allow for informed consent, a parent, guardian, carer, power of attorney or responsible person is responsible. This requirement may vary between countries, and local laws should be adhered to.

### Professional practice and ethical requirements

Most countries provide clear guidance regarding professional practice and ethical requirements for ultrasound practitioners. These requirements include (ACSQHC NODS, 2017; APHRA SCoC, 2022):

- Act in accordance with the law
- Clear and appropriate communication
- Respectful and empathetic of the patient and carers
- Maintaining the necessary skill competence to perform the examination requested
- Not work outside the scope on which qualified or competent to perform
- Recognise the limitations of the ultrasound examination performed, whether impacted by the healthcare professional, patient, equipment or environment
- Understand the potential for harm and mitigate risk
- Uphold the duty of care for the patient and do not discriminate on grounds such as race, religion, sex, disability or those listed in the anti-discrimination legislation.
- Culturally sensitive service requires a genuine effort to understand cultural needs (AHPRA, 2022)
- Good working relationships with other healthcare practitioners

### Cultural, religious and spiritual needs

Working with diverse patients requires a specific emphasis on respectful practice, an open mind, and understanding of the ultrasound practitioner's own bias. To provide a sensitive approach, ultrasound practitioners must be open to listening, critically reflect on their knowledge and attitudes, and consider their expectations of roles if reversed.

It is essential to:

- Acknowledge and address the ultrasound practitioner's bias and assumptions

- Acknowledge, respect and embrace our differences without judgement
- Recognise the need to communicate with patients and carers to determine what is culturally safe for them. Partner with patients to deliver the best outcome for them, and you
- Understand your own culture, values and expectations to allow better consideration of others' needs
- Lead by example, reflecting on how you expect to be treated

The practice may benefit from diversity, equity and inclusion policy or training for patients and internal staff to establish and maintain a positive culture.

Cultural and language considerations may impact the patient's understanding during communication, mainly where English is not the first language. Potential misinterpretation of instructions and information impacts the patient's understanding and expectation of the ultrasound services and outcomes. Make every effort to explain the examination so that informed consent may be forthcoming (ACSQHC, 2014). When patients provide verbal or explicit consent, be sure to document it.

### A clean environment and sterile equipment

*Physical Comfort* encompasses the patient's comfort physically, psychologically and in the environment. This principle requires ultrasound practitioners to consider the patient's physical and psychological well-being, including demonstrating empathy to patients who feel vulnerable or are immunocompromised. Physical comfort extends to patients feeling comfortable wearing a mask at any time, no matter the environment and includes the ultrasound practitioner and other staff offering to wear a mask for the comfort of the patient should be encouraged.

In addition, ultrasound practitioners and personnel have a duty of care to ensure a safe and clean workplace for all patients. Practices must adhere to relevant guidelines and are encouraged to set clear expectations for employees as part of their induction and ongoing education, as well as for patients. Guidelines must incorporate country or state requirements and align with the patients' charter of healthcare rights, describing what patients, or someone they care for, can expect when receiving health care (ACSQHC, 2019) and patient expectations. Ideally, these guidelines should incorporate communicable diseases such as influenza and COVID-19. Close and sustained personal contact defines the two critical circumstances by the Communicable Disease Network of Australia (CDNA) that increase the risk of diseases such as COVID-19 in the ultrasound setting (Basseal JM et al., 2020).

Ultrasound practitioners are required to adopt standard precautions for infection control. These standards range from hand hygiene to the need for low, intermediate or high-level disinfection of transducers and equipment according to the spectrum of activity and guidelines. Infection control processes should include the following:

- Environmental surface cleaning and disinfection. Examples include workbenches and the patient examination table.
- Ultrasound machine disinfection. Examples include high-touch areas such as the keyboard, screens and transducer cables.
- Ultrasound transducer disinfection. For example, a transducer in contact with intact skin may only require cleaning and low-level disinfection. Intracavity transducers, in contact with mucous membranes, require cleaning and high-level disinfection per the Guidelines for Reprocessing Ultrasound Transducers – ASUM and ACIPC (Basseal JM et al., 2017). High-level disinfection requires traceability as per the Standards AS/NZS4187:2014 for recall if and when needed (AS/NZS, 2006).
- Transducer covers. Single-use transducer covers offer a barrier for protection and a level of confidence for the patient but cannot be relied upon 100% (Basseal JM et al., 2019). Every effort should be made to explain the process used in the practice to inform the patient of the protection measures. In particular, offer each patient the option of latex or non-latex to avoid allergic reactions.
- Workstations. For example, the keyboard and mouse (Basseal JM et al., 2020).
- Ultrasound gel. While the warm gel is more appealing and comfortable for the patient, ensuring the gel bottle is clean and appropriately washed and dried if reused is required to reduce the risk of bacteria.

### Safety and bioeffects

Ultrasound practitioners must be knowledgeable of the bioeffects and safe acoustic output settings in ultrasound, similar to a radiographer's knowledge of radiation exposure. The ALARA (as low as reasonably achievable) principle is essential in clinical practice (Abramowicz J, 2015). Ultrasound practitioners must adjust to the lowest possible thermal and mechanical indexes for each patient depending on the anatomical area and the patient's size, particularly in obstetrics, paediatrics and small parts such as an eye. Machine settings must not be adjusted to a higher level than is legally allowable, keeping the thermal index below 1.0, except for peripheral vascular ultrasound examination, which is a maximum of 3.0 (WFUM, 2013).

### Recognising patients needs during the ultrasound experience

*Emotional Support and Alleviation of Fear And Anxiety* consider the patient's fear, anxiety, stress and frustrations related to their diagnosis, prognosis, and impact of the illness on the patient and family. This principle requires ultrasound practitioners to pay particular attention to stress and anxiety related to the impact being unwell has on the patient.

### Involving formal and informal support people – parents, guardians, carers, family, interpreters

*Involvement of Family and Friends* recognises the vital contribution of family, friends and support people (this can include employed support workers) in many patient experiences. For ultrasound practitioners, the focus is to involve family and close friends in decision-making to the extent the patient chooses. At all times, be guided by the patient.

In addition, consider cultural, language or cognitive needs, and recognise the role family and friends may play in providing mental health and social support and interpreting for patients. For many patients, it is incredibly comforting to have a support person with them, particularly young children, the elderly, and people with a disability. They may also help the ultrasound practitioner by assisting the patient with undressing and moving onto the bed. Consider an older person with dementia; a family member may provide medical history and context for the ultrasound examination request.

In keeping with a person-centred approach, patients who do not speak the local language require a translator to ensure they are fully informed. The need for a translator must be established early and can help the ultrasound practitioner understand the patient's needs and obtain clinical information relevant to the examination.

*Continuity and Transition* recognise any concerns expressed by patients about their ability to care for themselves after the ultrasound examination. For ultrasound practitioners, this requirement must extend to consider any patient challenges related to pre-ultrasound preparation requirements, such as fasting or cessation of medication prior to the examination. During and after the examination, patients' needs will vary based on many factors, including the patient's ability to manage their care, the clinical reason for the ultrasound, the type of ultrasound examination, the result or outcome, and post-ultrasound requirements.

*Access To Care* reminds ultrasound practitioners that an ultrasound examination is not an isolated experience. Instead, it is a necessary experience in a patient's continuous healthcare journey. As such, the patient needs context about how the ultrasound examination outcomes enhance or change their existing care, what additional care they may require, and how to access that care beyond the ultrasound examination. While ultrasound practitioners may not know the answers to all these questions, they must consider what they can do to meet, or exceed, patients' expectations as they respond.

Remember, our patients are trying to navigate a complex healthcare system and follow unfamiliar processes while trying to achieve the outcomes that matter to them. We need to remind ourselves that every patient's journey includes highs and lows, with family, friends and support people that are part of their journey. Often overwhelmingly, patients will interact with many healthcare professionals and organisations and navigate many and various touchpoints and channels along the way.

Importantly for patients, ultrasound practitioners must recognise that their journey starts before and ends after the ultrasound service. There are many supportive ways for us to help our patients navigate and access care along the way.

## Summary

As a result of reading the information shared in this chapter, we hope ultrasound practitioners embrace and implement a person-centred approach in their practices and organisations. The patient insights shared throughout the chapter provide an opportunity to recognise the diversity of ultrasound experiences today, some good and some not so good. Looking to the future, we have shared a framework to recognise what great person-centred care looks like and hope ultrasound practitioners will grasp the knowledge with both hands. We hope we have inspired ultrasound practitioners to build respectful and trusting relationships with every patient. We can consistently deliver great ultrasound experiences and better outcomes by working as partners.

## Checklist: Considerations when adopting person-centred care

The following checklist supports ultrasound practitioners adopting person-centred care. It is not intended to replace existing ultrasound protocols or checklists.

### Before commencing the ultrasound examination

Before escorting the patient to the ultrasound room, ultrasound practitioners should:

- Review the patient's past imaging, making every effort to retrieve the report and review it in the context of the current ultrasound examination if it is available and relevant to the current symptoms.
- Provide a comparison of condition progression or regression gives patients context, as they may need to become more familiar with medical terms or the nuances of staging or protocols.
- Consider the clinical question, patient concerns and desired outcomes and adjust the imaging protocol accordingly.
- Explain that a brief scan of the anatomy of interest will be performed before providing any commentary. This explanation reduces stress and anxiety for both parties. In addition, the overview scan gives the ultrasound practitioner time to plan well-thought-out responses to patient questions.

### In the ultrasound room

Ultrasound practitioners should allow patients to:

- Use a gown to avoid the unnecessary gel on the patient's clothes, so patients are comfortable at work or during other commitments throughout the day. The presence of gel on clothing may alert others to something out of the ordinary that the patient may not be willing to discuss. However, if not essential to perform the examination, give the patient a choice. Gowns in an appropriate range of sizes must be available to cater for all sizes and shapes.
- Change in privacy and only expose the examined area, using a towel to cover their body.
- Speak up if they are cold so a sheet or blanket can be provided to keep them warm.
- Feel comfortable by using warm gel, particularly in the young and elderly age groups.
- Easily communicate the result with others by providing context, i.e. the calf tear is improving – it is 0.5cm, compared to 1cm at the previous ultrasound examination. Sharing progress can also motivate patients to continue their healthcare journey, i.e. my tumour was approximately the size of a golf ball, and has shrunk by a third since the last ultrasound examination.

### After the ultrasound examination

Ultrasound practitioners should:

- Provide adequate towels to clean off the gel.
- Give patients privacy to clean and dress.
- Give patients total attention if they ask questions.
- Refrain from multitasking, writing a report or tidying the room, as the patient may interpret this as a lack of care.
- Help patients who need assistance. For example, help put shoes back on. A shoehorn within the room is helpful for patients with swollen feet or limited mobility. Helping with a shirt, jumper or belt, or assisting a lady in fastening her bra helps patients with shoulder injuries or who cannot extend behind their back.

### Beyond the ultrasound room

Ultrasound practitioners should:

- Escort the patient back to the reception area.
- Arrange transportation for an inpatient.

- Be aware of delays and check in with patients frequently so they know you have remembered them.
- Consider the patient's physical comfort, paying particular attention to stress and anxiety levels. Offer tea, coffee or a glass of water (if appropriate).
- Explain when and how to access results and images and how long till they and their referring doctor receives an official report.
- Notify the referrer or reporting doctor to ensure continuity of care, mainly when there are urgent findings.

### Paediatric patients

Paediatric patients require extreme patience and skill to capture images without movement. This outcome is only achievable if the ultrasound practitioners can:

- Prepare the room and the machine before commencing the examination, so there is no need to do this once the child is ready for an ultrasound examination.
- Seek to build a good rapport with the child and parent(s) or guardian to keep them still and calm. Suggestions include practising breathing techniques, so the child understands what is required. Imaging with quiet respiration is more manageable for toddlers and babies than if they are crying.
- Reassure the child and parent that the ultrasound examination will not hurt. They may wish to touch the transducer or gel with their finger before the examination.
- Distract the child to achieve compliance. Examples include a favourite toy, watching a show on the phone or iPad, singing or reading a book.
- Ask the parent to restrain the child safely or hold the child in different positions to provide access to anatomy, such as the kidneys.
- With babies, asking the mother to breastfeed or bottle feed the baby in a decubitus position may allow access to the upper abdominal organs.
- Recognise a 'textbook' position is not always achievable, and adapt to the needs of each patient.
- If young children have a full bladder apply minimal pressure with the transducer.
- Save a cine loop to save time instead of capturing still images. The cine loop technique is fundamental if there is little time before the patient needs to empty their bladder.

### Teenagers/adolescent patients

Teenagers may not want their parents in the room. Ultrasound practitioners should take into consideration the following:

- If a boy has a scrotal examination, it is often helpful to facilitate an open discussion with the parent present, giving the option for privacy without the parent present.
- Teenagers may share additional information relevant to the ultrasound examination when taking charge of their care, such as how they broke their limb, sexual activity, or habits such as vaping that they were unwilling to share in front of a parent.
- Be sensitive in approaching why the patient needs the ultrasound examination, as they may be embarrassed about their symptoms.
- The need for privacy and coverage of the body is essential.
- There are legal age requirements that will vary in different jurisdictions.

### Older patients

Among older patients, there is a wide variation in physical, mental health and emotional capability. A 90-year-old patient may be cognitive, fit and mobile, whilst a 70-year-old may suffer from advanced dementia, be physically deteriorating and require constant support.

Ultrasound practitioners should:

- Take the time to understand the mobility limitations of each patient and prepare the room before bringing patients into the room.
- Turn up the light when the older patients move around the room and bed, pointing out any equipment that could be hazardous to them as they walk into the room.
- Put the bed down as low as possible or use a step to assist the elderly or disabled in climbing onto the bed (Figures 18.2 and 18.3). An adjustable rail on the side of the bed is helpful to keep the patient feeling safe and secure (Figure 18.1).
- Consider the benefits of a mechanical bed that raises erect, particularly for venous studies, as it supports the patient, and the ultrasound practitioner does not need to worry about the patient collapsing (Figure 18.4).
- Complete the ultrasound examination in a wheelchair or on a chair, preferably with arms, to prevent the patient from falling. A carer or support person sitting close by can help steady them throughout the ultrasound examination.
- Be conscious of incontinence challenges or patients who may need a catheter clamped. Offer assistance and provide privacy for the patient.
- Be mindful that some patients will need new dressings on ulcers following the ultrasound examination. Failing to redress wounds adequately may be a challenge for patients to solve themselves, particularly those who are immobile and rely on others for support and transportation.
- Escort older patients to the bathroom if required and stay close if they need assistance.

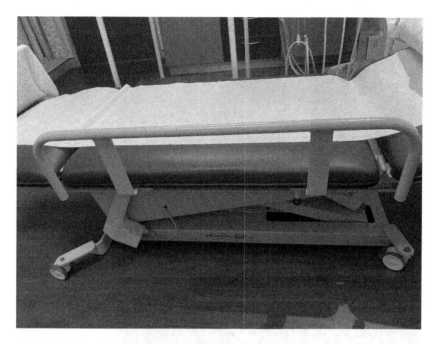

*Figure 18.1* A rail on the bed.

## Obstetric examinations

Unlike most other ultrasound examinations, the obstetric ultrasound examination brings excited parents ready to view their new family addition. As ultrasound practitioners, we are in a very privileged position to be a part of the pregnancy journey and should:

- Make the experience as comfortable and enjoyable as possible whilst remaining cognisant of our diagnostic requirements.
- Ensure the mother does not have an over-distended bladder and there is a chair for the support person, and both can view the ultrasound examination in comfort.
- Offer to perform the ultrasound examination in a decubitus position in the third trimester, as the mother may not be comfortable lying supine. Never lie the mother flat, as this will instigate a vasovagal reaction.
- Use towels to protect the mother's clothing from the gel and offer to adjust the bedhead higher or extra pillows behind her head or under her knees.
- Consider patient comfort and use warm gel.
- Push lightly with the transducer

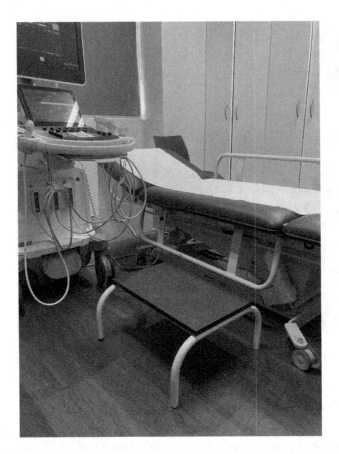

*Figure 18.2* A stable step.

- Accept that maternal body habitus will often impact the image quality, which may not be optimal. Consider asking the mother to roll onto her side to enable scanning from different windows and through less adipose tissue.

### Gynaecology examinations

Gynaecology examinations are incredibly intimate and sensitive. Patients expose their pelvic region to the symphysis pubis for a transabdominal ultrasound examination. For a transvaginal ultrasound examination, patients expose their perineal region. Ultrasound practitioners should consider the following:

- The patient may have experienced sexual violence and should be alert to any history triggering these traumatic events.

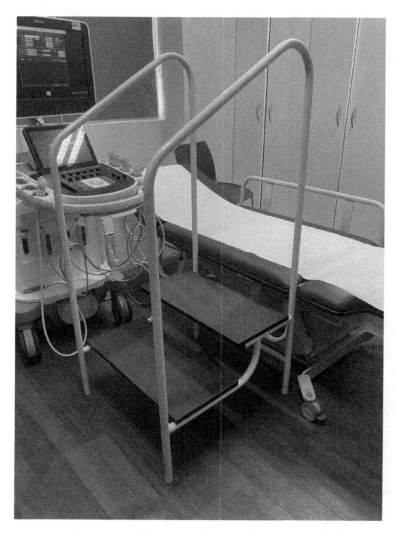

*Figure 18.3* A step with handrails.

- The patient may not have had intercourse or are underage (depending on the local legal age limits), so performing a transvaginal ultrasound may not be appropriate.
- A transvaginal examination's appropriateness may depend on the transabdominal findings.
- Allow the patient time to consider if they consent to the examination. Never rush this decision.

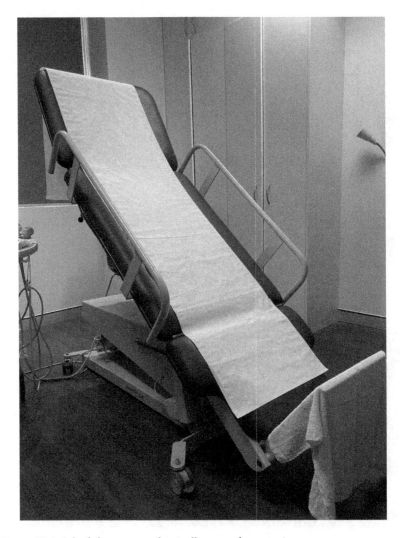

*Figure 18.4* A bed that can mechanically move from supine to erect.

- Privacy when undressing and a gown and sheet to cover the lower abdomen are essential.
- Before inserting the probe, explain and demonstrate how the probe is used and manipulated when asking for consent. If using a latex probe cover, ensure the patient has no latex allergies.
- The benefits of a dedicated gynaecology bed include an adjustable platform that lowers from the abdominal position, drops at the foot end, and is more comfortable than a pillow under the patient's pelvis (Figure 18.5).

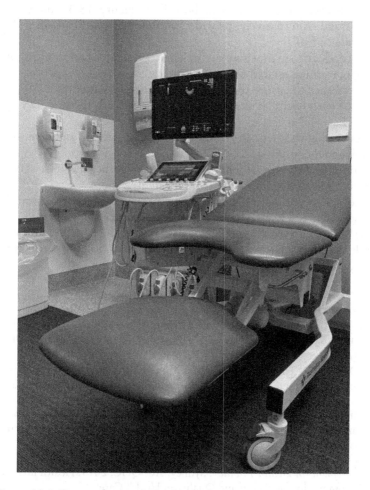

*Figure 18.5* Gynaecology examination bed.

- Explain what is happening whilst performing the ultrasound examination and checking that the patient is comfortable.

## Vascular Examinations

During vascular examinations, ultrasound practitioners must know the impact of standing for lengthy periods. Notably, chronic venous insufficiency (CVI) patients are susceptible to vasovagal attacks. Ultrasound practitioners must consider the following:

- Patients who cannot stand for CVI examinations can sit on the edge of the bed, raising the bed to a height where the patient's foot can rest on the ultrasound practitioner's lap.

- A dedicated vascular bed that can be raised (Figure 18.4) allows the patient to stand. Alternatively, provide a mobile hoist to lift them if they are immobile or frail.
- A room big enough to manoeuvre a wheelchair around the bed comfortably.
- Appropriate training related to lifting patients.
- Ask for assistance to ensure the appropriate lifting of patients.
- Be aware of early signs; if the patient gets fidgety, ask them if they feel lightheaded.
- If patients are lightheaded, sit them on the bed or a chair and give them water and a cold compress.
- After recommencing, the patient will often get lightheaded in a shorter time frame, so keep sitting them down repeatedly.
- Ideally, complete one leg only in a single appointment and ensure the patient has eaten and is well-hydrated for the second visit.
- When performing an ultrasound examination on young patients, they may try to hide that they are starting to feel dizzy, so keep checking throughout the ultrasound examination.

## References

Abramowicz J, 2015. *ALARA: The Clinical View. Ultrasound in Medicine and Biology*, 41(4), S102. www.umbjournal.org/article/S0301-5629(15)00162-3/fulltext#relatedArticles

ACSQHC NODS, 2017. The Australian Commission on Safety and Quality in Health Care. National Safety and Quality Health Service Standards (NSQHS), Standard 2: Partnering with Consumers, Partnering with Patients in Their Own Care, Action 2.05, Healthcare Rights and Informed Consent. www.safetyandquality.gov.au/standards/nsqhs-standards/partnering-consumers-standard/partnering-patients-their-own-care/action-205

ACSQHC, 2014. *Australian Commission on Safety and Quality in Health Care. National Statement on Health Literacy*. Sydney: ACSQHC; 2014. Health literacy I Australian Commission on Safety and Quality in Health Care

ACSQHC, 2017. *The Australian Commission on Safety and Quality in Health Care*. National Safety and Quality Health Service Standards (NSQHS), Standard 2: Partnering with Consumers. www.safetyandquality.gov.au/standards/nsqhs-standards

ACSQHC, 2019. Australian Commission on Safety and Quality in Health Care. *Australian Charter of Healthcare Rights*. www.safetyandquality.gov.au/consumers/working-your-healthcare-provider/australian-charter-healthcare-rights

AHPRA SCoC, 2022. *The Australian Health Practitioner Regulation Agency (AHPRA) Shared Code of Conduct Australian Health Practitioner Regulation Agency – Shared Code of Conduct (ahpra.gov.au)*

AHPRA, 2022. The Australian Health Practitioner Regulation Agency (AHPRA) *Shared Code of Conduct, Case Study – Culturally Safe Practice*. www.ahpra.gov.au/Resources/Code-of-conduct/Resources-to-help-health-practitioners/Case-studies-Shared-Code-of-conduct.aspx

AS/NZS, 2006. Australian/New Zealand Standard AS/NZS 4815:2006 Office-Based Health Care Facilities – Reprocessing of Reusable Medical and Surgical

Instruments and Equipment, and Maintenance of the Associated Environment, Standards Australia and Standards New Zealand. https://scholar.google.com/scho lar?hl=en&q=+Australian%2FNew+Zealand+Standard+AS%2FNZS+4815%3A2 006+Office%E2%80%90based+health+care+facilities+%E2%80%93+Reprocess ing+of+reusable+medical+and+surgical+instruments+and+equipment%2C+and+ maintenance+of+the+associated+environment%2C+Standards+Australia+and+ Standards+New+Zealand

ASUM Consent, 2018. The Australasian Society for Ultrasound in Medicine (ASUM). Discussion Paper on Consent Discussion-Paper-on-Consent.pdf (asum.com.au)

Basseal JM et al., 2017. Basseal JM, Westerway SC, Juraja M, van de Mortel T, McAuley TE, Rippey J, et al. *Guidelines for Reprocessing Ultrasound Transducers. Australasian Journal of Ultrasound in Medicine* 2017; 20: 30–40. https://onlinelibr ary.wiley.com/doi/full/10.1002/ajum.12042

Basseal JM et al., 2019. Basseal JM, Westerway SC, Hyett J. *Analysis of the Integrity of Ultrasound Probe Covers Used for Transvaginal Examinations* December 2019 http://dx.doi.org/10.1016/j.idh.2019.11.003

Basseal JM et al., 2020. Basseal JM, Westerway SC, McAuley TE. Covid-19: Infection Prevention and Control Guidance for All Ultrasound Practitioners. https://online library.wiley.com/doi/10.1002/ajum.12210

Copeland WE, Keeler G, Angold A, Costello EJ. (2007). *Traumatic Events and Posttraumatic Stress in Childhood. Archives of General Psychiatry,* 2007;64(5):577– 584. https://pubmed.ncbi.nlm.nih.gov/17485609/

Edwards, L., & Hui, L. (2018). *First and Second Trimester Screening for Fetal Structural Anomalies. Paper Presented at the Seminars in Fetal and Neonatal Medicine.*

IOM, 2001. The Institute of Medicine (US) Committee on Quality of Health Care in America. *Crossing the Quality Chasm: A New Health System for the 21st Century.* https://pubmed.ncbi.nlm.nih.gov/25057539/

Kezelman C, 2014. Dr Cathy Kezelman, President, Blue Knot Foundation. Trauma Informed Practice. Mental Health Australia website. https://mhaustralia.org/gene ral/trauma-informed-practice

McLaughlin KA, Koenen KC, Hill ED, Petukhova M, Sampson NA, Zaslavsky AM, Kessler RC. 2013. Trauma Exposure and Posttraumatic Stress Disorder in a National Sample of Adolescents. *Journal of the American Academy of Child & Adolescent Psychiatry,* 52(8):815–830. doi:https://doi.org/10.1016/j.jaac.2013.05.011 www. ncbi.nlm.nih.gov/pmc/articles/PMC3724231/

Moore, L., Britten, N., Lydahl, D., Naldemirci, Ö., Elam, M. and Wolf, A. 2017. Barriers and facilitators to the implementation of person-centred care in different healthcare contexts. *Scandinavian Journal of Caring Sciences,* 31(4), pp.662–673.

Nicholls et al., 2014. Nicholls, D., Sweet, L., Hyett, J. (2014). Psychomotor skills in medical ultrasound imaging. *Journal of Ultrasound in Medicine,* 33(8), 1349– 1352. doi:10.7863/ultra.33.8.1349

Picker Institute, 1987. The Picker Institute. *The Eight Picker Principles of Person-Centred Care.* https://picker.org/who-we-are/the-picker-principles-of-person-cent red-care/

RANZCR, 2020. Royal Australian and New Zealand College of Radiologists (2020). *Standards of Practice for Diagnostic and Interventional Radiology,* version 11.1. Retrieved from www.ranzcr.edu.au/documents-download/document-library-22/ document-library-25/510-ranzcr-standards-of-practice-for-diagnostic-and-interve ntional-radiology/fileCached

Salomon et al., 2011. Salomon, L. J., Alfirevic, Z., Berghella, V., Bilardo, C., Hernandez-Andrade, E., Johnsen, S. L., Kalache, K., Leung, K.Y., Malinger, G., Munoz, H., Prefumo, F., Toi, A., Lee, W., on behalf of the ICSC. (2011). Practice Guidelines for Performance of the Routine Mid-Trimester Fetal Ultrasound Scan. *Ultrasound in Obstetrics & Gynecology*, 37(1), 116–126. doi:10.1002/uog.8831

Teisberg E, Wallace S, O'Hara S. 2020. Teisberg, Elizabeth PhD; Wallace, Scott JD, MBA; O'Hara, Sarah MPH. 2020. Defining and Implementing Value-Based Health Care: A Strategic Framework. Academic Medicine. Academic Medicine: May 2020 –95(5), 682–685. doi: 10.1097/ACM.0000000000003122 https://journals. lww.com/academicmedicine/Fulltext/2020/05000/Defining_and_Implementing_V alue_Based_Health_Care_.14.aspx

The Health Foundation, UK, 2016. The Health Foundation. *Person-Centred Care Made Simple: What Everyone Should Know about Person-Centred Care.* www.hea lth.org.uk/publications/person-centred-care-made-simple

WFUM, 2013. World Federation of Ultrasound and Medicine. (2013). WFUMB Policy and Statements on the Safety of Ultrasound. *Ultrasound in Medicine and Biology*, 39(5), 926–929. doi:10.1016/j.ultrasmedbio.2013.02.007. www.umbjour nal.org/article/S0301-5629(13)00080-X/fulltext

# 19 Person-centred ultrasound communication

## Shelley Thomson, Lyndal Macpherson, Samantha Thomas and Anja Christoffersen

### Introduction

Person-centred care in ultrasound starts before the person or patient enters the ultrasound department. A clear appointment letter with understandable instructions regarding any preparation required and why is essential. Creating such resources with patients using co-creation can ensure that instructions are clear, understandable and easy to follow.

Ultrasound scans require close contact between the practitioner and the person undergoing the ultrasound scan and for the person, the day of an ultrasound scan is often associated with powerful emotion due to the insights it may provide, and their understanding and expectations are high. Patients expect ultrasound practitioners to support their needs and deliver a person-centred ultrasound experience. A critical component of the experience is the quality of the ultrasound practitioner's verbal and non-verbal communication skills, as they can make or break a patient's experience. The ultrasound practitioner's communication may also impact the patient's confidence in the examination and the findings. Therefore, ultrasound examinations require the patient and ultrasound practitioners to rapidly develop rapport and trust due to the proximity of both parties and the often invasive nature, including interventional techniques. An ultrasound practitioner's role differs from other imaging specialties due to the extended time the sonographer spends in close contact with the patient.

At its most fundamental, person-centred communication starts with both parties respecting each other and building trust and rapport. At its most aspirational, person-centred communication sees ultrasound practitioners evolve conversations beyond exchanging information to support patients to actively participate and take responsibility for their care.

Expectations of communication between the ultrasound practitioner and patient have evolved to include before, during and after the scan. With the advent of real-time imaging, societal and technological changes, and readily accessible information, many patients now expect the ultrasound practitioner to communicate the findings at the examination time. This expectation is particularly critical for obstetric patients, where non-disclosure has been

DOI: 10.1201/9781003310143-26

proven to negatively impacts parents. (Johnson et al., 2020; Morse, 2011; NSW Gov, 2022; Boyle et al., 2019; Perinatal Society of Australia and New Zealand, 2018).

Historically, ultrasound practitioners did not perform real-time imaging. Doctors generated reports from still images the ultrasound practitioner acquired after the patient left the department. As a result, patients did not expect the ultrasound practitioner to discuss the results at the time of the scan. Increasing consumer expectations of person-centred care require ultrasound practitioners to adapt and evolve their communication practices to ensure patients feel supported before and during their scan and understand the next step in their healthcare journey. This expectation can be stressful for some ultrasound practitioners, particularly if they are prohibited from openly communicating with their patients because of the company or reporting doctors' policies (Thomas et al., 2020; Thomas et al., 2017).

## Person-centred communication

The Picker Institute's Eight Principles For Delivering Person-Centred Ultrasound Care were introduced in the chapter titled Person-centred ultrasound experiences. These principles embody the conviction that all patients deserve high-quality healthcare and that patients' views and experiences are integral to healthcare service delivery (The Picker Institute, 1987).

Vague or dismissive communication can harm the patient, shape their beliefs, and influence future behaviour when accessing the health system. Person-centred care relies on both parties respecting each other and building trust and rapport. Effective and efficient face-to-face communication is crucial. It allows all people involved to interpret what is about to happen and enables them to take an active role throughout the examination. The body language and facial expressions provide vital information, helping to understand the meaning behind and intention of the words. Furthermore, once embedded, high-performing practitioners extend person-centred communication beyond merely exchanging information to empowering patients to actively participate and improve the quality of their care (Moore et al., 2017).

Person-centred communication includes describing the ultrasound process as clearly as possible, using simple patient-specific language and active listening techniques to ensure understanding from the patient. Patients have a variety of expectations related to the ultrasound examination, so expectation management is essential. Managing expectations includes communicating the goals of the scan and discussing limitations of the scan, i.e. a stomach ulcer during an abdominal scan.

Throughout the conversation, dismissing the patient's concerns, questions or reason for being there can detrimentally impact the patient experience. Personal beliefs or biases on what a positive and negative result is should not be discussed. For some, not finding anything on the scan can mean subsequent invasive tests and prolonged suffering.

Communication also should consider other aspects of person-centred care, including respect for patients' values, preferences and expressed needs. The experience of a young patient, shared below, demonstrates the critical role of person-centred communication and its subsequent impact on the patient's behaviour and participation in her care:

A 12-year-old indigenous girl presented to a community health clinic in Australia with significant acute rheumatic fever symptoms. She was sent via Careflight to a tertiary hospital for rheumatic heart disease diagnosis and treatment. She found this stay very distressing, partly because she only had one family member with her (her mother was not in the community at the time, so she was not present with her in the hospital). She did not have all the tests explained to her appropriately (English is her third language), and she found the echocardiogram a scary experience and the probe painful. The test and its results were not well explained. She was diagnosed with significant valvular disease, commenced treatment, and then travelled home within two weeks. Due to the significance of the disease, she was scheduled for a repeat echocardiogram in three months during the next community cardiology visit.

Due to her prior scary experience in the hospital, she did not attend her first two echocardiography appointments in the community. Fourteen months later, on the third attempt, she was persuaded to come to the appointment. However, she refused the echocardiogram as she felt uncomfortable with a male sonographer and was not offered a female chaperone.

Her follow-up echocardiogram was finally performed sixteen months after the initial hospital admission. A community rheumatic heart disease nurse and clinic nurse had developed a trusting relationship with the girl and worked on explaining the test and its importance. Before the appointment, the test was explained in language. Two female chaperones were in the room with her for the echocardiogram. She was comfortable with the test and even enjoyed seeing her heart as it was explained.

The young girl now regularly attends clinics for her monthly penicillin injections and is first at the door on her echocardiography appointment day.

This example shares the importance of the fundamentals of patient communication and empathy. It is also a reminder that people can experience cultural and language barriers in any environment. To ensure the patient's individual needs are considered, communicate with the patient in their language, and ask how to include and communicate with support people if present.

While ultrasound practitioners are widely recognised for their specialised diagnostic skills, they may need to be aware of if and when they could be more effective communicators, as this impacts patient outcomes. Good person-centred communication aligns with favourable health outcomes, increasing patient satisfaction, compliance and overall health status. Conversely, poor communication results in adverse outcomes, such as decreased adherence to treatment, patient and health professional dissatisfaction, inefficient use of resources and poor patient outcomes.

In clinical curriculums, communication skills are often taught, but frequently receive less attention than clinical skills. Research on physician and patient communication training in general practice indicate that both parties need practical communication training to improve participation and satisfaction with their experiences. In the ultrasound environment, where ultrasound practitioners provide a medical examination, the fundamentals of this research apply. The key message is that improving communication is a partnership that requires both parties to be actively involved. Healthcare practitioners are only half the equation.

According to Dr Kelly Haskard-Zolnierek, whose research is published in *Health Psychology*, health professionals and patients bring different perspectives, skills and levels of communication (Haskard-Zolnierek, 2008). Only focusing on one of the two people may be missing an important piece. Applying just a few minutes of coaching can encourage patients to ask questions and seek out information. For ultrasound practitioners, this research outlines a simple way to implement person-centred communication by coaching the patient to ask questions and actively participate in the examination. In addition, when practitioners receive communication training, they report lower stress and greater satisfaction with the relationship than when just one of the pair receives training.

### The patient perspective

While patient perspectives indicate that quality communication during ultrasound examinations is diverse, there is widespread agreement that establishing rapport via good communication is a fundamental requirement. Moreover, once rapport is lost, it is hard to recover. Rapport is lost when communication includes inappropriate or invalidating comments or questions, a lack of disclosure of information, or non-verbal cues creating an uncomfortable experience. The following patient experience shared by Charlie, a patient advocate, highlights the importance of ultrasound practitioners communicating with patients about what to expect during the scan.

> *The technician told me he would simply put the wand on my tummy. He failed to tell me that he would stick it up underneath my ribs and poke it into my ribs and side. He failed to tell me that this might be uncomfortable or painful (it was both for me; I am extremely touch-sensitive and have chronic pain). What he did with me was not fully informed consent. As a patient, I have the right to know what will be done to my body before it happens. It is important to fully explain procedures to patients (it could be written down beforehand, for example). This explanation needs to be done extra thoroughly with disabled and English as a Second Language (ESL) patients, as there is the potential for misunderstanding. It would be helpful to show as well as tell. When I told the technician he was hurting me, he made an awkward joke and continued. This is not ongoing consent.*
>
> (Charlie, Patient Advocate)

## Communication guidelines for ultrasound practitioners

As noted at the start of the chapter, patients' communication expectations are evolving, and patients expect to be informed of their scan results. In Australia, New Zealand, and the United Kingdom, the shift in patient expectations has led to the development of person-centred communication guidelines for ultrasound practitioners, specifically for obstetric examinations (ASUM, 2022; Johnson et al., 2020). These guidelines were developed in collaboration with parent groups, interdisciplinary medical professional groups, researchers, parents with lived experience and ultrasound practitioners. This approach was critical in ensuring a better understanding of the support needed for both parents and ultrasound practitioners when dealing with unexpected and often devastating findings. Ultrasound practitioners are encouraged to adopt these best practice guidelines and co-design protocols with patients to enhance the delivery of obstetric news aligned with patient and community expectations.

Education providers must also incorporate communication guidelines into their curriculum to help ultrasound practitioners master person-centred communication. This person-centred focus will ensure that the next generation of ultrasound practitioners have the skills to adapt their approach to each patient's needs and expectations. Essential person-centred communication and behavioural skills for ultrasound practitioners include respect, empathy, active listening and emotional intelligence. Furthermore, while these behaviours are classified as 'soft' skills, they are fundamental to delivering person-centred care.

All ultrasound practitioners and clinical practices are encouraged to develop and implement communication guidelines for patient-facing professionals, including administrative staff, to ensure a consistent, person-centred experience.

## Enhancing person-centred communication

By collaborating with patients, we have identified four critical areas for ultrasound practitioners to improve person-centred communication. They are:

### Enhancing person-centred communication: Patient introduction

Ultrasound practitioners may question the importance of an introduction. It sounds simple and is often dismissed as a minor behavioural requirement. However, this is the first opportunity for ultrasound practitioners to humanise healthcare by making a human connection and instantly building trust. To illustrate the importance of the introduction, below is a patient experience by Dr Kate Granger:

---

#### #hellomynameis campaign

Dr Kate Granger, MBE, was a doctor in the UK living with terminal cancer. As a doctor and terminally ill cancer patient, Kate made a stark observation during a hospital stay with post-operative sepsis; many staff looking after her did not introduce themselves before delivering her care. To Kate, it felt incredibly wrong that such a fundamental step in communication was missing.

After ranting at her husband Chris during visiting time one evening, he encouraged Kate to 'stop whining and do something!'. As a result, Kate and Chris co-founded a national #hellomynameis social media campaign [www.hellomynameis.org.uk/] to encourage all staff to introduce themselves by name and profession when meeting a new patient. In Kate's words, this is *"... the first rung on the ladder to providing compassionate care"*.

Following Kate's death in 2016, Kate's husband and co-founder, Chris, is keeping the campaign alive through conference talks worldwide, book writing, presenting awards and social media.

Kate's message is simple yet incredibly powerful *"... it is not just about the common courtesy, but it runs much deeper. Introductions are about a human connection between one human being who is suffering and vulnerable and another human being who wishes to help. Two people begin a therapeutic relationship and can instantly build trust in difficult circumstances"*.

(#hellomynameis)

The #hellomynameis campaign is implementable in any healthcare environment – hospital, emergency services, private practices, aged care facilities, home care, day surgery, outpatient services, diagnostic imaging and ultrasound examinations.

> As an ultrasound professional, when you meet a patient for the first time, stop and introduce yourself with *"Hello, my name is..."*

Introductions help ultrasound practitioners see patients as people, first and foremost, ahead of their examination or condition. The introduction includes asking the patient, *"What would you like to be called"* or *"What would you like me to call you?"*. This question demonstrates respect for people, particularly indigenous, culturally diverse and LBGTQIA+ patients. Additionally, consider the patient's relationship with a support person. Confirm whether the patients have any support people accompanying them and how to communicate with them. By asking, *"Who have you brought in with you?"* the patient can explain, rather than the ultrasound practitioner assuming the relationship between the patient and the support person.

### Enhancing person-centred communication: During the examination

For patients, an ultrasound scan can be a confronting and traumatic experience. Interactions between patients and ultrasound professionals require an empathetic approach to develop trust and compassion. After introductions and before beginning the examination, patients and ultrasound practitioners should discuss and agree on expectations about what the ultrasound can and cannot achieve and the process that will happen. Otherwise, a mismatch of expectations can occur.

In the experiences described below by Charlie, a patient advocate, she details two contrasting ultrasound practitioners' communication approaches and the resulting outcomes she experienced when having two different scans five years apart.

> Experience 1: *I was not told beforehand that the ultrasound was internal – they just pulled out the wand, lubed it up, and said they were about to insert it. They continued banging the wand against my cervix and pushing their hand into the painful spot well after I said it was painful, and I writhed in pain. They were determined to tick off a list of organs to photograph, no matter my experience. I felt sexually violated. Nothing of note was found on the scans. They did not mention my pain in the report.*

> Experience 2: *My referring specialist, the receptionist and the sonographer explained that it would be an internal ultrasound. The sonographer told me to tell her if it hurt, which I did. The sonographer offered to stop that part, which I took up. The sonographer wrote in the report that there was pain on the left side, and they had to stop. I believe that this note provided evidence for my gynaecologist to proceed with a laparoscopy. I was diagnosed and treated for endometriosis, which was found worse on the side with the pain during the ultrasound.*
>
> *I had similar pain in both ultrasounds, approximately five years apart, yet very different approaches and outcomes for me.*
>
> (Charlie, Patient Advocate)

Creating a comfortable environment for the patient during the scan includes refraining from judgemental remarks and addressing people by their preferred names. For ultrasound professionals, adapting to meet each patient's needs is about asking how much the person wants to know about their scan. This information includes explaining what the ultrasound practitioner is doing, what happens next, what is being scanned and measured, and whether the patient would like to view the screen. These are all requirements of person-centred care delivery.

Stephanie, a patient advocate, has had many ultrasounds throughout her life and shares the positive behaviours demonstrated by the ultrasound practitioners she has met during her experiences.

> *I always feel more comfortable with staff members who are empathetic and make an effort to personalise the experience and make conversation. Even though they would see many people every day, the interaction during the scan may not be important to them, but the patient is the one having their important scan, and that scan may mean a lot to them. It could mean answers to long-term pain, or it may change their life in the form of a cancer diagnosis!*
>
> (Stephanie, Patient Advocate)

During the ultrasound examination, patients will likely experience heightened emotions. The ultrasound practitioner's verbal and non-verbal communication behaviours, tone of voice, pauses, silence and length of time scanning an area all impact. For patients, the behaviours displayed by the ultrasound professional throughout the examination determine their scan experience.

As the scan progresses, ultrasound practitioners must continue to check with the patient, permitting them to ask questions and communicate their comfort levels, pain, concerns, needs and requests throughout the scan.

Many patients are eager to understand more about their bodies. Some have complex or rare anatomy, which they can explain to help the ultrasound practitioners navigate during the examination. Patients have extensive knowledge of their symptoms, pain, discomfort, and lived experiences; this is often overlooked because patients are not encouraged to contribute.

Moreover, as patients of all ages increasingly use the internet to understand their health needs, the nature of the ultrasound practitioner–patient relationship continues to change.

The ultrasound examination may identify a way to improve a patient's quality of life rather than just a clinical outcome. Furthermore, an ultrasound result that does not identify a cause or abnormality may provide relief or cause anxiety. Regardless of the outcome, the scan provides valuable information for the patient and referring clinician. The result should be communicated with care to ensure the patient does not feel invalidated or dismissed because nothing is identified.

Throughout this section, many patients have shared experiences to help ultrasound practitioners increase their awareness of the impact of communication on the patient's experience and outcomes. Taking the time to understand patient concerns and adapt to each patient's needs can make a significant difference and improve outcomes for all.

### Enhancing person-centred communication: Concluding the scan

There will be times when the results of the scan require urgent action. In this situation, how ultrasound practitioners explain the findings impacts the patient's experience.

Below, Tanya, a patient advocate, shares that a poor diagnosis due to the scan does not necessarily translate to a poor ultrasound experience. When the ultrasound practitioner uses clear, non-clinical language to explain the findings and an empathic approach, this demonstrates a person-centred approach.

> *The best ultrasound I ever had was when they found a 21cm blood clot in my arm. The sonographer scanned my arm and showed me on the screen, then scanned her arm to show me the difference between the two. The way she delivered terrible news that was potentially life-threatening was amazing because she took the time to explain it so well. I am a visual learner, so that was great. My support worker was with me and commented on how impressed she was too.*
>
> (Tanya, Patient Advocate)

Ultrasound practitioners often perceive that what matters most to patients is to reassure them that 'everything is okay'. However, patients have different perspectives. Patients indicate that when they experience symptoms or something different from usual, their priority is to undergo the examination to discover why. The patients 'why' may extend beyond a clinical outcome to consider their quality of life. The outcomes that matter most to patients are more likely related to improving functional capability, reducing pain, relieving emotional suffering, or working or living while receiving care. Ultrasound practitioners can help patients achieve the outcomes that matter to them by ensuring they ask for and document the information shared by the patient and include it in the report.

When a patient schedules an ultrasound examination, language and context are also important considerations. For example, when describing ultrasound findings to patients, ultrasound practitioners often use phrases such as '*everything is okay*' or '*you are all good*' to describe findings where nothing is relevant to the patient's condition. These phrases need to be clarified. How should the patient interpret this information? How should the patient interpret this information? Does the phrase refer to the images being technically okay or good, or is the patient themself okay or good? Incorrect interpretation can break the trust between the patient and the ultrasound practitioner, particularly if the patient has additional diagnostic tests and the outcome of these tests are not '*okay*'. Below, patient advocates, Wendy and Brittany share their experiences relating to the ultrasound practitioners' delivery of the scan results:

> *When I had my Mammogram in August 2015, it did not go to plan, and I required further testing. At the end of a very stressful Friday, the doctor doing the biopsy and ultrasound told me she had seen many malignant tumours, and she was sure mine was benign. On Monday, I received the call to say that I not only had Breast Cancer, but it was Triple Negative.*
>
> Wendy, Patient Advocate

> *At a recent scan, I was told nothing was found, and there is no point returning to my GP (referring doctor) as nothing was wrong. It is important for sonographers to be aware that a person is there seeking answers and that when you cannot provide them, not deliver that in a way that is an answer in itself as it is not. There is always more investigation and conversation that can be had with the medical team, which should be communicated.*
>
> (Brittany, Patient Advocate)

Regardless of the outcome, patients expect to be informed of the results after the ultrasound examination. When discussing results, a person-centred approach is to engage patients by explaining the process of interpretation, the limitations of the examination (if any) and what the report will likely include. In addition, ultrasound practitioners must communicate what happens next to ensure continuity of care with the patient's referring doctor.

When patients do not receive feedback, they describe it as a feeling of "falling through the cracks" and feeling "lost" (Olsen et al., 2014) as they do not have a clear understanding of their next steps, which reduces the continuity of care.

### Enhancing person-centred communication: Patient access and interpretation of results

When patients are given access to their ultrasound images and report, most immediately access them to discover the outcome. Without an explanation of the findings, unexpected news can have a detrimental impact on all involved.

Changing patient expectations have driven the widespread adoption of patient apps and digital portals, allowing patients to access results immediately. Person-centred care requires ultrasound practitioners to respond to this phenomenon by discussing the result ahead of the patient accessing it. Patients expect the results conversation to encompass the process of interpretation, limitations of the examination (if any), the report, and a next steps pathway to ensure continuity of care. When these expectations are not met, patients will likely read the report out of context and search for information using keywords within the report to gain understanding. Without explanations and the support of healthcare professionals, the patient's findings can lead to misinformation, panic, and a feeling of being "lost" with no support.

The experience below, shared by an accredited sonographer and radiographer, highlights the impact for all involved;

> *My closest friend contacted me about her brother's PET scan results which he had received via the radiology app the day after he had the scan and a few days before he was due to have a follow-up appointment with his referrer. The report stated there was metastatic spread due to prostate cancer throughout his entire spine, pelvis, and upper and lower limbs. She said he has read the report and is googling what it all means. I could not believe that this report would be sent out directly to the patient before his referrer was notified to give them time to hopefully bring forward his appointment to discuss the results and his prognosis and treatment. I contacted the radiology practice to complain, and they have since changed their policy, which delays sending the images and report for seven days. When they discover these findings, all reporting doctors must notify the referrer immediately.*
>
> Accredited sonographer and radiographer

Some ultrasound practitioners may have limited autonomy. As a result, they may experience frustration in their quest to adopt a person-centred communication approach. The most common limitations are created by

(1) work policies preventing any disclosure of results or (2) a referrer who has requested the results not be discussed with their patient. This limitation may be overcome with robust company policies that empower ultrasound practitioners to discuss findings transparently within their knowledge base, particularly in obstetrics. In the United Kingdom and Australia, guidelines have been developed to guide ultrasound practitioners on best practices in unexpected obstetric findings and communicating these findings with parents and referrers (Johnson et al., 2020). Similar guidelines are expected to be developed for additional ultrasound specialties.

### Patient complaints

Inevitably, when communicating, individual communication styles differ. As a result, the speaker's information may be misunderstood or misinterpreted by the receiver. When patients have an experience that is not aligned with their expectations, it may result in a complaint.

As patient expectations change, ultrasound practitioners can quickly recognise this shift and evolve by listening to patients' needs. Conversely, if expectations are not aligned with the patient's needs, the number of patient complaints arising from communication deficiencies will likely escalate, signalling a mismatch in expectations.

While complaints may not seem like a good outcome at face value, ultrasound practitioners should shift to a more modern mindset and see complaints as opportunities to learn and improve.

A patient complaint is a gift given freely by a patient to help us improve. When asked, most patients share that the motivation for complaining is that they do not want other patients to experience what they did. Keeping this in mind, ultrasound practitioners should recognise the stressful nature of lodging a complaint and thank the patient for bringing this to their attention. With patient feedback, ultrasound practitioners would know how or where to focus our improvement efforts (Thomson, 2018).

In addition, patients provide a unique perspective as they are sensitive to and able to recognise a range of problems, many of which may not be identified by internal incident reporting or quality and safety monitoring. Thus, patient complaints provide essential and additional information to ultrasound practitioners on improving patient experiences. Furthermore, analysing data on negative communication experiences strengthens the ability of ultrasound organisations to detect systematic problems in ultrasound care delivery.

### Online patient reviews

Like other industries, healthcare is subject to the same market forces as cafes, restaurants or accommodations. Until recently, healthcare had largely avoided the comparison-website movement, but the marketplace has responded to patients' expectations to understand more about their healthcare professionals (Thomson, 2018). Like it or not, ultrasound practitioners and practices are subject to online patient reviews, which increasingly influence patients' choices.

Many insights are gained by browsing ratings and review sites. The most common references relate to poor quality of communication and include lack of communication, quality of reports, and avoiding discussion of results following the examination.

The online review below clearly outlines a patient's dissatisfaction with their ultrasound experience, including the financial, time and emotional impacts.

---

★☆☆☆☆ 2 years ago

██████████ is by far the biggest joke!

I am a person with high stress and anxiety levels and was given an ultrasound by ████████████████████████████████████

████ and was told I had a double ruptured breast implants, I then proceeded to contact the doctors whom I received the surgery from and already booking appointments to see what my next move was and taken days off work.

I was ready to fly back to where I had them done.

I spent $1000 on specialist appointments and MRI scans and only to be told that they are intact no rupture and it was not true!

The fact that there are inexperienced staff members providing ultrasounds is beyond me and I can not believe that after me phoning in tears 3 days ago (part relief and part confused) and asking to talk with someone in charge to get to the bottom of their "diagnosis" I was told no one was in to talk to me and I gave them my name and number and absolutely no one phoned me back.

Very very poor customer service knowledge from ████████████ and this is the last time I will ever be visiting one of their branches along with recommending them to anyone.

If I could give less than one star I would!

---

The patient's ability to share online information continues beyond there. Patients also use social media to connect with others with similar medical conditions. This rapid spread of review-style websites, online communities and disruptive business models indicates that patients demand online information and services (Thomson, 2018). Ultrasound practitioners must recognise that a component of future success is their ability to provide person-centred care to improve ultrasound experiences and rapidly address shortcomings.

## Aspiring to person-centred communication mastery

To excel at person-centred communication requires several mindset shifts. First, ultrasound practitioners need to see beyond the clinical question to acknowledge the patient as a person. The second fundamental shift is to adopt person-centred affirming communication (verbal, non-verbal) to create a partnership between ultrasound practitioner and patient, where both parties share expectations of needs. The third critical person-centred care communication shift is discovering what matters most to the patient rather than being satisfied by answering the clinical question.

In **ultrasound practices**, one of the most significant challenges is translating discussion about improving person-centred communication to tangible actions to enable this new approach to be embedded as the new way of working. Leadership support is critical to implementing person-centred care, as it requires coaching to help people evolve beyond their current state.

**Patients** require education about their essential role in shared decision-making about ultrasound examinations and their role in managing conditions, including practical tools to help them understand their options and the consequences of their decisions. Patients should also receive the emotional support they need to express their values and preferences and be able to ask questions with censure from their ultrasound practitioners.

In turn, **ultrasound practitioners** need to evolve their patient relationship from paternalistic authority to become effective coaches, learning to ask, *"What matters most to you?"*. Recognition of the shift from a diagnostic-oriented way of thinking to a person-centred approach is critical to widespread adoption (Thomson, 2018).

Aspiring to ask patients, "What Matters to You (#WMTY)?" recognises ultrasound practitioners are part of an international person-centred care movement inspired by the article *Shared Decision-Making: The Pinnacle of Patient-Centered Care* (Barry et al., 2012). WMTY conversations help healthcare teams understand what is "most important" to patients, leading to better care partnerships and improved patient experience (What Matters To You website).

More than 15 countries are supporting the movement, including the National Health Service (NHS) in the United Kingdom, which hosts an annual '*What Matters To You?*' day that focuses on improving meaningful discussions with patients and their care partners as routine practice, with the promise of how asking a patient what matters to them can shape the future of healthcare. Focusing on the outcomes that matter most to patients, this movement aligns all healthcare professionals with the essential purpose of healthcare: improving patients' health outcomes.

## Summary

Aspiring to achieve high-performance person-centred communication requires a shift from a diagnostic-oriented focus to a person-centred approach, focussing on health outcomes that matter most to the patient. Person-centred care respects the needs and preferences of individuals receiving care and empowers patients to take responsibility for their care. For ultrasound practitioners, person-centred communication reduces stress and provides greater satisfaction as it aligns with their purpose of improving health. Aspiring to achieve person-centred communication sees ultrasound practitioners evolve conversations beyond exchanging information to support patients to actively participate and take responsibility for their care.

### Checklist: Considerations when adopting person-centred communication

The following checklist supports ultrasound practitioners adopting person-centred communication. It is not intended to replace existing ultrasound protocols or checklists.

### Before entering the ultrasound room

Before escorting the patient to the ultrasound room, ultrasound practitioners should:

- Introduce yourself by saying, "Hello, my name is …". This is the first step towards providing compassionate care.
- Ask the patient, "What would you like to be called" or "What would you like me to call you?". This question demonstrates respect for people, particularly indigenous, culturally diverse and LBGTQIA+ patients.
- Ask the patient if they have any support people accompanying them and how openly they would like you to communicate with/in front of them.
- If the patient has a support person, ask "Who have you brought in with you?" rather than assuming the relationship between the patient and the support person.

### In the ultrasound room

Complying with patient privacy requirements, ultrasound practitioners should:

- Determine the patient's understanding of the examination they have been referred for and what is expected of them.
- Ask the patient about their expectations for the ultrasound examination and '*What matters to you?*', which establishes a foundation of trust.
- Ask the patient about any needs with the environment; temperature, lights, music, fast access to the bathroom, and accessibility.
- Explain the ultrasound examination and process before asking the patient to change, including:

- How long it will take – this should include any additional time at the end to have images reviewed by the doctor, write a worksheet etc. Whatever is relevant to your workplace.
- What to expect – change into a gown, gel, latex gloves for the ultrasound provider (and therefore any potential allergies for the patient), lots of clicking of the keyboard to identify the position relative to the patient, change the position of the patient, breath holding, any discomfort or pain, area to be covered, barriers to protect the patient, e.g. towel or sheet.
- Confirm the patient's understanding before starting the examination.
- Ask the patient if they have any questions before commencing.
- Make every effort to explain the examination, so informed consent may be forthcoming.
- When patients provide verbal or explicit consent, be sure to document it.
- Ask the patient to get changed if required.
- Explain to the patient that they can speak up anytime if they are unsure, anxious or uncomfortable, and the examination can stop anytime.
- Ask the patient to share if something is uncomfortable during the scan, as this is a useful clinical sign for the ultrasound practitioner.
- If performing a Doppler ultrasound, let the patient know up front that there will be times they are asked not to speak due to interference.

Example 1: A thyroid ultrasound examination requires the neck to be exposed and examined. However, in the course of this procedure, the ultrasound professional's arm often brushes over the chest. Communicating this up front and providing a towel over the chest as a barrier ensures the patient does not need to be concerned by this action.

Example 2: A patient who is experiencing calf pain is referred for a Deep Vein Thrombosis (DVT) ultrasound examination. As the patient has calf pain, they wore shorts to make it easier for the examination. However, they were asked to change into a gown, and the examination began at the groin, which made the patient uncomfortable and confused. A simple explanation of the examination itself and why it was necessary to investigate the vessels at their starting point in the groin would have informed the patient and avoided a complaint.

**During the ultrasound examination**

Ultrasound practitioners should:

- check in with the patient, any questions or concerns, and pain points; and
- ask if the patient is comfortable.

Example: During a breast examination, the arm is positioned above the head and rotated, often leading to pain or discomfort. While this is necessary, the patient may need a short break from this position.

## After the ultrasound examination

Ultrasound practitioners should explain:

- how images will be sent, to whom they are sent, the timing of the report; and
- the importance of the patient contacting or returning to the referrer, especially if there is an immediate need.

## Vascular examinations

Example: During a vascular ultrasound, the patient was quite talkative. However, the ultrasound practitioner had communicated the need for silence during specific parts of the examination due to interference of the voice on the signal. When the patient was asked to be quiet, they were not offended and could hear the machine noise, as explained. The result was that the examination was completed efficiently due to the compliance and understanding of the patient.

## Paediatric patients

Paediatric examinations may be challenging as the patient may be anxious or not understand. Ultrasound practitioners should:

- involve the parents and carers;
- take time for the child to touch the transducer with a finger, touch the gel, put some gel on the parent's hand/skin and ensure the experience is positive; and
- offer a sticker or certificate at the end of the examination as a reward

Example: Paediatric hip examinations can be difficult when the baby is stressed or upset. Ideally, this exam could be timed to occur during or immediately after feeding when the baby is content. A great example of a sonographer and parent working together involved the baby being breastfed while the ultrasound was in progress. This was not easy for either but together, they completed the examination successfully without undue stress on everyone.

## References

ASUM. 2022. *The Australasian Society for Ultrasound in Medicine (ASUM). Parent-Centred Communication in Obstetric Ultrasound ASUM Guidelines.* www.asum. com.au/files/public/SoP/curver/Obs-Gynae/Parent-centred-communication-in-obstetric-ultrasound.pdf

Barry. et al., 2012. New England journal of medicine. Shared Decision-Making: The Pinnacle of Patient-Centered Care. Michael Barry and Susan Edgman-Levitan www.nejm.org/doi/full/10.1056/nejmp1109283

Boyle, F. M., Horey, D., Middleton, P. F., Flenady, V. 2019. Clinical practice guidelines for perinatal bereavement care – an overview. *Women and Birth*, 33(2), 107–110.

Haskard-Zolnierek, 2008. Haskard, K.B., Williams, S.L., DiMatteo, M.R. Rosenthal, R., White, M.K., & Goldstein, M.G. 2008. Physician and patient communication training in primary care: Effects on participation and satisfaction. *Health Psychology*, 27(5), 513–522. https://pubmed.ncbi.nlm.nih.gov/18823177/

#hellomynameis website. *A Campaign for More Compassionate Care*. www.hellomynameis.org.uk/

Johnson et al., 2020. Johnson, J, Dunning, A, Sattar, R, et al. Delivering unexpected news via obstetric ultrasound: A systematic review and meta-ethnographic synthesis of expectant parent and staff experiences. *Sonography*, 2020; 7, 61–77. https://onlinelibrary.wiley.com/doi/10.1002/sono.12213

Moore, L., Britten, N., Lydahl, D., Naldemirci, O., Elam M., Wolf A. 2017. Barriers and facilitators to the implementation of person-centred care in different healthcare contexts. *Scandinavian Journal of Caring Sciences*. 2017.

Morse, J. 2011. Hearing bad news. *Journal of Medical Humanities*, 32, 187–211. doi10.1007/s10912-011-9138-4); https://pubmed.ncbi.nlm.nih.gov/21503780/

*NSW Government*. 2022. Health. centre for genetics education.www.genetics.edu.au/SitePages/Pregnancy-support-resources.aspx

Olsen, 2014. Olsen, R. and Bialocerkowski, A. Interprofessional education in allied health:A systematic review. *Medical Education*, 48(3), 236–246. https://pubmed.ncbi.nlm.nih.gov/24528458/

Perinatal Society of Australia and New Zealand. 2018. Clinical practice guideline for care around stillbirth and neonatal death (3rd ed.). Retrieved from www.stillbirthcre.org.au/assets/Uploads/Respectful-and-Supportive-Perinatal-Bereavement-Care.pdf

Picker Institute. (1987). Principles of patient-centered care. Retrieved from http://pickerinstitute.org/about/picker-principles/

Thomas S; O'Loughlin K; Clarke J, 2020, 'Sonographers' communication in obstetrics: Challenges to their professional role and practice in Australia', *Australasian Journal of Ultrasound in Medicine*, 23, 129–139, http://dx.doi.org/10.1002/ajum.12184

Thomas S; O'Loughlin K; Clarke J, 2017. 'The 21st century sonographer: Role ambiguity in communicating an adverse outcome in obstetric ultrasound', *Cogent Medicine*, 4, 1373903–1373903. http://dx.doi.org/10.1080/2331205x.2017.1373903

Thomson, 2018. Thomson, Shelley. *Patients for Life*. 2018. https://patientsforlifebook.com.au/

What Matters To You website. n.d. *"What Matters to You (WMTY)?" Is an International Person-Centered Care Movement Inspired by a 2012 New England Journal of Medicine Article, Shared Decision-Making: The Pinnacle of Patient-Centered Care*, written by Michael Barry and Susan Edgman-Levitan. https://wmty.world/

# Section eight

# Person-centred care in nuclear medicine

# 20 Nuclear medicine

## Addressing the myths and improving the patient journey

*Aruna Jago-Brown*

### The department of nuclear medicine

Many, but not all, patients are referred to nuclear medicine (NM), also sometimes referred to as radionuclide imaging. After having previous procedures in radiography that have detected abnormalities patients can be already anxious. The title of the department is already a little intimidating for a lot of patients; we are often told after all, that it has the word 'nuclear' in it.

Whilst NM provides an outline of anatomy, the major strength of this modality is the great insight into physiological changes within body systems. Although the radiation doses can be high and the procedures longer compared to other radiology modalities, the information gained certainly justifies the dose to the patient and the length of the procedures. However, both the high doses, long scan times and the need to return some hours after injections in many cases can all be concerning for patients undergoing NM investigations. NM has a diagnostic role as well as a therapeutic role. Focusing on one of many definitions it is a: *'Medical specialty that uses radioactive tracers (radiopharmaceuticals) to assess bodily functions and to diagnose and treat disease. Diagnostic nuclear medicine relies heavily on imaging techniques that measure cellular function and physiology'* (NIH, 2023).

The clinical information is gained by administering a radioactive isotope or a radiopharmaceutical tracer into a patient via intravenous access, orally or by inhalation. Then either following the tracer's pathway to the area of interest (dynamic phase) or waiting for the tracer to travel to the site of interest and observe its accumulation in the area of interest (early phase) and then be taken up by a specific organ and assessing the distribution of the tracer within the organ (late phase). This initial or delayed distribution of tracer informs us of the physiological uptake process in the organ (Sharp et al., 2005). These imaging procedures can take from a few seconds up to an hour, sometimes needing the patient to return to the department after 4, 24, 48 or 72 hours after the initial administration of the tracer. It is essential that the patient returns for their scan(s) following their injection and staff must therefore ensure the patients understand the instructions and the importance of returning for their scans. The staff without fail will connect with the

DOI: 10.1201/9781003310143-28

patient, and much time is invested in preparing, reassuring and explaining the process to the patient.

## The patient journey

Many patients are facing a potentially serious diagnosis when coming to the NM department for a procedure, the majority of procedures in the department are linked to a diagnosis of cancer or other serious illnesses, resulting in many patients feeling vulnerable and afraid. It is important that the initial NM experience a patient has, is a positive and high quality one, where they feel listened to and welcomed. This experience can be continuously improved upon, using the patient experience improvement framework, and by including co-design methods that address patient, families and carers views that are inclusive of all patient population. Integrating policy guidance with the most frequent reasons the Care Quality Commission in the UK gives for rating acute trusts 'outstanding', is both important and essential. All imaging processes should include the clients' needs and points of view, including their feedback and feedforward improves future care allowing previous patients to co-create the service, ensuring it includes the patients' perspective (Health Education England, 2021). The department would wish for them to have experienced a bespoke, individual, high-quality care environment (Inova Fair Oaks Hospital, 2011). Taking into consideration their individual needs and greeting them with reasonable adjustments, accounting for their safety and risk management while in our care, that would make all the difference to their patient journey.

### Prior to arrival in the department

Initially, patients will need to be prepared prior to their procedure appointment time, they may be sent a letter, a text or receive a phone call with instructions on what to do prior to the scan. They may have a choice of which NM department they can attend, geographically within their catchment area, the opportunity to choose from days when their procedure is offered and a time and date for their attendance. Is important that the patient does not feel overloaded by information and can access a 'real person' if they have any questions or queries or worries about the information they received. The appointment invitations are constructed with patient accessibility in mind, friendly font type and size, colours, and language used. Some instructions are not well accepted, e.g., to stop consuming caffeine-containing foods and drinks, such as chocolate, coffee and tea. Clear justification from the staff for these restrictions is paramount to patient compliance, satisfaction and experience. Medication may also need to be stopped, often for a period of days, which can result in patient anxiety, side effects and stress (Abreu et al., 2017).

*The nuclear medicine department – the injection*

On arrival to the department, patients will encounter rules and regulations like no other department, which may involve for example using a specific toilet prior to their radioactive injection and a different one after. These are frequently known as the 'cold' and 'hot' toilets, it is courteous and responsible to explain the need for this to them. These communications must be done with understanding, patience, empathy and authority. Parents can feel alarmed when preparing their child for the procedure, to consent to their child receiving a radiopharmaceutical, i.e., a radioactive dose to their baby or young child, can be a hard decision to make. Being asked to handover a soiled radioactive nappy can also be unnerving if the reasoning is not made clear (Garcia et al., 2007). Furthermore, some adult patients can feel reluctant to consent to having a radioactive tracer administered into them. It is important not only to give the correct information, but the right amount of information, so a patient is not overwhelmed (Kaya et al., 2010), but also has sufficient information to make an informed decision. On occasions radiopharmaceutical doses will need to be calculated according to the weight of a bariatric or paediatric patient, gathering of data must be carried out in a professional and respectful manner. To gain a good quality image in NM, the radioactive tracer dose in the UK is dictated is by the Administration of Radioactive Substances Advisory Committee (ARSAC). ARSAC recommend and established doses, which are all set for a patient weighing 70 kg. Patients who weigh heavier, will have their dose adjusted to gain the optimum image. This will be calculated according to their weight in kilograms, something that must be gathered with sensitivity and respect. Similarly, patients under the age of 18 who weigh less than 70 kg have a lower radiopharmaceutical dose administered, in this case the child is weighed, and dose reduced accordingly. Communicating this clearly to the parents, carers and young patient has been reported to be reassuring and well received. The doses are calculated precisely with the patients' weight; once more, understanding and kindness is essential for that patient's positive journey (Uppot, 2018).

NM equipment is adapted with both Lead and Tungsten shielding, including sharps bins and syringe shields, these can appear unusual to patients and may need justifying and explanations to reassure them. Due to the nature of the radioactive tracers administered, radiation protection pre- and post-injection, local rules must be followed (Royal College of Radiologists, 2023; UK Government, 2017). A willingness to explain the unusual environment to a patient is essential to easing their fears, particularly relating to being administered with a radiopharmaceutical.

Pregnancy checks are essential for patients undergoing NM examinations and ensuring privacy for pregnancy checks, in a safe environment is essential. Patients from the age of 12 to 55 are asked to confirm their pregnancy or breast-feeding status (Abushouk et al., 2017; UK Government, 2017) and this must be inclusive for anyone with a uterus. If a child is being imaged, the staff in NM will need to be sensitive to the nature of these questions, as the patient in some cases may feel unable to answer this in front of a parent or carer.

In NM, each procedure can be divided into the administration of tracer, possible dynamic, early imaging and delayed imaging. When considering the administration of the radiopharmaceutical or radioisotope, room preparation must be precisely considered.

Oral administration – prior to oral administration of capsules and drinks radiation protection considerations should be put in place in case of spills or side effects, due to the nature of symptoms such as vomiting and diarrhoea e.g., SeHCAT Se $^{75}$ Tauroselcholic (Fani et al., 2018).

Intravenous administration – prior to cannulation, needle-phobic patients can be offered anaesthetic cream to the potential cannulation sites; good aseptic technique and aftercare prevents pain and bruising and possibilities of infection. The nervous patient should be offered the opportunity to lie down, and potential side effects should be explained prior to administration. An explanation of what is being administered may help the educated patient understand better why this particular tracer was chosen, e.g. Tc $^{99m}$ HDP Hydroxydiphosphonate for a bone scan or simply a tracer that 'goes' to your bones for others, it is important to read your audience (Adams and Banks, 2020).

Administration by inhalation – this administration method is simple but can be daunting to patients. A simple explanation 'pretend you are scuba diving', seal your lips around the mouthpiece, allowing the patient to place the nose piece on themselves and practice, is reassuring and gives the patient control in the administration of Tc $^{99m}$ Technegas or other inhaled isotopes. Certain manufacturers offer videos, which can be offered to the patient prior to their scan (Cyclomedica, 2023).

Following the injection, clear instructions are required so that the patient knows when to return to the department for their scan, what they can and cannot do and any specific instructions relating to hydration. It is essential that patients understand that they are radioactive and must avoid prolonged contact with children and pregnant people. These instructions should also be provided in writing prior to their appointment so they can make arrangements for children in their care as necessary.

For inpatients who are returning to the ward, it's essential that the staff caring for the patient are made aware of radiation safety aspects and how to manage visitors to the patient. Many hospitals use wrist bands on patients to signpost staff and carers to alerts regarding their patients. In certain NM departments it is common to use a banding system for all patients who are radioactive after a NM procedure. Depending on the half-life of the radioactive isotope, the patient or ward staff are instructed both verbally and in writing (on the wristband itself), when their wristband can be removed, in most cases the wrist band is yellow and worn in addition to an inpatient ID wrist band.

## The scan

There is a compromise between administering the lowest possible radioactive dose to the patient and the quality of the scan results. This negotiation is between the length of scan time, which can be on average 20–30 minutes and the dose administered to the patient, which must be As Low As Reasonably Practicable (ALARP) for the patients radiation protection (UK Government, 2017). The duration of a scan can be a long time for a person to remain still whether sat up or lying supine on an imaging table.

The Gamma cameras can be single, double or triple headed and may seem intimidating and large, and the detectors are positioned close to the patients during the scan, so prior warning of the scan set up is essential. Even with prior warning and guidance the proximity of the detectors can leave patients who are claustrophobic or in pain, feeling unable to tolerate the procedure. Positron Emission Tomography (PET) Computed Tomography (CT) scanners may feel like entering a tunnel for patients. Clear signage, guidance and effective, adaptable preparation and communication is essential to a successful scan outcome, both in NM and in PET. The tunnel opening of the gamma camera can be up to 1 m in diameter; however, it is still a large and overpowering piece of machinery in the imaging room Siemens Healthineers, 2023).

The removal of metal objects to eliminate artefacts on the images, can result in patients needing to wear a hospital gown, having interfaith gowns in a range of sizes is both inclusive and culturally sensitive, helping the patient feel comfortable and in turn relaxed enough to remain still during their scan (Hawkins, 1993; Syed et al., 2022; Frankel et al., 2021; Roseman et al., 2021). The initial contact with the patient is so essential, to their NM experience, availability of language interpreters prior to their visit or informing them to bring a DVD or audio piece of equipment of their choice to listen to music or watch a film during the scan can improve their involvement and overall experience. This is particularly relevant to children – if they are distracted, they may not fidget as much and feel more settled. Play specialists can be useful to work with children prior to the scan to ensure the experience is as positive as possible and to help both the patient and carers to understand what to expect.

Overall, a NM department's pleasant and inclusive physical environment must be considered for all patients, in particular for elderly and paediatric patients, patients who are neurodiverse, have dementia, or have special needs. Limitations can be both mental and physical; it is our responsibility to plan ahead and create the environment that addresses the patients' needs, through suitable lighting, décor, accessibility, cultural competence, noise levels, smells and temperature control. The NM scanning rooms are maintained colder than an average room temperature due to the crystals in the Gamma cameras, which are sensitive to temperature variations. Therefore, having clean blankets for patients is essential. Considering using blankets infused with essential oils, such as lavender, can be calming, but patients should be given a choice because some people dislike certain smells or may be allergic to some essential oils. Soft lighting and background music can assist the positive experience of the patient further and this is particularly useful prior to Positron Emission Tomography (PET) scans, where the mind must also be still (Lee et al., 2017).

Patients undergoing PET/CT scans will frequently be required to wait for up to an hour in an uptake room on their own and clear explanation of why this is required is important. Making the room an inviting place with calming artwork and prior notification of this waiting time is important so that patients can mentally prepare for this. Often neurological scans such as brain scans, for the diagnosis of Parkinsonian syndromes or dementia will require the patient's mind to be in a 'calm' state and their body to be supported, comfortable, and immobile. A darkened room can help achieve this, however once again, patients must be informed and choose to consent to be left in the darkened room for what may feel a long time to them. Furthermore, scanning rooms are kept cool for the imaging equipment, so it is helpful if the patient is dressed according to their temperature needs, as it can become hard to keep still if a person is too cold, and for PET scans this can also cause activation of brown fat, which increases uptake and can confound images.

Imaging – once the tracer is administered, the patient needs to lie still and breathe normally. Ensuring the patient is comfortable enables them to remain still; providing foam supports for arms and legs, clear instructions for positioning changes and reassurance you are in the room with them will ensure your patient feels confident. Communication in way of a countdown of the scan time and reassurance of the tracers exit from their body naturally also helps the patient (Pandit and Vinjamuri, 2014). There is a balance between talking to your patient to reassure them and chatting so much that they move during the scan; ensuring professionalism and boundaries are also important, but sharing commonalties such as a love for sport only further humanises us to the patient, making their NM experience agreeable.

If a diuretic is necessary to be administered during a NM scan, pre-warning of the process, the feeling during its administration and side effects must be communicated both verbally and non-verbally. Preparation of a commode or toilet immediately after the scan is considerate to patient dignity, as the effects of diuretics can be very intense (Matiti et al., 2007). Finally, monitoring a patient getting on and off the bed, the use of steps, monitoring dizziness and/or other symptoms and communicating this to the patient ensures that they are feel you are aware of their needs before they tell you (King and Hoppe, 2013).

## Positron Emission Tomography (PET)

PET imaging is in principle very different to NM; as it is classed as molecular imaging, however, the patient experience, fears, questions and concerns are similar to those in NM. PET is mainly used to diagnose malignancies, epilepsy, myocardial ischemia, inflammatory conditions, Alzheimer disease and the monitoring of therapy responses (Gambhir, 2002). Radioactive tracers such as [18]F Fluorodeoxyglucose (FDG) are administered to the patient intravenously and scanning must be carried out with the patient's consent and understanding that they must not talk (Ashraf and Goyal, 2022). The radioactive doses are on average higher than in NM, so staff may be wearing more monitoring devices that may have audible alarms, which may unsettle a patient. Taking the time to explain the differences between PET scans and NM scans is important as they are often used by patients who need both types of scans. The half-lives are much shorter in PET radioisotopes, hence the importance of setting up the patient correctly and ensuring that they can comply with the scanning procedure prior to administration of the radiopharmaceutical is crucial. In all PET scan procedures, the patient's mind and body must be quietened prior to the scan. Restriction on food consumption sometimes up to 6 hours prior to the appointment, avoidance of strenuous exercise for 24 hours before the scan, all contribute to good images (Bailey et al., 2005). As mentioned previously, the overall radioactive dose is higher and the time between injection and scan is lower, giving the patient less interaction within the department and in many cases, scans are performed in a mobile van. Accessibility, suitability for the scan, ability to remain still and quiet require clear instructions and that ability and cooperation from the patient (Ducharme et al., 2009) is essential for a successful scan. Effective communication and organisation skills in this modality is essential; PET combined with CT and MRI for image co-registration result in effective diagnosis and fundamentally can result in a satisfactory patient journey.

### The fond farewell

As the scan process ends, the aftercare instructions for the patient, families and carers can be extensive depending on the procedure the patient has undergone. Limiting time with children and pregnant persons after leaving the department, ensuring good hydration to aid 'flushing' the tracer out of their bodies through the urinary system, specific instructions for incontinent patients and in-patients can be detailed in various formats. If this were to be communicated by the same professional who greeted, treated and cared for the patient during their visit, trust and respect has already been established between them. This continuity is important to patient-centred care; it ensures the patient is exposed to only one person whom they can establish a short professional relationship with. Focusing on the patient journey, it demonstrates consistency in communication and ensures the patient forms a trusting bond with a specific professional within the department. Relating back to a bespoke and personal service, mentioned earlier (O'Brien et al., 2023), this can be particularly helpful for those patients who are neurodivergent or with dementia and should be implemented in the department when the resources are available (Holmér et al., 2023). Helpful leaflets, departmental contact methods and transparency is important at this stage, honesty about the limitations such as not using public transport or returning to work, if working with children is essential. The offer of further support and ensuring there is time after the procedure gives the patient a sense of not being rushed out of the department, as their worries and questions may make a difference to how they comply with the instructions they have received. Thanking the patient for allowing students to observe or assist with their procedure, apologising for any delays, or errors, ensuring carers and family members are waiting or transport is ordered result in a satisfactory experience, a positive holistic one in what started as a complex and unknown journey. Requesting patient feedback and in turn feed forward in form of a survey or focus group would help improve future patient-centred care, reassuring the patient of their value. Investigating further the needs of patients in particular who are sight impaired, hearing impaired, physically immobile, and those with other limitations (Nolan–Bryant and Lockwood, 2023; Aipanda et al., 2023). Reflecting and adapting practice when possible and safe, then informing the patients of the changes that are being implemented due to their suggestions and experiences. Taking into account patients' values, reassuring them that the department is sustainable of as environmentally friendly as possible, are all considerations that ensure patients feel respected and appreciated (Perkins, 2013; Sullivan et al., 2023).

Finally, the last and one of the most important considerations when contemplating patient centred care, is how the patient will receive their results. Many NM and PET/CT scans will require data processing, image co-registration with the CT scan and possibly a diagnosis and management plan only after a multidisciplinary team meeting. Ensuring the patient is informed of how they will receive the results, the length of time this will take and how they may need to return to the department in the future should be communicated with empathy and respect.

Lastly, post-scan care and regulations should be explained distinctly to both the patient and their family or carers. In some cases, if the patient is travelling abroad soon after the scan, they may need a letter declaring what type of scan they had, to explain the airport detector readings to the authorities.

Finally, NM and PET are modalities in radiology that require staff to be excellent, empathic and reassuring communicators; on occasion a patient will refuse the scan – this is their human right and it is our duty to respect their choices.

## References

Abreu, C., Grilo, A., Lucena, F. & Carolino, E. 2017. Oncological patient anxiety in imaging studies: The PET/CT example. *Journal of Cancer Education*, 32, 820–826.

Abushouk, A. I., Taheri, M. S., Pooransari, P., Mirbaha, S., Rouhipour, A. & Baratloo, A. 2017. Pregnancy screening before diagnostic radiography in emergency department; An educational review. *Emergency*, 5.

Adams, C. & Banks, K. 2020. *Bone Scan*. StatPearls website.

Aipanda, C., Karera, A., Kalondo, L. & Amkongo, M. 2023. Radiation risk-benefit communication during paediatric CT imaging: Experiences of radiographers at two public hospitals. *Radiography*, 29, 301–306.

Ashraf, M. A. & Goyal, A. 2022. *Fludeoxyglucose (18F). StatPearls [Internet].* StatPearls Publishing.

Bailey, D. L., Karp, J. S. & Surti, S. 2005. *Physics and Instrumentation in PET. Positron Emission Tomography: Basic Sciences.* Springer.

Cyclomedica Clinical Guidelines. 2023. Technegas clinical guidelines. www.cyclomed ica.eu/clinical/ (Accessed on 21.05.23)

Ducharme, J., Goertzen, A. L., Patterson, J. & Demeter, S. 2009. Practical aspects of 18F-FDG PET when receiving 18F-FDG from a distant supplier. *Journal of Nuclear Medicine Technology*, 37, 164–169.

Fani, B., Bertani, L., Paglianiti, I., Fantechi, L., De Bortoli, N., Costa, F., Volterrani, D., Marchi, S. & Bellini, M. 2018. *Pros and Cons of the SeHCAT Test in Bile Acid Diarrhea: A More Appropriate Use of an Old Nuclear Medicine Technique.* Gastroenterology Research and Practice, 2018.

Frankel, R., Peyser, A., Farner, K. & Rabin, J. M. 2021. Healing by leaps and gowns: A novel patient gowning system to the rescue. *Journal of Patient Experience*, 8, 23743735211033152.

Gambhir, S. S. 2002. Molecular imaging of cancer with positron emission tomography. *Nature Reviews Cancer*, 2, 683–693.

Garcia, P., Biggs, H., Hunter, J. & Pearson, T. 2007. Evaluation of a paediatric preparation booklet in reducing patient anxiety in nuclear medicine. *ANZ Nuclear Medicine*, 38, 27–29.

Hawkins, D. G. 1993. Transcultural hospital pastoral care. *Journal of Religion and Health*, 32, 291–298.

Health Education England. 2021 Patient experience improvement framework health education England (27.04.21) www.england.nhs.uk/publication/patient-experie nce-improvement-framework/

Holmér, S., Nedlund, A.-C., Thomas, K. & Krevers, B. 2023. How health care professionals handle limited resources in primary care–an interview study. *BMC Health Services Research*, 23, 1–12.

Inova Fair Oaks Hosptial. 2011. The patient experience journey https://cdn.ymaws. com/www.theberylinstitute.org/resource/resmgr/OTR-PDFs/Inova_Final.pdf (Accessed on 25.05.23)

Kaya, E., Ciftci, I., Demirel, R., Cigerci, Y. & Gecici, O. 2010. The effect of giving detailed information about intravenous radiopharmaceutical administration on the anxiety level of patients who request more information. *Annals of Nuclear Medicine*, 24, 67–76.

King, A. & Hoppe, R. B. 2013. 'Best practice' for patient-centered communication: a narrative review. *Journal of Graduate Medical Education*, 5, 385–393.

Lee, W.-L., Sung, H.-C., Liu, S.-H. & Chang, S.-M. 2017. Meditative music listening to reduce state anxiety in patients during the uptake phase before positron emission tomography (PET) scans. *The British Journal of Radiology*, 90, 20160466.

Matiti, M., Cotrel-Gibbons, E. & Teasdale, K. 2007. Promoting patient dignity in healthcare settings. *Nursing Standard (through 2013)*, 21, 46.

NIH. 2023. *Nuclear Medicine* [Online]. Available: www.nibib.nih.gov/science-educat ion/science-topics/nuclear-medicine [Accessed 14/07/2023].

Nolan–Bryant, A. & Lockwood, P. 2023. Diagnostic radiography students' perceptions towards communication with service users who are deaf or hearing impaired. *Radiography*, 29, 207–214.

O'brien, E. C., Ford, C. B., Sorenson, C., Jutkowitz, E., Shepherd-Banigan, M. & Van Houtven, C. 2023. Continuity of care (COC) and amyloid-β PET scan: The CARE-IDEAS study. *Alzheimer's Research & Therapy*, 15, 6.

Pandit, M. & Vinjamuri, S. 2014. Communication of radiation risk in nuclear medicine: Are we saying the right thing? *Indian Journal of Nuclear Medicine: IJNM: The Official Journal of the Society of Nuclear Medicine, India*, 29, 131.

Perkins, A. C. 2013. *Ethical, Green and Sustainable Nuclear Medicine*. Springer.

Roseman, A., Morton, L. & Kovacs, A. H. 2021. Health anxiety among adults with congenital heart disease. *Current Opinion in Cardiology*, 36, 98–104.

Royal College of Radiologists. 2023. IR(ME)R Implications for clinical practice in diagnostic imaging, interventional radiology and diagnostic nuclear medicine BFCR(20)3. www.rcr.ac.uk/system/files/publication/field_publication_files/bfcr21.pdf (Accessed on 25.05.23)

Sharp, P. F., Gemmell, H. G., Murray, A. D. & Sharp, P. F. 2005. *Practical Nuclear Medicine*, Springer.

Siemens Healthineers. 2023. Technical Specifications. www.siemens-healthineers.com/en-uk/molecular-imaging/spect-and-spect-ct/symbia-evo (Accessed on 25.05.23)

Sullivan, M. D., Vowles, K. E., Powelson, E. B., Patel, K. V. & Reid, M. C. 2023. Prioritizing Patient Values for Chronic Pain Care: A Path Out of the Pain Reduction Regime?. Oxford University Press UK.

Syed, S., Stilwell, P., Chevrier, J., Adair, C., Markle, G. & Rockwood, K. 2022. Comprehensive design considerations for a new hospital gown: A patient-oriented qualitative study. *Canadian Medical Association Open Access Journal*, 10, E1079–E1087.

UK Government. 2017. The ionising radiation (Medical Exposure) regulaitons 2017: Guidance. www.gov.uk/government/publications/ionising-radiation-medical-exposure-regulations-2017-guidance

UK Government. 2017. The ionising radiations regulations 2017. www.legislation.gov.uk/uksi/2018/121/contents/made

Uppot, R. N. 2018. Technical challenges of imaging & image-guided interventions in obese patients. *The British Journal of Radiology*, 91, 20170931.

# 21 The role of artificial intelligence in supporting person-centred care

*Geoff Currie, Eric Rohren and
K. Elizabeth Hawk*

## Artificial intelligence and person-centred care

The emergence of artificial intelligence (AI) in radiology and nuclear medicine represents the most significant disruptive technology since the developments of Roentgen, Becquerel and Curie. The term "artificial intelligence" is very broad, inclusive of an array of concepts from physical AI (e.g. robotic devices) to purely algorithmic processes, with boundless applications across industries. Indeed, AI as a general concept is so broad that its use as an unrefined term creates ambiguity and misunderstanding, and can even build a sense of mistrust in the user. To illustrate, AI is not new in radiology and nuclear medicine, with long-standing applications in business analytics, voice-to-text, computer added detection (CAD), automated software and predictive software. In radiology and nuclear medicine, the recent hype around AI relates to emerging applications of machine learning (ML) and deep learning (DL). Both the principles (1–4) and applications (2,5) of ML and DL have been previously detailed so only a brief clarification of concepts is provided here. Collectively, ML, DL and AI when applied in this manner in radiology and nuclear medicine might be better referred to as "intelligent imaging" and "engineered learning" to be more explicit about what technology is being referred to.

In radiology and nuclear medicine, ML is a sub-type of AI that uses algorithms to improve the prediction of outcomes. In this way, ML is consistent with the principles of precision medicine but harnessed appropriately, ML recognises the unique aspects of individuals to incorporate person-centred features into algorithmic decision making. While more abstract aspects of identity are difficult to distil quantitatively for inclusion in algorithms (e.g. aspects of cultural like faith and belief), a wide array of features not typically included in human decision making could be incorporated into algorithms (e.g. cultural aspects of diet, social aspects of disease risk) alongside image-based features (e.g. radiomics) and patient demographics (e.g. age and gender). A limitation of the use of ML in health is the use of historical data that may lack diversity and included bias (6). Nonetheless, ML provides a powerful tool to deliver person-centred care.

DOI: 10.1201/9781003310143-29

Patient-centred care and personalised medicine focus on the individual relationships with patients, their choices, beliefs, and preferences to improve overall health care quality one individual at a time. The differentiator of patient-centred and person-centred care lies in identifying the whole person rather than a patient. Person-centred care adapts the patient-centred approach based on individuals needs and expectations.

While the artificial neural network (ANN) is the backbone of newer applications of ML and AI, ANNs are not the only algorithms used. ML includes, without being limited to:

- Ensemble learning (e.g. random forest).
- Clustering (e.g. K-means).
- Classification and regression (e.g. decision trees, Naïve Bayes K-nearest neighbour [KNN] and support vector machine [SVM]).
- Reinforcement learning (e.g. deep Q-network).
- Neural networks (e.g. ANN, convolutional neural networks [CNNs] and generative adversarial networks [GANs]).

An ANN is an analysis algorithm composed of layers of connected nodes. The inputs may be radiomic features combined with other patient data that have been extracted from the image files or, if using a CNN, the input is the images itself (input tensor). A CNN is synonymous with DL but the terms have very different meanings. A CNN is a neural network that includes a convolution step to allow the image itself to be the input (rather than features extracted from the image in the form of data, e.g. SUV). DL refers to the depth of the neural network, the number of nodes and layers. More complex problems need a more complex neural network with outputs associated with the deep layers of the neural network, hence DL. ML with an ANN typically do not require deep neural networks unless the data is particularly dense while a CNN produces data density that requires a neural network of greater depth. As a result, DL is almost always associated with a CNN and seldom an ANN based ML solution (Figure 21.1). Clearly, over-fitting of simple data with DL or underfitting complex data with a shallow ANN produce poor outcomes.

ML applies algorithmic approaches to data analysis where complex, person-centred data is inputted and the neural network is trained (adjusting weightings on nodes to maximise the number of correct responses) against a grounded truth for individual cases. The trained network can then be used to identify unique features in individual patients to predict outcomes and optimise patient care. Not all ML applications apply directly to person-centred health care. There exist valuable AI applications in business analytics which could be used to optimise scheduling and staffing in a manner that benefits patients. In this way, such an application is patient-centred without being person-centred (Figure 21.2). ML can also identify from among a large number of patient-specific variables, the individual or aggregate weighted and

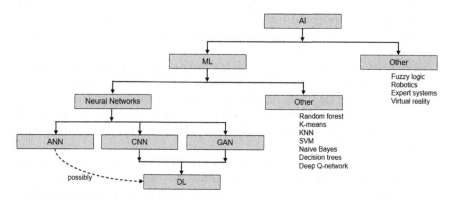

*Figure 21.1* Flow chart delineating types of AI and ML.

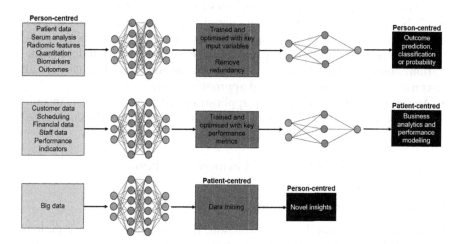

*Figure 21.2* Schematic representation of several roles of ML in radiology and nuclear medicine that provide patient-centred or person-centred care.

Source: Adapted with permission 7.

scaled combination (algorithm) of variables most predictive of the outcome of interest (Figure 21.2). Since this is focused on both the unique features of individual patients and the optimised outcome for that single patient, it is consistent with person-centred care. ML is also a powerful tool for data mining of big data (Figure 21.2). This would allow learning from large packets of data to be applied to and benefit the individual patient: person-centred care.

The most widely reported applications of AI in radiology and nuclear medicine are associated with DL and CNNs. These applications could be simplified into several novel approaches:

- Use of DL and CNN for object detection and segmentation. While these approaches could integrate into classification and interpretation applications, the key application is in auto-segmentation to allow radiomic feature extraction. Hand-crafted regions across multiple modalities and time series in complex tomographic imaging data is time consuming and potentially susceptible to inter- and intra-operator variability. The use of CNN based DL improves accuracy and creates efficiency. This allows improved patient outcomes and, in theory, more time for the health professionals to provide care. This is consistent with patient-centred care.
- Use of DL and CNN to extract abstract features directly from image tensors. These features are not intuitive to the human observer and represent unique features within images for individual patients that identify patient management needs. This is consistent with person-centred care.
- A number of CNN-based algorithms are designed for triaging urgent cases for priority reporting which contributes to patient-centred care.
- CNN approaches to attenuation correction, image co-registration, optimisation of patient positioning, optimisation of acquisition parameters, enhancement of image quality and noise reduction, radiation dose reduction, and detection of radiomic feature change over serial images are all consistent with a patient-centred approach to care.
- An important nuclear medicine application of CNNs and DL is in radiation dosimetry for radionuclide therapy which provides patient-centred and person-centred molecular theranostics.

These principles will be discussed in more detail below.

### Radiomics and person-centred precision medicine

Radiomics or radiomic feature extraction, like AI, is not new in radiology and nuclear medicine, with radiomic features being routinely incorporated into semantic reporting (Figure 21.3) consistent with traditional or system-centred care. The depth of insight associated with radiomics has evolved with improved imaging, data science and analysis capabilities. AI has played an important role in the increasing accuracy of radiomic features and the increasing depth. Radiomics relates to radiological images (including scintigraphy) but with the emergence of higher order imaging associated with computed tomography (CT), magnetic resonance imaging (MRI), single photon emission computed tomography (SPECT), positron emission tomography (PET) and hybrid imaging (SPECT/CT, PET/CT and PET/MRI), a more appropriate term would be "molecular radiomics" (7). Radiomic feature extraction is a widely integrated application of AI in radiology and nuclear medicine. Conventional radiomics tend to be associated with AI augmented segmentation. Hand-crafted regions across multiple modalities and time series in complex tomographic imaging data is time consuming and potentially

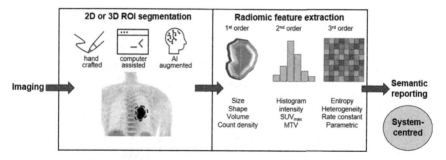

*Figure 21.3* The traditional semantic reporting incorporating radiomic features in a system-centred approach.

Source: Adapted with permission 7.

susceptible to inter- and intra-operator variability. The use of CNN-based DL improves accuracy and creates efficiency. This allows improved patient outcomes and, in theory, more time for the health professionals to provide care. This is consistent with patient-centred care (Figure 21.4). The use of CNNs and DL for efficient radiomic feature extraction directly from images (no segmentation) produces unique abstract radiomic features associated with the individual patient that are integrated into patient management; person-centred care (Figure 21.5).

Radiomics are used to improve decision making and use higher dimensional data to drive patient-centred or person-centred care. Specific radiomics can be classified into tiers or orders (*8–10*):

- First order or primary radiomics capture the spatial characteristics associated with signal intensity like size, shape, and count density.
- Second order or secondary radiomics capture the mathematical relationships between pixels or voxels like intensity histograms, mean intensity, maximum intensity and total intensity.
- Third order or tertiary radiomics examine texture (e.g. entropy and heterogeneity) or use kernels/filters to create layers of textural insights (e.g. wavelet transform and harmonisation).
- Fourth order or quaternary radiomics are more abstract features produced in the deeper layers of the CNN so they are referred to as deep radiomics.
- Multi-parametric radiomics integrate first through fourth order features extracted from multiple modalities (e.g. CT and MRI), hybrid imaging (e.g. SPECT/CT, PET/CT and PET/MRI), multiple phases from the same modality (e.g. MRI sequences, pre/post contrast, dual time-point PET, dynamic PET), or multi-tracer studies (e.g. $^{177}$Lu SPECT and $^{68}$Ga PET, or $^{18}$F FDG with $^{68}$Ga PSMA-617) to produce richer insights.

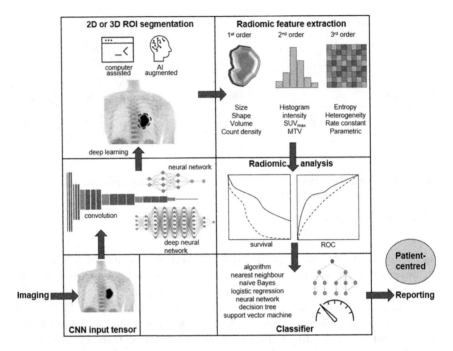

*Figure 21.4* Computer-aided or AI augmented segmentation and radiomic feature extration (first, second and third order) for reporting drives a patient-centred approach.

Source: Adapted with permission 7.

These radiomic features can be evaluated with tools or classifiers to generate predictive insights that drive patient-centred care and precision medicine. For example, intensity histogram scores from MRI and [18]F FDG PET/CT have been shown, in combination, to better describe tumour heterogeneity and, thus, predict tumour response to therapy (*11,12*). While fourth order radiomics have been used to predict survival in patients (*13*), a more important person-centred application of deep radiomics relates to molecular theranostics. The ability to predict response to therapy and radiation dosimetry allow improved patient outcomes by increasing target dose and decreasing collateral exposure. This is achieved by adjusting parameters based on person specific characteristics.

There remain a number of important challenges or barriers to not only application of radiomics in clincial decision making but also in ensuring radiomics provide patient-centred or person-centred care. Firstly, reliable and valid radiomics need ML and DL algorithms from highly curated, diverse data that is free from bias. Secondly, explainability of ML and DL algorithms is necessary to ensure a person-centred approach. This relates to lack of

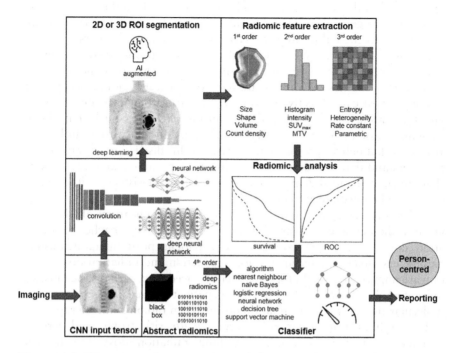

*Figure 21.5* AI augmented segmentation with first, second and third order radiomic feature extration integrated with fourth oprder deep radiomic feature extraction from the input tensor use DL informed reporting for person-centred care.

Source: Adapted with permission 7.

transparency of DL algorithms (black box or magic box) but the use of a grounded truth instead of a gold standard in algorithm training can undermine the value to individual patients. Thirdly, there is variability and lack of standardised guidelines for the way radiomics are integrated into reporting.

## Person-centred applications of artificial intelligence

It should be kept in mind that no single tool provides person-centred care. Person-centred care requires a holistic approach that is transdisciplinary and relies on skills and capabilities beyond the healthcare team (including the patient). For example, a theoretically appropriate person-centred care policy can be undermined by lack of cultural competence of the workforce. Conversely, a well-qualified workforce to deliver person-centred care can be shackled by lack of vertical consistency of policy. Moreover, external factors can conspire to undermine person-centred care. For example, lack of diversity and inclusivity drives policy making based on historical records containing social or cultural bias. While AI stands as a tool to address these social asymmetries from a general social and health benefit perspective, in parallel it also contributes to the culturally safe environment required for person-centred care (discussed below). While no single tool and attribute provides person-centred care, there are a raft of tools and attributes that are either necessary or provide necessities. AI and DL are such tools that may not always provide person-centred care directly but does deliver tools that provide the necessary foundations. Patient-centred tools in radiology and nuclear medicine, including those of AI, are essential foundations on which to scaffold other attributes that deliver person-centred care. In this discussion, it is important to understand that branding against patient- or person-centred care is not a claim of single-handed delivery but recognition of consistency with the philosophy and contributory to delivery.

Patient-centred care is an outcome of numerous applications of CNNs and DL in radiology and nuclear medicine. While CNN approaches to attenuation correction, image co-registration, patient positioning, acquisition parameters, image enhancement, noise reduction and radiation dose reduction are important contributors to patient-centred care, perhaps the most significant application of CNNs and DL is in patient-centred and person-centred molecular theranostics for radiation dosimetry in radionuclide therapy.

In accommodating a patient's specific health, risk profile, and personal preferences, it may be prudent to limit radiation dose and truncate standardised procedures. For hybrid imaging, this may involve the elimination of the CT (SPECT/CT and PET/CT) or replacement of CT (PET/MRI). Examples of particularly vulnerable patients for whom radiation dose reduction is vital include paediatric patients and women of childbearing age. AI can assist in overcoming the historical barriers to achieving some of these aims. For example, MRI avoids the use of ionizing radiation, but is limited when it comes to accurately estimating an attenuation map for SPECT or PET. A CNN approach has been developed to produce pseudo-CT scans which provide accurate attenuation maps superior to MRI approach and with only a 1–2% error compared to CT (*14–18*).

Radiation dose reduction strategies in radiology and nuclear medicine are not only consistent with regulatory and ethical principles, but also accommodate patient variations in understanding of and comfort with receiving radiation doses for medical purposes. The key is to improve or maintain image quality and diagnostic integrity while decreasing radiation doses administered. CNNs and DL can be used for dose reduction without image quality compromise. Doses as low as just 1% of the original dose have been used to produce diagnostic quality PET with the use of encoder-decoder GANs (*19–22*). These approaches can also be used for noise reduction in images (e.g. due to dose reduction or body habitus) (*23,24*).

In radiation therapy, CNNs and DL are used for auto-contouring and auto-planning to optimise radiation dosimetry and treatment outcomes. The emergence of theranostics and growth in radionuclide therapies demands similar tools. Optimisation of the dose for dosimetry that damages the target tissue and spares other tissues is essential in radionuclide therapy and is consistent with patient- and person-centred care. The CNN and DL can be used to predict dosimetry from the diagnostic scan (e.g. $^{68}$Ga DOTATATE image for $^{177}$Lu-lutate therapy). The CNN and DL can also continue to learn and evolve through input of the serial $^{177}$Lu images made possible by the two low abundance gamma emission that accompanies the low energy beta decay. For $^{177}$Lu-lutate (neuroendocrine tumour) and $^{177}$Lu-PSMA (prostate cancer) therapy, $^{177}$Lu allows gamma imaging can be used to assess tumour burden, non-target dose burden and optimisation of subsequent cycles of therapy. The accompanying $^{68}$Ga PET scan could be used to train a CNN to predict dosimetry metrics against the serial $^{177}$Lu scans. While challenging, this application of AI is strongly grounded in patient- and person-centred care (Figure 21.6).

GANs are an important emerging technology in personalised radiation dosimetry. In conventional neural networks there are challenges with identifying organic images outside the training set. A GAN uses two neural networks in tandem (Figure 21.7). The first neural network is a discriminator network that functions in the same way a typical CNN would for predicting an image classification and compare that to a grounded truth (*1,3,4*). The second neural network is the generator neural network which produces fake images and feeds those into the discriminator network in parallel to the real images. GANs could be used to produce accurate digital twins for radiation dosimetry. A digital twin is a digital model of the actual object. For example, a digital twin might provide an accurate model of a specific patient that incorporates all their individual characteristics. The digital twin could be treated with a variety of medications and doses to determine the best approach based on the patients' individual characteristics. This would optimise the actual medication and the dose while reducing adverse effects and potential toxicity. The person-centred approach to personalised medicine could be used to predict patient outcomes (target tissue and non-target tissues) by varying the therapeutic dose and dose frequency in the digital twin before implementing the best option in the actual patient. There is significant development to be undertaken in this space but it is an important next step in person-centred molecular theranostics.

### Artificial intelligence, social asymmetry and person-centred care

While AI has the potential to enhance the quality of healthcare provided, there is also potential associated with implementation of AI to deepen (or widen) healthcare inequities (*26,27*). Person-centred care is dependent on equity in access and opportunity for services. Social asymmetry threatens otherwise well-crafted person-centred strategies. AI offers a potentially valuable tool in addressing healthcare inequities, including radiology and nuclear medicine services.

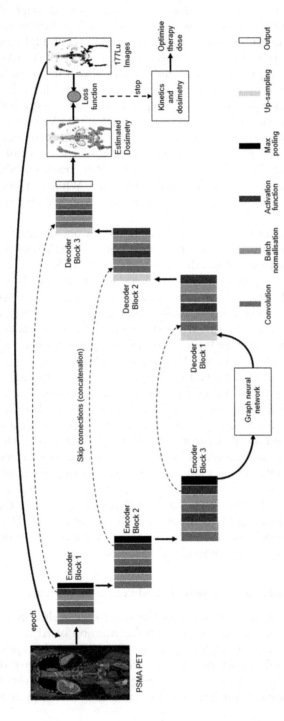

*Figure 21.6* Schematic representation of an encoder-decoder GNN in the U-net architecture that could be used to develop person-centred, dosimetry-based optimisation of therapy doses of 177Lu based on the 68Ga PET/CT.

Source: Adapted with permission 25.

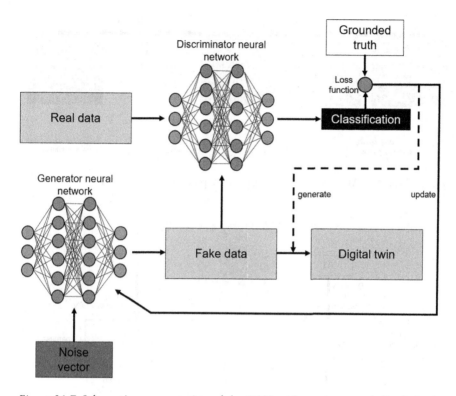

*Figure 21.7* Schematic representation of the GAN with generator and discriminator neural networks producing a digital twins.

First and foremost, AI algorithms implemented in radiology and nuclear medicine need to address potential bias and any resulting inequity (Figure 21.8). Lack of diversity in the teams of data scientists and clinicians developing the AI tools continues to be an issue but if resolved enables creative problem solving among diverse target patient populations (26). Lack of diversity in training data also creates bias and inequity in AI algorithms. To extend the advantage of AI to person-centred care, data sets need to reflect diversity.

AI in radiology and nuclear medicine has the potential to reduce social asymmetry by increasing accessibility to expertise and assets that may not otherwise be possible. For example, remote expert reporting using telemedicine platforms bring a high level of expertise to communities that would be otherwise unserviced. This could include expertise from developed health economies reaching into developing countries. Indeed, an image interpretation DL algorithm validated with good generalisability that is trained against expert radiologists or physicians can provide expert second reader systems into clinical departments with novice or developing support. The same

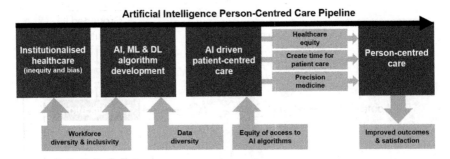

*Figure 21.8* The AI driven person-centred care pipeline.

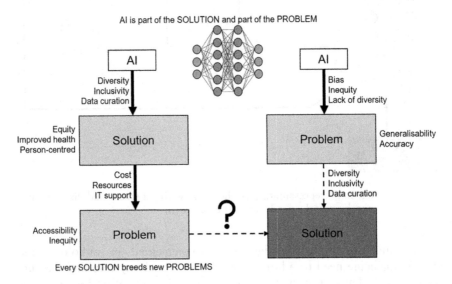

*Figure 21.9* AI is part of the solution for inequity and providing person-centred care but can also be part of the problem associated with bias, inequity and social asymmetry.

algorithms could be used as training tools for residents or for less experienced clinicians in developing health economies. While these types of AI could improve the delivery and quality of care, cost prohibitive for some communities could exacerbate inequity. There is significant debate about whether AI will improve or worsen social asymmetry (Figure 21.9). The application of AI in radiology and nuclear medicine occurs within heterogeneous global and national imaging environments. Will the costs and technology requirements of AI be prohibitive of homogenous access and application? The effectiveness of AI in addressing social asymmetry and health inequity will be challenged by this heterogeneity.

Lack of diversity in the AI workforce is often raised as a concern because it is a driver for perpetuating historical and institutional bias (*28*). This concern generally overlooks the bias in the absence of AI. If the risk of bias and inequity (and associated threat to person-centred care) is viewed through the lens of standard care in radiology and nuclear medicine, the problem becomes more obvious; AI learns the biases from humans. It is an opportunity, therefore, to use AI to address issues of diversity, bias and inequity so the foundations of person-centred care are supported. There are a number of risks associated with AI that could reinforce or perpetuate bias and inequity including, without being limited, to:

- AI algorithms targeting workflow, efficiency or cost containment could redirect resources away from minority socioeconomic, cultural or ethnic populations.
- Machine bias results from lack of diversity in training datasets and other sources of bias within training sets.
- Improvements in social equality and diversity continue to be undermined when AI development uses historical data that pre-dates social improvements.
- Improvements in social equality and diversity are also confounded by inherent bias associated with language that lacks neutrality typical in health (e.g. gender or ethnicity).
- The means to implement AI solutions to drive improved health outcomes, equity and person-centred care fail when data diversity undermines generalisability of AI algorithms for specific populations in need.
- Barriers to implementation of AI in clinical practice based on factors that influence social asymmetry (e.g. socioeconomic status of a community) are prohibitive of putting AI tools in the hands of those capable of applying them for the greater good.

AI can provide solutions to bias and inequity. AI algorithms can be designed to recognise hidden prejudice and correct the biases with careful curation of data. Availability and reliability of well curated data validated to be representative of broad groups, inclusive of individuals and populations who may be excluded, is essential in AI development. This may allow AI to inform policy that redirects radiology and nuclear medicine resources based on need rather than postcode. Recruitment of more diverse and inclusive teams enriches AI development and reduces bias and inequity. More diverse and inclusive data eliminates historical bias, improving accuracy of algorithms and positive health outcomes equitably. An important perspective is to ensure that AI algorithm development is undertaken from the context of the problem being solved. This requires problem selection and outcome definition that recognises inequity and the needs of individuals within a target population.

For example, AI could be a powerful tool to bring otherwise unavailable expertise into the rural or remote setting, through the collection of medical information which is used to provide real-time analytics, triage and remote evaluation by experts. Nonetheless, an AI solution built out of homogenous data may lack intrinsic objectivity and could exacerbate systemic or institutional bias. Compounding this may be variations in accessibility to AI solutions based on socioeconomic factors. Internet connectivity and bandwidth variability based on socioeconomic factors creates inequity of access and opportunity to AI solutions. Globally, 41% of people lack resources associated with information technologies such as a computer and internet access, while 16% live without electricity (29). This technology differential limits data diversity and inclusivity in training AI solutions but also decreases the capacity for AI solutions to be implemented among the most vulnerable populations (30,31). This digital divide is problematic in radiology and nuclear medicine because large and complex datasets used in AI for DL and CNNs are resource-intensive and expensive.

As stated previously, AI in radiology and nuclear medicine is far broader than image-based DL algorithms. AI systems supporting a person-centred approach to healthcare include scheduling systems, health informatics, hospital information systems, mobile and tele-medicine solutions, and cloud computing. The challenge for person-centred care is to ensure automation does not limit agility and flexibility. For example, AI driven scheduling systems need to include flexibility to accommodate specific needs of individual patients (e.g. cultural requirements).

An important application of AI to address inequity and provide person-centred care involves analytics and data mining of historical electronic medical records in an effort to identify and eliminate implicit bias; facilitating corrective policy and practice (32). Eliminating historical and institutional bias from policy and procedure is an important step toward equity and person-centred care but also potentially reduces healthcare costs and increases accessibility for those not generally having access to resources or expertise. The risk relates to creating an additional layer of inequity between communities serviced remotely by AI and those serviced in-person with the benefits of AI augmentation. The general premise 'electronic access is better than no access' remains true and sets a fluid benchmark that recognises a phase rather than final solution. Cost sensitivity is also likely to create greater inequity for some (e.g. developing economies, rural communities, those with complex morbidity and lower socioeconomic groups).

Perhaps the most significant impact of AI on person-centred care in radiology and nuclear medicine relates to the potential of AI to enhance the patient experience by enabling deeper, more meaningful interactions between the patient and their imaging team. AI augmented automation of menial, tedious and time-consuming tasks in medical imaging could allow more time for staff to interact with the patients. Moreover, that interaction has the *cognitive and emotional* space (33) to be more meaningful and impactful, and play a more direct role in person-centred care (*34*). While the patient experience is improved, such person-centred care also enhances staff satisfaction. AI is an important tool for driving a culture shift that emphasises the value of the interaction between the patient and the health professional. These tenets are in alignment with initiatives from many international radiology and nuclear medicine organisations advocating for greater engagement of the imaging team with patients and their caregivers, including "Imaging 3.0" by the American College of Radiology.

Developments in AI, ML and DL have also ignited discussion about ethical and legal challenges associated with the use of AI in radiology and nuclear medicine, and health more generally. In all of the AI discussion, one should remember that AI is just one of many tools, how it is used is a human choice. Indeed, how AI is developed and implemented, including data diversity and access, are human choices. That human choice determine whether AI addresses or causes inequity, and whether it supports a person-centred approach or undermines it. Adoption of transformative yet disruptive technologies such as AI brings with it professional responsibility for ethical, social and legal implementation. In radiology and nuclear medicine, the learning around ethical, social and legal issues in AI is occurring in parallel to innovation and implementation of AI (*35*). Under these conditions, the foundations of ethical practice (autonomy, beneficence, justice, and respect for knowledge) can be challenged (*36*) and this could significantly undermine person-centred care philosophy. There is also an ethical and social risk in radiology and nuclear medicine that decision making about patient management decision might become AI autonomous. This alters the accountability framework (*37*) which in turn may undermine patient-centred outcomes.

An interesting person-centred care consideration in AI is when a patient's personal preferences are to have care that is not AI augmented. A number of complex questions are raised:

- Can a patient refuse care associated with AI or where decision making has been influenced by AI (*38*)?
- Does the answer change if the refusal of AI influence threatens outcomes or the patient's health?
- Does a patient reserve the right to be informed when AI is part of the decision-making process?
- What constitutes AI if a patient must be notified?

There are no easy answers. Certainly a patient should reserve the right to have care without AI augmentation if that is their preference, consistent with person-centred care. Determining where the line is drawn is somewhat more difficult. It could simply eliminate AI augmentation in decision making but it is less clear where AI augmented segmentation of images would be positioned in the discussion. A simpler example might be a patient who refuses AI-based patient scheduling but human scheduling remains AI augmented. The AI process is not being changed, only the interface with the patient is different. Indeed, a human interface for AI augmentation of processes is consistent with a person-centred care environment. Consider AI augmented radiomic outputs for imaging that may form one of the tools used to influence decision making. The interpretation and reporting is human but the human may have varying influence from one or more AI tools that are not typically declared. Is it adequate, therefore, to only require patients to be informed when AI provides autonomous decision making (e.g. triage)? This debate circles back to issues of inequity. If AI extends a benefit to patient outcomes, there is not only the risk of harm associated with a patient refusing care associated with AI but there is also harm created by inequitable access to the AI.

## Artificial intelligence and healthcare team diversity

For some patients, inequity can be perceived as a lack of diversity in the healthcare team providing care. While the preceding discussion has focused on how implementation of AI tools can impact healthcare inequity in terms of the quality of care delivered, it is important to understand that the implementation of AI tools also has the ability to directly impact healthcare providers in different ways.

In the same manner that an AI tool built on homogeneous data has the ability to create bias and deepen healthcare inequities for underrepresented patient minority groups, AI tools designed by teams fundamentally lacking diversity risk producing underperforming tools that do not meet the needs associated with the diversity among healthcare providers.

There is great potential for new AI tools to contribute to healthcare provider burnout, as problems with adaptation and integration can occur. New AI tools, designed with creative diversity, have the ability to better support underrepresented minorities in the workforce and ultimately help to build a clinical environment that nurtures a more diverse healthcare provider team. Fostering a care environment, supported by AI tools designed by a diverse team, ultimately builds a place where a diverse team can thrive and practise patient-centred care that can meet the needs of our complex patient population.

## Summary

In radiology and nuclear medicine, the use of AI is expected to provide earlier and more accurate detection of disease, and provide enhanced personalised medicine and precision medicine (35,36) (Figure 21.10). While there are very real risks of AI augmenting inequity, there remains significant promise and excitement about the potential applications of AI in closing the inequity gap in healthcare and supporting person-centred approaches to care. AI risks

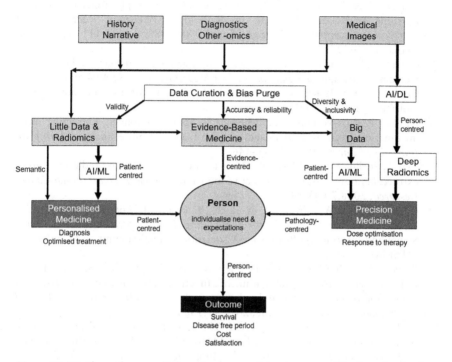

*Figure 21.10* Flow chart outlining the connectivity between AI, diversity in data, radiomics and person-centred care.

reinforcing societal, institutional and individual biases, but AI can also be used to correct bias and drive equity. The potential of AI to be harnessed to improve patient outcomes, inform optimised management based on patient specific characteristics, and the array of AI tools to individualise imaging and analysis approaches, is consistent with a person-centred approach to care. Efficiencies created by some AI applications can reduce costs and create the space required for more meaningful interactions with patients.

### References

1. Currie G, Hawk KE, Rohren E, Vial A, & Klein R. Machine learning and deep learning in medical imaging: Intelligent imaging, *Journal of Medical Imaging and Radiation Sciences*, 2019;50:477–487.
2. Currie G. Intelligent imaging: Artificial intelligence augmented nuclear medicine, *Journal of Nuclear Medicine Technology*, 2019;47:217–222.
3. Currie G. Intelligent imaging: Anatomy of machine learning and deep learning, *Journal of Nuclear Medicine Technology*, 2019;47:273–281.
4. Currie G, & Rohren E. Intelligent imaging in nuclear medicine: The principles of artificial intelligence, machine learning and deep learning, *Seminars in Nuclear Medicine*, 2021;51:102–111.

5. Currie, G & Rohren E 2022, Intelligent Imaging; Applications of Machine Learning and Deep Learning in Radiology, *Veterinary Radiology and Ultrasound*, accepted. 2022 Dec;63 Suppl 1:880–888. doi: 10.1111/vru.13144. PMID: 36514225

6. Currie G, Rohren E. Social asymmetry and artificial intelligence: The nuclear medicine landscape, *Seminars in Nuclear Medicine*, 2022; 52:498–503.

7. Currie, G, Hawk KE & Rohren E 2022, The transformation potential of molecular radiomics, *JMRS*, under review.

8. Lambin P, Rios-Velazquez E, Leijenaar R, et al. Radiomics: Extracting more information from medical images using advanced feature analysis. *European Journal of Cancer*, 2012;48:441–446.

9. Vial A, Stirling D, Field M et al. The role of deep learning and radiomic feature extraction in cancer-specific predictive modelling: a review. *Translational Cancer Research*, 2018;7:803–816.

10. Clifton H, Vial A, Stirling D, et al. 2019. *Using Machine Learning Applied to Radiomic Image Features for Segmenting Tumour Structures.* Asia-Pacific Signal and Information Processing Association Annual Summit and Conference (APSIPA).

11. Bowen SR, Yuh WTC, Hippe DS, Wu W, Partridge SC, Elias S, et al. Tumor radiomic heterogeneity: Multiparametric functional imaging to characterize variability and predict response following cervical cancer radiation therapy. *Journal of Magnetic Resonance Imaging*, 2018;47:1388–96.

12. Park C, Na KJ, Choi H, Ock CY, Ha S, Kim M, et al. Tumor immune profiles noninvasively estimated by FDG PET with deep learning correlate with immunotherapy response in lung adenocarcinoma. *Theranostics*, 2020;10:10838–48.

13. Arshad MA, Thornton A, Lu H, Tam H, Wallitt K, Rodgers N, et al. Discovery of pre-therapy 2-deoxy-2-18F-fluoro-D-glucose positron emission tomography-based radiomics classifiers of survival outcome in non-small-cell lung cancer patients. *European Journal of Nuclear Medicine and Molecular Imaging*, 2019;46:455–66.

14. Hwang et al 2018. Improving the accuracy of simultaneously reconstructed activity and attenuation maps using deep learning. *Journal of Nuclear Medicine*, 2018; 59:1624–1629.

15. Hwang D, Kang SK, Kim KY, Seo S, Paeng JC, Lee DS, Lee JS. Generation of PET attenuation map for whole-body time-of-flight [18]F-FDG PET/MRI using a deep neural network trained with simultaneously reconstructed activity and attenuation maps. *Journal of Nuclear Medicine*, 2019 Jan 25. pii: jnumed.118.219493. doi: 10.2967/jnumed.118.219493. [Epub ahead of print]

16. Torrado-Carvajal A, Vera-Olmos J, Izquierdo-Garcia D, Catalano OA, Morales MA, Margolin J, Soricelli A, Salvatore M, Malpica N, Catana C. Dixon-VIBE deep learning (divide) pseudo-ct synthesis for pelvis pet/mr attenuation correction. *Journal of Nuclear Medicine*, 2019 Mar;60(3):429–435. doi: 10.2967/jnumed.118.209288. Epub 2018 Aug 30.

17. Leynes A et al. Zero-Echo-Time and Dixon deep pseudo-ct (zedd ct): Direct generation of pseudo-ct images for pelvic pet/mri attenuation correction using deep convolutional neural networks with multiparametric mri. *Journal of Nuclear Medicine*, 2018; 59:852–858.

18. Liu H, Wu J, Lu W, Onofrey JA, Liu YH, Liu C. Noise reduction with cross-tracer and cross-protocol deep transfer learning for low-dose PET. Physics in Medicine and Biology, 2020 Sep 14;65(18):185006. doi: 10.1088/1361-6560/abae08. PMID: 32924973.

19. Xu J, Gong E, Pauly J, Zaharchuk G. 2017. *200x Low-Dose PET Reconstruction Using Deep Learning*. arXiv preprint arXiv:1712.04119.

20. Kaplan S, Zhu Y-M. Full-Dose PET Image Estimation from Low-Dose PET Image Using Deep Learning: a Pilot Study. *Journal of Digital Imaging*, 2019;32(5):773–778.

21. Ouyang J, Chen KT, Gong E, Pauly J, Zaharchuk G. Ultra-low-dose PET reconstruction using generative adversarial network with feature matching and task-specific perceptual loss. *Medical Physics*, 2019;46(8):3555–3564.

22. Lei Y, Dong X, Wang T, et al. Whole-body PET estimation from low count statistics using cycle consistent generative adversarial networks. *Physics in Medicine & Biology*, 2019;64(21):215017.

23. Cui JN, Gong K, Guo N, et al. PET image denoising using unsupervised deep learning. *European Journal of Nuclear Medicine and Molecular Imaging*, 2019;46(13):2780–2789.

24. Hashimoto F, Ote K, Tsukada H. Dynamic PET image denoising using deep convolutional neural network without training datasets. *Journal of Nuclear Medicine*, 2019;60:2.

25. Currie, G 2022, *Artificial Intelligence In Nuclear Medicine: A Primer for Scientists and Technologists, Snmmi Publishing*, Reston. (ISBN: 978-0-932004-56-7)

26. Currie G, Hawk KE. Ethical and legal challenges of artificial intelligence in nuclear medicine. *Seminars in Nuclear Medicine*, 2021;51:120–125.

27. Currie G, Hawk KE, Rohren E: Ethical Principles for the application of artificial intelligence (AI) in nuclear medicine and molecular imaging. *European Journal of Nuclear Medicine and Molecular Imaging*, 2020;47:748–752.

28. McCoy LG, Banja JD, Ghassemi M, et al. Ensuring machine learning for healthcare works for all. *BMJ Health Care Inform*, 2020;27:e100237.

29. Routley N: Mapped: The 1.2 billion people without access to electricity, www.visualcapitalist.com/mapped-billion-people-without-access-to-electricity/ 2019. accessed 16 March 2021.

30. Zhang X, Peˊ rez-Stable EJ, Bourne PE, et al. Big data science: Opportunities and challenges to address minority health and health disparities in the 21st century. Ethnicity & Disease, 2017;27:95–106.

31. Breen N, Berrigan D, Jackson JS, et al. Translational health disparities research in a data-rich world. *Health Equity*, 2019;3:588–600.

32. Ibrahim SA, Charlson ME, Neill DB. Big data analytics and the struggle for equity in health care: The promise and perils. *Health Equity*, 2020;4:99–101.

33. Recht M, Bryan RN. Artificial intelligence: Threat or boon to radiologists? *Journal of the American College of Radiology*, 2017 Nov;14(11):1476–1480.

34. Luh JY, Thompson RF, Lin S. Clinical documentation and patient care using artificial intelligence in radiation oncology. *Journal of the American College of Radiology*, 2019 Jun 22; 16(9):1343–1346.

35. Choy G, Khalilzadeh O, Michalski M et al. Current applications and future impact of machine learning in radiology. *Radiology*, 2018;288:318–328.

36. Ristevski B, Chen M. Big data analytics in medicine and healthcare. *Journal of Integrative Bioinformatics*, 2018. https://doi.org/10.1515/jib-2017-0030
37. Geis JR, Brady A, Wu C, et al. Ethics of artifical intelligence in radiology: Summary of the joint European and North Americam miltisociety statement. *Insights into Imaging*, 2019. https://doi.org/10.1186/s13244-019-0785-8
38. Jalal S, Nicolaou S, Parker W. Artificial intelligence, radiology, and the way forward. *Canadian Association of Radiologists Journal*, 2019;70:10–12.

# Person-centred care in forensic imaging

# 22 Forensic imaging

*Chrissie Eade and Charlotte Primeau*

## Introduction

This chapter should be read in conjunction with local/national Codes of Professional and Ethical Conduct and/or Scope of Practice and any other guidelines for radiographers relating to forensic topics.

The term forensic imaging is applied to diagnostic imaging when the acquired images form part of a forensic investigation and may be used in court proceedings within the criminal justice system. At times, it is not known from the onset of image acquisition that images may become part of an investigation, and only later become evidence within the criminal justice system and therefore imaging of a forensic nature. A typical example would be acquiring images of a child or an adult where the referral for diagnostic imaging is considered accidental in nature, and only later, either on the basis of the reporting of the images or a change in circumstance of the patient, such as death, that images are reviewed and may be requested as evidence.

With regard to patient care, the fundamental difference between radiography and radiology is that the radiographers undertaking the imaging have direct contact with the patient, carers and the wider multidisciplinary (MDT) team. The overseeing radiologist often never sees the patient face to face and is reliant predominantly on two things; the clinical information they have received from the referrer to justify and oversee the examination, and, the images that are acquired by the radiographers, to enable them to make a diagnosis. Patient care is therefore principally the radiographer's responsibility and it is important they have the skillset to undertake all the key challenges of looking after patients in forensic imaging.

Bear in mind, that in cases of forensic imaging and especially in cases of Suspected Physical Abuse (SPA), the radiology opinion may be used as legal evidence and on some occasions the radiologist themselves may be required to give an opinion in a court of law. It is therefore imperative that the procedure of image acquisition to image reporting, is optimised to assist in the correct outcomes for the patient.

DOI: 10.1201/9781003310143-31

This chapter, will first discuss forensic imaging of SPA of children, then shortly mention forensic imaging more generally in living patients, and lastly post-mortem forensic imaging is also included.

## Living patients

### Patient-focused care in preparation for forensic imaging

A large part of forensic radiography involves paediatrics and SPA. As part of the work-up for investigating cases of SPA in children, a skeletal survey maybe undertaken. This involves a series of approximately 30 images of the entire skeleton. Imaging includes an initial detailed survey and a limited follow up skeletal survey with specific views to maximise detection of occult fractures particularly in young infants. Studies indicate that additionally up to 12% of young siblings of children who have fractures, may also have inflicted injuries identified on a skeletal survey, with twins being particularly at risk (RCPCH, 2020). The radiographer has multiple streams of thought to contend with; trying to create a child-friendly environment, reassuring carers, and interacting with members of the multidisciplinary team, such as a paediatric nurse, that may accompany the child and take part in the examination. This is all going on whilst simultaneously thinking about the technical aspects of acquiring each image and minimising radiation exposure. Therefore, by its nature, the skeletal survey is a complex workflow opening opportunity for error. To safeguard against this, departments have developed their own protocols which are underpinned by guidance documents from the Royal College of Radiologists and Society of Radiographers (RCoR & SCoR, 2018).

### Communication

Not only does it require the technical knowledge of the radiographer to acquire the images, but also calls upon the softer interpersonal skills of good communication, active listening, emotional intelligence, and sensitivity. This is because both the child and the carers could present angry or distressed creating a volatile atmosphere in which to practise. It is important that the procedure is explained to carers and the reasoning for the referral. Written consent from the carers via the referral team needs to be obtained before the examination can proceed. It is at this point that excellent support of the carers is crucial so they can fully engage with the complex process. This will only be achieved through good communication. In the words of George Bernard Shaw, 'The single biggest problem in communication is the illusion that it has taken place'. The consent form detailing the investigation that both parties sign (carer and referrer) goes some way to negate any miscommunication, nonetheless, it is important that the radiographer(s) undertaking the skeletal survey never assume and confirm when the child and carers arrive in the department, that all parties are happy for the skeletal survey to proceed. It is often at this point that carers may voice any questions or anxieties they have, having had time to digest the information given to them on the ward. Conclusions drawn from meta-analyses of child protection processes are that often families feel misunderstood, blamed, mistrusted and threatened rather than helped, and the perspectives of parents, carers and children should be core during any investigation (Bywater et al., 2022).

It is essential that the radiographers practise active listening skills to understand and fully participate with the conversation. Aligning body language to reflect engagement is also imperative, using regular head nodding, maintaining eye contact, an empathic tone of voice and respect of personal space are all skills radiographers should display to optimise care of the patient and their carers (Walker, 2022). From personal experience, carers often say they feel judged during the process as one of the differential diagnoses of undertaking the skeletal survey is to investigate SPA and, as carers, they are within the pool of potential perpetrators.

### Patient-focused care during imaging acquisition

#### Immobilisation

It is essential that the radiographers involved in imaging SPA are trained in paediatric imaging and the heath professional accompanying the child should have experience in handling and comforting the child during immobilisation. In some hospitals, additional healthcare professionals such as the play specialist and/or the learning disability team may be able to assist as they use their understanding of child development and therapeutic play activities to support children to cope with the imaging experience (Malik et al., 2022). It is advisable beforehand to find out what the child likes playing with or listening to, as toys and/or music can be used to help keep the child content.

In further preparation to the forensic imaging process of a child with SPA, most departments will try the 'feed and wrap' technique with young babies, as they will generally fall asleep after a feed. Therefore, pre-scan preparation also includes liaising with carers or colleagues on the ward (if the baby is an in-patient) about keeping the baby awake and timing feeds to just before the imaging appointment to optimise the 'feed and wrap' method. Radiographers can then take advantage of this and undertake the imaging whilst the baby is asleep and still. This is particularly useful for the head scan where no clothes need to be removed and the baby can stay securely wrapped in the supine position for the procedure. If the baby is severely distressed and there is too much movement to acquire images, sedation may be given pre imaging, or on rare occasions, general anaesthesia. (GOSH, 2020)

Clothing of the child is also another factor to consider; as garments can cause artefacts on the image (Kim and Kweon, 2021), it is advisable to recommend the patient comes to the department in easily removable clothing so the child can be undressed with ease.

As many as 55% of young children who have been physically abused present with fractures, 18% of whom have multiple fractures. Therefore, diagnosing SPA in children is a broad and complex task often involving multiple body parts. Metaphyseal fractures, for example, which are a subtle fracture found on the end of long bones, are commonly described in fatal abuse, and most children with classic metaphyseal fractures have other associated injuries (RCPCH, 2020).

It is vital that handling and immobilisation of these children considers the trauma that the child may have been through, but also crucially acknowledges the importance of accurate positioning to identify such subtle injury.

Figure 22.1a shows a poor anterior-posterior (AP) view of a knee that does not display either the tibia or femur in the AP projection. Instead, the image is oblique with the leg externally rotated, the fibula is hidden behind the tibia, and therefore does not allow a metaphyseal fracture to be identified. The lateral view is adequate and shows a large knee joint effusion (elliptical outline) which is a soft tissue sign often associated with bony pathology. It is therefore vital that a second adequate AP view should be attempted to complement the lateral view for diagnosis.

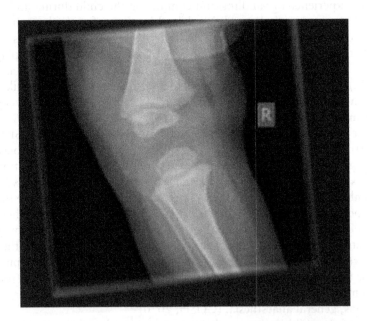

*Figure 22.1a* Oblique AP view of a knee of a 13-month-old child showing knee joint effusion.

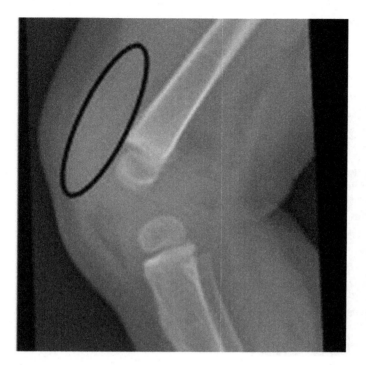

*Figure 22.1b* Lateral view of the same knee.

The complexity of the above diagnosis comes down to the initial positioning and care of the patient. In this scenario where the child has a painful leg, internal rotation of the hip to bring the knee into the AP projection is difficult. In such an instance, turning of the whole body toward the left side to counteract the external rotation of the right leg, will bring the knee into the AP projection. The knee can then be steadied on x-ray pads as necessary with immobilisation and close comfort of the child by the carer holding the patient (see Figure 22.2a for positioning and Figure 22.2b for resultant image). When the x-rays obtained in Figure 22.1a are compared to Figure 22.2b, it is apparent how the slight change in patient position provides a different projection, obtaining an optimum postion resulting in an improved image, allowing for a more accurate diagnosis.

This image displays the knee in the AP projection as all the metaphyses of the distal femur and proximal tibia are projected AP (medial proximal tibial metaphyseal region indicated with the white pointer), the proximal fibula is also visualised which was obscured by tibia on the previous image (Figure 22.1a).

The image in Figure 22.3 demonstrates patient positioning for a whole spine x-ray in the lateral projection. The carer holding the child has close

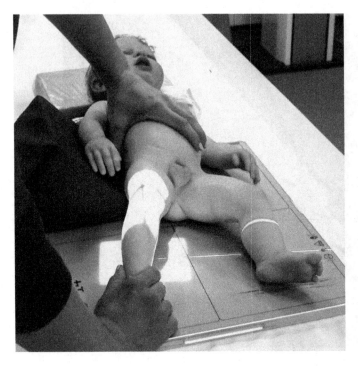

*Figure 22.2a*  Patient positioning using pads to rotate the child to the left and bring the right knee into the AP position whilst being held securely by the carer. *(A silicon manikin was used in the above image.)*

physical interaction, making eye to eye contact trying to attract the child's attention to reduce stress and movement. Clothing has been removed, and the child is completely lateral with both arms and legs securely held in front of the child to ensure no rotation.

### Gender, ethnicity, and health inequality

There is a tendency for boys to undergo varying types of childhood abuse more so than girls, with rates of physical abuse showing the greatest difference, which is 28% higher in boys (Holbrook and Hudziak, 2020). There are higher rates of physical abuse amongst black children compared to their white peers (Kim & Drake, 2018). It is also widely acknowledged that poverty is a contributory causal factor in childhood abuse and neglect, and has been for a number of decades (Leventhal et al., 1993; Bywaters et al., 2022).

These considerations should be given context when imaging children as to both acknowledge the evidence base and importantly not to show bias against certain characteristics which may jeopardise outcomes for the child.

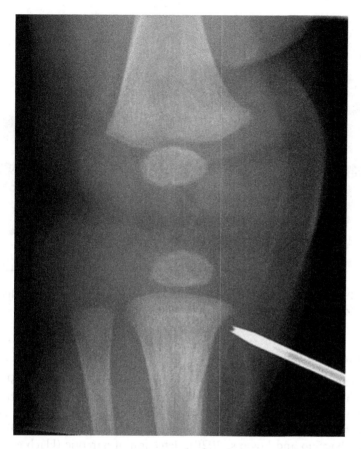

*Figure 22.2b* Resultant image from the postioning illustrated in 2a.

There are few studies, for example, on child protection concerns in families of average or higher levels of income and therefore caution should be employed to ensure practitioners do not stereotype against children and their carers (Bywater et al., 2022). Deprivation (the consequence of a lack of financial income) is strongly associated with lower educational attainment, and 5.1% of some local populations have no qualifications. Approximately 71,000 people live in the 20% most deprived areas in England (Ministries of Housing, Communities & Local Government, 2019). Therefore, factors such as being able to financially afford transport to hospital and the ability to read and understand hospital referral letters should be a consideration as to why some patients may not attend appointments. Other modes of communication such as telephone confirmation should be considered for some patients/carers, and the use of hospital services that can include advice on how to obtain financial support.

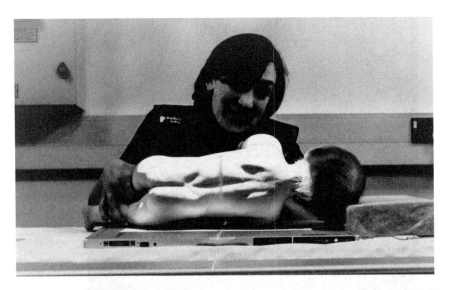

*Figure 22.3* Patient positioning for a whole lateral spine x-ray of a young child. *(A silicon manikin was used in the above image.)*

### Living patients (other than SPA)

While the majority of forensic imaging of living patients is for cases of SPA, there are other situations where imaging of living individuals become part of a forensic investigation. Such situations are much rarer but may involve documentation of violence for cases such as domestic (Kharoshah et al., 2020) or elder abuse (Baccino and Lossois, 2020), detection of narcotic (Flach et al., 2011; Pinto et al., 2014; Reginelli et al., 2020) and ballistic cases (Puentes et al., 2009; Reginelli et al., 2020). One such case was of a 22-year-old male asylum seeker who feared repatriation to Eritrea. He alleged that he had been shot and that returning to his country of origin would put him at risk of further harm. The United Kingdom Border Control when considering the eligibility of the claimant for entry into the UK, required a skull x-ray to corroborate the claimant's story. The skull x-rays (AP and Lateral) confirmed multiple metallic fragments consistent with bullet pellets, embedded within the forehead soft tissues and skull of the victim; this helped authorise the claimant's request to remain in the UK for humanitarian protection.

Imaging for age assessment for asylum seekers, which is only undertaken in some countries, should only be carried out where there is significant reason to doubt that the applicant is a child (Schmeling and Rudolf, 2022).

Section 17 of the Children Act 1989 enforces a duty on local authorities to safeguard and promote the welfare of children within their area who are in need. Unaccompanied Children Seeking Asylum (UASC) who have no responsible adult to care for them or are separated, are by definition 'in need'. They account for 7% of total asylum applications (approximately 5,152 children in 2022). On occasion, these children may present for a chest x-ray to rule out infectious diseases such as tuberculosis (Home Office, 2022). In countries, such as the UK, where ionised imaging for purposes of age estimation is not allowed, images taken for diagnostic purposes, such as dental x-rays for, can be used. Additional issues are likely to arise in relation to this cohort of children, as they may also be victims of trafficking or modern slavery. Some of these children may have witnessed or been subject to horrendous acts of violence. Assessing the child's needs is only possible if their background events and culture are considered, including the societal shock of arrival in the UK. It is vital that the services of an interpreter are utilised in the patient's first language and that care is taken to confirm that the interpreter knows the correct dialect. Particular sensitivities which may be present include: fears of repatriation; anxiety raised by yet another professional asking similar questions to ones previously and potential lack of understanding of the need for the x-ray; previous experience of being asked questions under threat or torture, or seeing that happen to someone else. Past trauma can impact upon the child's mental and physical health and needs to be considered when asking the patient to undress for the x-ray for example. The journey into the UK as well as the previous living situation may have been the source of trauma; the surprise of arrival – the strange culture, and language may cause shock and uncertainty, and may affect behaviour and presentation; the patient may have also been subject to frequent changes of address or location within the UK and may be living with the fear of sudden further unexplained moves. Radiographers need to be mindful of this when asking seemingly trivial questions to confirm identity, such as name, date of birth and address.

## Deceased patients

Some of the concepts described for live patients also applies to deceased individuals who may need to undergo imaging to establish a cause of death. Considerations of artefacts and positioning remain the same. Strict infection control measures should be practised when handling a deceased body. Dependant on the circumstances of death, stages of decomposition and underlying disease processes, both the radiographer and equipment may have increased risk of contamination from blood or other bodily fluids.

Rigor mortis is a post-mortem change causing stiffening of the body due to chemical changes in muscle myofibrils (Madea et al., 2022). Consequently, imaging of deceased patients with rigor mortis can be challenging. Adapted technique to work around abnormal body position is needed to optimise the resultant images. Working closely with mortuary colleagues, who are experienced at handling deceased patients, is often of benefit.

Forensic imaging of the deceased may be for case work such as suspected abuse of vulnerable (e.g., children, elderly, and those with disabilities, including reduced mental capacity), sharp or blunt force trauma (e.g., homicide cases), road traffic accidents, suicide, medical negligence, custodial injury, industrial injuries, mass fatality incidents, atrocity/genocide or for identification purposes of unknown remains (Doyle et al., 2020).

Forensic imaging of deceased individuals is less regulated than living individuals as there is no longer a need to keep radiation doses 'as low as reasonably practicable' (ALARP) or 'as low as reasonably achievable' (ALARA) as radiation exposure dose to a deceased is not of concern. Therefore, imaging work involving ionising radiation of deceased individuals does not requiring the radiographer to be a fully trained and licenced radiographer. Hence, imaging of deceased individuals can be conducted by employees such as mortuary technicians/anatomical pathology technicians (APTs), academics or researchers, although these should be trained in the regulations concerning ionising regulations relevant to the country of practice to understand the minimum legal duties, requirements and standards required to protect workers, the public and others that may be exposed, against the dangers arsing with working with ionising radiation (HSE, 2018).

Unlimited images and the utilisation of any imaging equipment available (such as a Faxitron or micro-CT) is therefore feasible when imaging deceased. Conversely, there is also less guidance available for conducting forensic imaging of deceased individuals (Primeau et al., 2020). Currently, there are no international established protocols for forensic post-mortem imaging, although guidance documents have been published, though not all guidance documents include specifically the forensic aspect of post-mortem imaging (Ruder et al., 2014; Lucraft, 2019; Offiah and Dean, 2016; Schmelding and Arthurs, 2022; RCT/RCPath, 2021; Doyle et al., 2020). Imaging requests involving ionising radiation of deceased individuals, unlike those of living individuals, do not fall under IR(ME)R 2017; however, the request does need to be from a Legally Designated Responsible Person (e.g., coroner, medical examiner, procurator fiscal, police official) as an authorised referrer (Doyle et al., 2020). Further, it should be mentioned that forensic imaging of a deceased individual upon the request of an authorise referrer as part of a forensic autopsy does not require consent from family of the deceased (Peres, 2017). However, keeping relatives informed of the post-mortem events is evidence of good, open, and honest practice.

Post-mortem imaging may be acquired at the hospital where the individual died or was initially transferred to or may be acquired at the facility where the post-mortem examination is conducted. If the post-mortem examination is conducted where there are no imaging facilities, imaging may be acquired at a third location, or additional imaging may be acquired at a specialised imaging facility. An example for the UK, is the Forensic Centre for Digital Scanning and 3D Imaging at the University of Warwick or the University of Leicester, that provides specialised high resolution micro-CT imaging of samples from Home Office post-mortem examinations on the request of the police.

In later years forensic imaging has seen an increasing use of a wider set of modalities such as specimen radiography system (also known as an x-ray cabinet), 2D radiographs, CT, MRI, ventilated angiography and more recently also micro-CT (Grabherr et al., 2016; Grabherr et al., 2017; Baier, 2019; Rutty et al., 2013; Franhetti et al., 2022; Guareschi, 2021). However, current post-mortem imaging modalities for forensic imaging are highly dependent on availability and access to facilities (Primeau et al., 2020; Ruder et al., 2014).

### Post-imaging considerations in forensic imaging of deceased patients

Even when imaging a deceased individual, patient-centred care is still of the utmost importance for the justice of the individual itself, but also for others who may also be endangered as potential victims from the same source. Forensic imaging can lead to perpetrators being apprehended, thus avoiding further cases (e.g., knife crime), or avoiding multiple cases of elder abuse or otherwise valuable individuals in care or at home.

Particularly with deceased children, patient-centred care is still vital, as there are other important concerns of a child who has died from suspected child abuse. The foremost concern is that other siblings may remain in the care of those responsible of fatal abuse, be it from carers, parents, or siblings. Child abuse court cases can take many years before reaching a final verdict, and hence siblings of a deceased child remain in immediate danger if not removed from the home. Or those in charge of the deceased child may have more children, either as parents, carers or adopters, and hence not diagnosing abuse through forensic imaging may endanger other children. Conversely, if suspected child abuse was incorrectly diagnosed, there is risk of removing siblings or other cared for children from a safe home and unnecessarily entering these children into the care system, placing undo strain on an overstretched system and risking traumatising a child and their immediate family. Lastly, patient-centred care in forensic imaging involving deceased individuals is also protecting the dignity of the individual through maintaining a respectful approach around the deceased person.

### Mental health considerations in forensic imaging

It is essential to appreciate that the handling and appearance of the deceased patient may be distressing to the radiographers undertaking forensic imaging. Hence, radiographers need to be mentally prepared for undertaking such tasks. Further, it is important to understand that there are different ways that radiographers may process the psychological aspects, as some professionals may need to articulate their feelings to colleagues while others preferer to process in private. Additionally, what is needed, is to have support mechanisms in place to reduce psychological harm and develop professional resilience and that radiographers undertaking forensic imaging know their welfare is of concern at their place of work and what help is available to them, if needed.

Consideration of the deceased patient and those affected by the death of the individual, should be a priority for the radiographer(s) who handle the body to ensure dignity and respect are demonstrated at all times.

## Conclusion

Forensic imaging is a broad and complex task that involves care from the radiographers of both the patient, whether they be deceased or living, and of the wider support group/relatives of the patient. Accurate imaging is of paramount importance as it may be used not only for diagnosis but as evidence in a court of law. Optimum imaging can only be obtained with full co-operation from the patient and from those that care for/have responsibility for the patient. In cases of deceased patients, adaptive techniques for abnormal body positions is a necessary skill for optimum image quality. It is important that the differences in ethnicity, gender, religious, social and ethical circumstances of the patient are taken into account, when treatment and imaging of the patient is undertaken.

## References

Baccino, E. and Lossois, M. (2020). Imaging and elder abuses. In: Lo Re, G. Argo, A., Midiri, M. and Cattaneo., C. (eds.) *Radiology in Forensic Medicine*. Springer, pp. 145–156.

Baier, V. Mangham, C., Warnett, J.M., Payne, M., Painter, M. and Williams, M. (2019). Using histology to evaluate micro-CT findings of trauma in three post-mortem samples – First steps towards method validation. *Forensic Science International*, 297, pp. 27–34.

Bywaters, P., Skinner, G., Cooper, A., Kennedy, E. and Malik, A. (2022). *Final Report: The Relationship Between Poverty and Child Abuse and Neglect: New Evidence*. March 2022. Available at: www.nuffieldfoundation.org/wp-content/uplo ads/2022/03/Full-report-relationship-between-poverty-child-abuse-and-neglect.pdf (Accessed: 9 January 2023).

Doyle, E., Hunter, P., Viner, M.D., Kroll, J.J.K., Pedersen, C.C. and Gerrard, C.Y. (2020). IAFR Guidelines for best practice: Principles for radiographers and imaging practitioners providing forensic imaging services. *Forensic Imaging*, 22:200400.

Flach, P., Ross, S., and Thali, M. (2011). Forensic and clinical usage of X-rays in body packing. In: Thali, M.J. (ed.) *Brogdons Forensic Radiology*. Boca Raton: CRC Press, pp. 311–34.

Franchetti, G., Viel, G., Fais, P., Fichera, G., Cecchin, D., Cecchetto, G. and Giraudo, C. (2022). Forensic applications of micro-computed tomography: A systematic review. *Clinical and Translational Imaging*, 10, pp. 597–610.

GOSH. (2020). Great ormond street hospital. Your child is having an MRI scan using ' feed and wrap' technique. Available at: www.gosh.nhs.uk/conditions-and-treatme nts/procedures-and-treatments/your-child-having-mri-scan-using-feed-and-wrap-technique (Accessed 12 December 2022).

Grabherr, S., Baumann, P.m Minoiu, C., Fahrni, S. and Mangin, P. (2016). Post-mortem imaging in forensic investigations: Current utility, limitations, and ongoing developments. *Research and Reports in Forensic Medical Science*, 6, pp. 25–37.

Grabherr, S., Egger, C., Vilarino, R., Campana, L., Jotterand, M. and Dedouit, F. (2017). Modern post-mortem imaging: An update on recent developments. *Forensic Science Research*, 2(2), pp. 52–64.

Guareschi, E.E. (2021). Postmortem imaging in forensic cases. In: Guaresschi, E.E. (ed) *Forensic Pathology Case Study*. Academic Press, pp. 79–93.

Holbrook, HM., and Hudziak, J.J. (2020). 'Risk factors that predict longitudinal patterns of substantiated and unsubstantiated maltreatment reports', *Child Abuse & Neglect*, 99. doi: 10.1016/j.chiabu.2019.104279

Home Office. (2022). National satistics. How many people do we grant protection to? Availible at: www.gov.uk/government/statistics/immigration-statistics-year-end ing-september-2022/how-many-people-do-we-grant-protection-to (Accessed 10 January).

HSE. (2018). Health and safety executive. Work with ionising radiation. Ionising Radiations Regulations 2017. Approved Code of Practice and guidance. Available at: www.hse.gov.uk/pubns/books/l121.htm (Accessed 5 December 2022).

Kharoshah, M., Gaballah, M., Bamousa, M., Alsowayigh, K. (2020). Domestic vio-lence. In: Lo Re, G. Argo, A., Midiri, M. and Cattaneo, C. *Radiology in Forensic Medicine*. Springer. pp. 133–144.

Kim, H., and Drake, B. (2018). 'Child maltreatment risk as a function of poverty and race/ethnicity in the USA', *International Journal of Epidemiology*, 47(3), pp. 780–787. doi: 10.1093/ije/dyx280

Kim, M.S. and Kweon, D.C (2021). Comparison of the conventional reusable cotton and disposable nonwoven fabric patient gowns in digital chest radiography, *Radiation Effects and Defects in Solids*, 176(3–4), pp. 358–367.

Leventhal, J.M., Thomas, S.A., Rosenfield N.S. and Markowitz, R.I. (1993). Fractures in young children. Distinguishing child abuse from unintentional injuries. *The American Journal of Diseases of Children*, 147(1), pp. 87–92.

Lucraft, M. (2019). Guidance No. 32. Post-mortem examinations including second post-mortem examinations. Guidance document published by the Chief Coroner of England and Wales, 23 September 2019. Available at: www.coronersociety.org.uk/_img/pics/pdf_1569321731-72.pdf (Accessed 12 December 2022).

Madea, B., Henßge, C., Reibe, S. and Tsokos, M. (2022). Postmortem changes and time since death. In: Madea, B. *Handbook of Forensic Medicine*. Second edition. Wiley, John Wiley & Sons ltd. pp. 91–133.

Malik, M.M.U.D., Majumdar, G., Kaushik, U. and Gupta, S. (2022). Pediatric imaging: A challenging task in radiography. *Current Pediatric Research*, 26(9), pp. 1627–1632.

Ministry of Housing, Communities & Local Government. (2019). Th English indices of deprivation 2019 (IoD2019). Available at: https://assets.publishing.service.gov.uk/government/uploads/system/uploads/attachment_data/file/835115/IoD2 019_Statistical_Release.pdf (Accessed 12 January 2023).

Offiah, C.E. and Dean, J. (2016). Post-mortem CT and MRI: Appropriate post-mortem imaging apperences and changes related to cardiopulmonary resuscitation. *The British Journal of Radiologyand*, 89(1058), 20150851.

Peres, L.C. (2017) Post-mortem examination in the United Kingdom: Present and future. [editorial]. *Autops and Case Reports*, 7(2), 1–3. Available at: www.ncbi. nlm.nih.gov/pmc/articles/PMC5507562/pdf/autopsy-07-02001.pdf (Accessed 21 December 2022).

Pinto, A., Reginelli, A., Pinto, F., Sica, G., Scaglione, M., Berger, F.H., Romano, L., Brunese, L. (2014). Radiological and practical aspects of body packing. *The British Journal of Radiology*, 87(1036), 20130500.

Primeau, C., Marttinen, F., Pedersen, C.C.E. (2020). The status of forensic radiography in the Nordic countries: Results from the 2020 IAFR questionnaire. *Forensic Imaging*, 29, 200502.

Puentes, K., Taveira, F., Madureira, A.J., Santos, A. and Magalhães T. (2009). Three-dimensional reconstitution of bullet trajectory in gunshot wounds: A case report. *Journal of Forensic Legal Medicine*, 16, 407–410.

RCoR & SCoR. (2018). Royal college of radiologists & society and college of radiographers. The radiological investigation of suspected physical abuse in children. Revised 1st edition. November 2018. Available at: www.rcr.ac.uk/system/files/publication/field_publication_files/bfcr174_suspected_physical_abuse.pdf (Accessed: 19 December 2022).

RCPCH. (2020). Royal college of paediatrics and child health. Child protection evidence. Systematic review on Fractures. September 2020. Available at: https://chil dprotection.rcpch.ac.uk/child-protection-evidence/fractures-systematic-review/ (Accessed: 5 January 2023).

RCT/RCPath. (2021). RCR/RCPath statement on standard for medico-legal post-mortem cross-sectional imaging in adults. Published in 2012, reviewed in 2021. Available at: www.rcr.ac.uk/publication/rcrrcpath-statement-standards-medico-legal-post-mortem-cross-sectional-imaging-adults (Accessed 21 November).

Reginelli, A., Russo, A., Micheletti, E., Picascia, R., Pinto, A., Giovine, S., Cappabianca, S. and Grassi, R. (2020). Imaging techniques for forensic radiology in living individuals. In: Lo Re, G. Argo, A., Midiri, M. and Cattaneo., C. *Radiology in Forensic Medicine*. Springer. pp. 19–28.

Ruder, T.D., Thali, M.J. and Hatch, G.M. (2014). Essentials of forensic post-mortem MR imaging in adults. *The British Journal of Radiology*, 87(1036), pp. 20130567.

Rutty, G.N., Brough, A., Biggs, M.J.P., Robinson, C., Lawes, S.D.S. and Hainsworth, S.V. (2013). The role of micro-computed tomography in forensic investigations. *Forensic Science International*, 10, 225(1–3), pp. 60–66.

Schmelding, S.C. and Arthurs, O. (2022). Post-mortem perinatal imaging: What is the evidence? The British Institute of Radiology, published online 5 May 2022. Available at: www.birpublications.org/doi/10.1259/bjr.20211078 (Accessed 17 December 2023).

Schmeling, A. and Rudolf, E. (2022). Medical age assessment in living individuals. In: Madea, B. *Handbook of Forensic Medicine*. Second edition. Wiley, John Wiley & Sons ltd.

Walker, S. (2022). *Quality Improvement Partner (QIP) Tips*. Synergy, July, pp. 29.

# Section ten

# Person-centred care in therapeutic radiography

# 23 Compassion as the heart of person-centred care in diagnostic and therapeutic radiography

*Amy Hancock and Jill Bleiker*

## Introduction

The research which underpins this chapter was qualitative; that means it sought to explore, understand and explain compassion rather than to produce compassion score charts for radiographers attending their annual reviews or generalise what compassion means to most, if not all, patients. As you can imagine, any attempt to generalise would be the very antithesis to the individual nature of person-centred care. Two radiographers, one diagnostic and one therapeutic, each working independently of the other conducted doctoral research projects exploring what compassion meant to patients, their carers, radiographers, and student radiographers (Bleiker, 2020; Taylor, 2020). In this chapter we will show you how that research has contributed to an understanding of person-centred care in radiography. There were remarkable similarities in the findings between the two disciplines in many areas, and some key differences which we will highlight. The chapter begins with a brief overview of the methods employed by the authors, a summary of the key findings, followed by the practical applications of these to the therapeutic treatment or diagnostic imaging procedure.

The research projects were conducted independently of each other over the period of 2016–2020, using a combination of interviews, focus groups, co-production workshops and online discussion forums. Both projects explored the experiences and opinions of patients and their carers, radiographers, and student radiographers with regard to compassion in diagnostic imaging and radiotherapy. Table 23.1 shows the number of participants for each project and the employed methods of data collection for each independent study.

For each study, the data were analysed independently, both inductively and deductively using Thematic Analysis (Braun & Clarke 2006, 2013, 2017). The subset of themes for the participant groups were then mapped by the authors.

DOI: 10.1201/9781003310143-33

*Table 23.1* Participant numbers and methods employed for each study

| Study | Patients +/- carers | Student radiographers | Radiographers | Methods |
|---|---|---|---|---|
| Author 1 Therapeutic | 16 | 24 | 27 | 3 x focus groups patients and carers |
| | | | | 5 x focus groups radiographers |
| | | | | 3 x focus groups student radiographers (Years/ stages: 1–3) |
| | | | | 4 co-production workshops |
| Author 2 Diagnostic | 34 | 30 | 85 | 34 x semi-structured interviews |
| | | | | patients 5 x focus groups |
| | | | | student radiographers |
| | | | | (years/stages: 1–3) |
| | | | | Online journal club discussion |
| | | | | radiographers |

Source: Hancock and Bleiker, 2023.

## Summary of the findings of the research

Caring with compassion for patients undergoing diagnostic medical imaging or radiotherapy treatment involves establishing how to engage and interact with each individual patient, and perhaps those accompanying them, on a number of levels. Physical interactions involve mainly undertaking the sometimes complex yet mechanical and technical components of the imaging or treatment task, but these take place at the same time as communicating both verbally and non-verbally with the patient. Adapting interpersonal and communication styles to meet the needs of the patient is as valuable a skill as adapting radiographic technique in diagnostic imaging or modifying your clinical approach in radiotherapy. It is also an important part of the patient–practitioner relationship in terms of patients' emotional wellbeing. Identifying and applying those behaviours that communicate compassion is arguably more important than authentic feelings of compassion in the practitioner, although a radiographer who feels no compassion towards their patient could be said by some to be in the wrong profession.

Whereas previously good communication skills have been the most highly valued, now patients also want to be seen promptly and imaged or treated quickly and proficiently but without feeling that they have been rushed or hurried through the examination (Hancock and Bleiker, 2023). We found that patients appreciated a radiographer who observes and interprets their nonverbal cues and body language, seeing them as empathic and caring (Taylor, Bleiker and Hodgson, 2021). This reduced feelings of fear and anxiety, which patients reported that they make deliberate efforts to hide in the interests of expediting the procedure, or else of not appearing weak or uncooperative. Recognition and acknowledgement of these cues signifies the radiographer is seeing the person, not just the patient which promotes engagement and rapport. We also found that appearing competent and proficient also calms the patient and appearing unhurried prevents feelings of being just another item on a conveyor belt or production line.

Person-centredness can also be achieved by means of the dialogue between radiographer and patient. Merely giving instructions whilst focusing only on the imaging or treatment task depersonalises the patient and contributes to perceptions of radiographers as technicians or, worse, simply buttonpushers. Whereas, asking questions relevant only to that patient, including the circumstances which brought them for imaging or about their levels of pain or discomfort, ability to move and turn on the examination or treatment table communicate interest in that patient specifically, and making eye contact and smiling convey warmth, allowing the patient to feel safe and calm. We found that questions as exemplified above, together with explanations about the procedure and what to expect confer these and other benefits including a swifter procedure as patients are better able to understand what to expect and co-operate with the radiographer.

Key values identified from our radiography research were those of kindness and an emotion-orientated caring about rather than a task-focused caring for the patient (this is particularly important because we talk about patient care without differentiating between the two). Other values included respect, acceptance, sensitivity and human connection between patient and the radiographer.

Although there was considerable accord between diagnostic and therapeutic radiography with regards to these values, the role and characteristics of a radiographer were not considered by patients to be a component of compassionate person-centred care within radiotherapy. There was, however, the acceptance that as a healthcare professional, the therapeutic radiographer should be compassionate as part of their role. This is perhaps due to the frequency of patient interactions with therapeutic radiographers, who in general will see patients repeatedly and for prolonged periods during their treatment regime compared to most commonly only a single occurrence for patients who attend for medical imaging. Patients perceive they are being cared for and subsequently cared about due to the frequency and continuity of their interactions with the radiographers. This sense of continuity supports the formation of a patient–practitioner relationship.

In diagnostic radiography, however, patients regarded the role and the person as a radiographer as synonymous and an important component of compassionate care. Patients used terms such as "natural", seeing compassion as "where your heart should lie", with "being caring" a basic minimum quality when appraising a student's suitability for their chosen profession. It could be that in the brief and highly technical diagnostic encounter there is no time for a radiographer to develop caring feelings towards that particular patient and so a caring nature is essential. Otherwise, there is a risk that the interaction becomes a perfunctory and a somewhat cold technical procedure. This was reported by patients during our research, who unhappily recalled their experiences. In both disciplines, however, and whether innate or otherwise, these values were shown to be at odds with the "hidden" values of the NHS which are not seen in policy and professional documents. These include speed, efficiency, time and cost-saving and throughput, all of which are in direct conflict with giving person-centred care.

The research identified specific behaviours which, when exhibited towards a patient, was regarded as demonstrating compassion. Often these were "little things" and small gestures such as checking whether a patient was warm enough or needed a blanket, an extra pillow perhaps or larger, more time-consuming acts such as halting the procedure to allow the patient time to calm down or talk. Such acts often cause radiographers to argue that "we haven't got time!" but compassion demands that we ignore the call of the waiting room and the pressures of workload and make the time for that particular patient in that particular moment.

In total, thirteen practices, or observable interactions, were identified through our research that were felt by the patients and carers, radiographers, and radiography students to illustrate a person-centred approach was being displayed to the patient (Figure 23.1). These, when witnessed by patients could be recognised and appreciated as being compassionate behaviour by the radiographer.

**Bedside manner:** Linked to communication, the way in which radiographers not only talk to, but also interact with patients, can promote, or hinder a person-centred relationship. As you would expect, patients would prefer calm, polite and friendly interactions, rather than those which are abrupt, hostile or directive. We know this can be challenging with the workload demands faced by clinical departments. But take a moment to consider when you are about to enter the waiting, imaging or treatment room. Is this considered, observant of the patient(s) in front of you, or is it rushed, purposeful and simply focused on the task at hand?

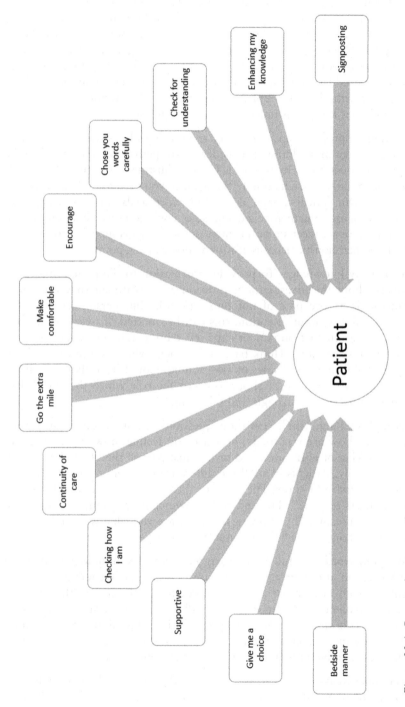

*Figure 23.1* Compassionate practices.

**Check for understanding:** A simple checkback signifies interest through the recognition that understanding can be influenced by many personal factors which can include prior knowledge and experience, emotional states and cognitive ability. By identifying a patient's understanding at this stage provides you with the opportunity to modify the approach you are taking, ensuring it is personalised to them and their individual needs. This may be breaking down the information or supplementing it with some written information, but you will not know this unless you ask the question.

**Signposting:** Appropriate and considered signposting allows you to think about the whole patient and address their wider needs, perhaps even external to the radiographic situation or environment. This can illustrate to patients a "genuine" desire to help, rather than its simply being a task you need to undertake as part of your role. Our patients hold a plethora of wider needs, some more apparent than others. Signposting them to a diagnosis or condition specific support group (for example, local cancer support centres or osteoporosis support groups) who can address their holistic needs is always a good starting point.

**Enhancing my knowledge:** To provide information to the patient and their caregivers about the investigation, procedure or treatment they are about to undergo is a core part of not only our role, but person-centred care. Sometimes, and without intention, this can be omitted or overlooked. This not only has an impact on patient experience but can hinder your ability to achieve the task at hand. Think back to a patient where you have faced difficulty immobilising, localising, or positioning. Could it be that you didn't fully explain what you were about to do or the role which they needed to play that caused you to struggle and take longer?

**Checking how I am:** You'll hear these questions echoing round radiography departments: how are you today, how are you feeling, how you doing? Are we really asking or even interested or is it just part of the blurb that we say to patients daily. Patients can tell the difference between a routine question and genuine desire to know. The latter is obviously preferred, and by asking "What matters to you?" (Healthcare Improvement Scotland, n.d.) can help us to be person-centred by knowing how they are and what we can do to help if the answer they give is "No, I'm not ok".

**Choose your words carefully:** Using jargon, medical terminology or pitching the information at the wrong level illustrates a lack of consideration for the patient (person) in front of you. It also reinforces the notion that you are only telling patients information or even talking to them because you must as part of your role and professional responsibilities. This dilutes the interaction and removes any chance of the patients thinking you are doing this with a genuine desire to help them to manage their current situation.

**Reassurance:** Reassurance can be displayed through words, body language and behaviours. We should aim to put patients at ease, and hopefully help to reduce their anxiety. Understandably, we can't assure patients that everything is going to be ok in terms of their health or outcomes. But we can reassure patients that we care about them and the situation they face, their well-being and that we take any concerns they may have seriously.

**Continuity of care:** It is important that we remember how the end of the interaction is just as important as the start. By engaging with patients in a way that is meaningful to them, i.e. displaying an understanding of their individuality throughout will connote how they are at the heart of your practice for the entirety of their appointment. This can be achieved in modest but effective ways. Simply maintaining eye contact can signify that you are still engaged and invested in them and how they are just as important to you after the procedure as they were before it began.

**Extra mile:** As we have already established, compassionate person-centred care is considered to be a core part of a radiographer's role. To go the extra mile is to do something with or for the patient which is beyond their expectations (but not beyond our scope of practice) and illustrates a personal touch to what could ordinarily become routine. Recall of personal events, for instance remembering a previous conversation and then enquiring about its outcome at a later appointment or a chance meeting in another department, were frequent examples given by patients during our research.

**Give me a choice:** Involving patients in their care personifies the concept of person-centred care and aligns with the drive for shared decision making within the UK National Health System (NHS) (NHS England and NHS Improvement, 2019; The Health Foundation, 2016; Department of Health and Social Care, 2012; NHS England, 2019a, 2019b). The decisions behind patient attendance to imaging or radiotherapy departments are part of their referral, request or prescription and are not ours to influence. But decisions around the process or procedure which we do control, for example where patients get changed, what they wear during the procedure, should be made together after discussing with the patient and hearing their preferences.

**Make comfortable:** You have a patient on the scanner or treatment couch, and they are crying out in pain. The clinical constraints, including the immobilisation devices, protocols and equipment parameters will, in this situation, dictate how we can adapt the position or technique. Although we are limited, we can ensure we adapt as much as possible to make the patient more comfortable in this situation. We can take this further, for example when we help them off the couch, transfer them across to their hospital bed or wheelchair, we can do all we can. Fluffing up or getting an extra pillow to help patients be more comfortable, or arranging for additional pain relief, can make a big difference. Those small things that attempt to alleviate pain and suffering exemplify compassion, illustrating a desire to help by doing what we can within the constraints.

**Supportive:** This is a collective term for those behaviours and practices which, by their presence, illustrate our purpose is to help guide and support patients through the procedure or treatment, alongside image acquisition or treatment delivery. Supportive actions personalise our role, reinforcing the caring element, and stops us being seen as technicians who simply operate the equipment. Supportive practices can include using words and gestures that convey encouragement to continue with their treatment or an imaging position which the patient may find a challenging. But you must also consider that your tone and body language are important for making your words sound supportive and must match the genuine intent behind them otherwise you may appear forceful or disgruntled (Taylor, Bleiker and Hodgson, 2021).

**Meeting needs:** We may consider the immediate needs of the patient are the measurable activities; for example, MRI acquisition or delivery of a radiotherapy treatment fraction. To be person-centred, you must consider and attempt to address the independent, or holistic, needs of patients. Behaviours that represent personalisation of your approach, for example using local phrases, changing the language or the detail of information provided (whilst making sure essential information is communicated) based on how the patient responds to you, illustrates that you appreciate not everyone is the same, and their needs at that time are different to the next patient who walks through the doors of the department.

The rest of this chapter will take you through a compassionate therapeutic treatment or imaging examination from start to finish using the evidence from our research to support the advice we offer. Although some patients did not necessarily perceive compassion in terms of observable acts and behaviours during their time in radiology or radiotherapy, compassion is nevertheless there, latent in the desire of the radiographer to minimise or alleviate their patient's suffering, be that physical or emotional.

### Before the imaging examination or radiotherapy treatment

Even before meeting them, you have an opportunity to start getting to know your patient by examining the request or referral form, clinical history, or patient notes. This can give you information about the patient's age, gender, and some (possibly brief) information about their clinical condition. This in turn may give clues as to their mobility and ability to co-operate with you, in addition to how unwell they may be feeling. Bear in mind that from the moment you meet your patient they will be taking in your facial expression and body language, so a friendly smile and open body language will go a long way to start the process of building rapport and engagement with your patient. Likewise, you can be observing your patient with a clinician's gaze to assess posture, movement and facial expressions that might indicate pain or emotions such as worry or fear, all essential to understanding your patient and the building of a positive relationship.

Once you come face to face with them, it is good practice to introduce yourself to your patient by telling them your name and role. Introducing yourself is an important initiative that was born of the "Hello my name is…" campaign of Dr Kate Granger (2014) which has improved immeasurably the patient–practitioner relationship. This begins the process of establishing rapport with your patient. When patients and radiographers talk about rapport there is often little detail about what this means and how it is established. Research has shown that introducing yourself is a key way to increase rapport, help radiographers to involve patients in their procedure, and move away from notions of passivity, of feelings of having something done to them (Dang et al., 2017; Price, 2017). Following on from this is the gaining of consent. For most projection imaging, this does not require a formal process, rather one or two questions regarding patients' understanding of what is about to happen, and explanations about what will follow. Consent is often implied through the co-operation of the patient with the radiographer during the procedure. In therapeutic treatment, informed consent is given prior to radiotherapy, during a wider discussion and explanation of the treatment plan. It is also sought again (including acquisition of a further countersignature) on the patient's first day of treatment. In our research, patients sometimes commented that they were brought into the room and told what to do without any form of welcome, leading to feelings of unease and a lack of trust and confidence. It's also worth knowing that our research indicates that a radiographer (i.e., you) is more likely to be viewed as a technician or button-pusher than a professional practitioner if the interaction is perfunctory and rushed. Establishing rapport in these ways means that your patient is "on board" with what is about to happen. They may not be terribly happy about it, as it may be uncomfortable or have side effects, but they will at least feel some element of choice and control. This facilitates the next thing you will both benefit from, their co-operation.

### During the examination or treatment

Early questions as you begin the procedure are designed not only to continue the process of rapport building, but now also to add in the technical aspects which form the basis of imaging or radiotherapy procedures. Our research suggests there is a stark differentiation between imaging and radiotherapy interactions, and those in other hospital departments. At this stage your patient's co-operation will make the difference between a procedure which goes smoothly and efficiently and one which is less so, irrespective of your own technical skills and abilities. You may require your patient to change into a radiolucent hospital gown. Maintaining dignity is essential to patients' wellbeing. If they do have to disrobe, you can well imagine the feelings of vulnerability this can engender. Hyde and Hardy (2021) found that patients preferred hospital scrubs or tracksuits to hospital gowns, as they provide fuller coverage of the body and therefore increased dignity. There will also be practical issues such as storing valuables during the procedure, and the potential reluctance of patients to be separated from their personal possessions for perhaps religious or emotional reasons. By noting and taking any concerns seriously you can display understanding of the meaning they hold, help to allay fears about the security of their belongings, and will continue to build those all-important relationships.

Once the imaging examination or radiotherapy treatment begins, and you become increasingly absorbed in acquiring the image or delivering the treatment, remember to keep looking at your patient wherever possible so that you can observe any non-verbal cues which may suggest something different to the verbal message the patient is attempting to communicate. This is particularly important when monitoring for pain or emotional distress; patients can be remarkably stoical and tolerant of even the most unpleasant experiences if it means they achieve what they have come for, be that a diagnosis or another treatment fraction completed. As well as eye contact, continuing a dialogue with your patient maintains rapport and calms and reassures them. Regular interactions, such as letting them know they are doing well, or how much longer the procedure will take, are particularly helpful. This is especially important when radiographers need to leave the room during a patients' treatment or move behind a screen to take an image. Patients may feel anxious due to fear of the unknown, being left alone or of the unfamiliarity of the environment (and that's without the big, noisy, and sometimes scary machines that we operate).

### After the imaging examination or radiotherapy treatment

As the imaging examination or radiotherapy treatment reaches its conclusion, the temptation can be to consider it finished and to hurry on to "the next one". However, this is a good time to remember that this is a person, not a procedure and that for them, this is only a part of an often-lengthy sequence of events, the timing and outcome or end of which may be uncertain. So, rather than ushering them out of the room, time must be allowed to check that your patient is as well as can be expected given what they have just endured. You should also ensure that they know what happens next, including the all-important results of their visit or the later stages of their treatment. You may have this information to hand, or you may not. Either way it is incumbent upon you to signpost them as well as you can to their onward journey, including – obvious to you, but not necessarily to them – the way out of the department.

### Summary

In this chapter we have provided you with concrete examples of ways in which you can care with compassion for your patient whilst you perform diagnostic imaging or therapeutic treatment. Some of these may appear universal to all, and indeed may seem like little more than common sense. But the use of tailored and individualised questions and explanations about your patient's particular procedure, as well as careful attention to their facial expressions, body language and responses, will personalise and individualise their experience. This will help your patient to leave the clinical department feeling cared about, as well as achieving both your and their aims; successful diagnostic imaging or completed treatment.

## References

Bleiker, J. (2020) *An Inquiry into Compassion in Diagnostic Radiography.* Unpublished thesis: http://hdl.handle.net/10871/121267

Braun, V. & Clarke, V., (2006) Using thematic analysis in psychology. *Qualitative Research in Psychology.* 3:77–101.

Braun, V. & Clarke, V., (2013) *Successful Qualitative Research.* London: Sage Publications.

Braun, V. & Clarke, V. (2017) *Evaluating and Reviewing Thematic Analysis Research: A Checklist for Editors and Reviewers.* University of Auckland Website. https://cdn.auckland.ac.nz/assets/psych/about/our-research/documents/TA website update 10.8.17 review checklist.pdf.

Dang, B.N., Westbrook, R.A., Njue, S.M. & Giordano, T.P. (2017) Building trust and rapport early in the new doctor-patient relationship: a longitudinal qualitative study. *BMC Medical Education.* 2;17(1):32.

Department of Health and Social Care. (2012) *Liberating the NHS: No Decision about Me, without Me – Government Response to the Consultation.* London: Department of Health and Social Care.

Granger K. (2014) YouTube – Dr Kate Granger talks about compassionate care and #hellomynameis. www.youtube.com/watch?v=Be_nIItj8bs

Hancock, A. & Bleiker, J. (2023) *But What Does It Mean to Us, Radiographic Patients and Carer Perceptions of Compassion. Radiography,* 29(1):S74–S80.

Healthcare Improvement Scotland. (n.d.) What matters to you? www.whatmattersto you.scot/why-ask/

The Health Foundation (2016) *Person-Centred Care Made Simple: What Everyone Should Know about Person-Centred Care.* www.health.org.uk/sites/default/files/Person CentredCareMadeSimple_0.pdf

Hyde, E. & Hardy, M. (2021) Patient centred care in diagnostic radiography (Part 1): Perceptions of service users and service deliverers. *Radiography.* 27(1):8–13.

NHS England. (2019a) *The NHS Long Term Plan.* London: NHS England. www.longtermplan.nhs.uk/

NHS England. (2019b). *Universal Personalised Care: Implementing the Comprehensive Model.* London: NHS England. www.england.nhs.uk/wp-content/uploads/2019/01/universal-personalised-care.pdf

NHS England and NHS Improvement. (2019) *Shared Decision-Making Summary Guide.* www.england.nhs.uk/publication/shared-decision-making-summary-guide/

Price, B. (2017) Developing patient rapport, trust and therapeutic relationships. *Nursing Standard.* 9;31(50):52–63.

Taylor, A. (2020) *Defining Compassion and Compassionate Behaviour in Radiotherapy.* Unpublished thesis: https://doi.org/10.7190/shu-thesis-00373

Taylor, A, Bleiker, J. & Hodgson, D. (2021) Compassionate communication: Keeping patients at the heart of practice in an advancing radiographic workforce. *Radiography.* 27(1):S43–S49.

# 24 Person-centred caring in therapeutic radiography

*Julie Hendry*

## Introduction

Person-centred and patient-centred caring appear frequently within healthcare policy and professional guidance. The terms may be used interchangeably alongside compassionate care and holistic care. All are nebulous phrases that may be hard to define and deconstruct making it more difficult for practitioners to deliver the concepts underpinning such approaches to their patients (Hyde and Hardy, 2020). Indeed educators who are required to help foster the concepts of centredness in students would benefit from such definitions. Since the latter part of the 1990s, an earnest move from the biomedical to a more holistic model of health has taken place. Person-centredness has grown in popularity and a more informed public, with access to internet resources, has expectations of inclusion, autonomy and shared decision-making.

The NHS Constitution and Values (Department of Health and Social Care [DHSC], 2021) has influenced practice in the United Kingdom by describing the need for all healthcare practitioners to have a 'commitment to quality of care'. Every patient has the right to appropriate individualised care, compassion, dignity and respect, which is promoted within professional codes such as the Society and College of Radiographers Code of Conduct and Ethics (2013). All healthcare practitioners are required to meet the expectations of relevant professional codes in addition to evidencing certain skills and attributes such as caring by their regulatory body, the Health and Care Professions Council [HCPC] (HCPC, 2023).

Despite caring practice being at the forefront of policy and a key focus of professionalism (Dewar, 2013), recent high-profile reports have highlighted a greater need for 'care to be patient-centred, compassionate and well informed' (Willis, 2015, p.3). Although primarily focused upon the nursing profession, the content of these reports is arguably of relevance to the multi-disciplinary healthcare teams responsible for all patient care, including therapeutic radiographers. There are clear expectations upon Allied Health Professionals (AHPs) to demonstrate high levels of caring, as an underpinning principle

DOI: 10.1201/9781003310143-34

of the NHS (DHSC, 2021). Therapeutic radiographers are part of the wider healthcare team, being one of the fifteen AHPs registered in the UK, who 'make a significant contribution to the care of people affected by cancer' (Health Education England [HEE], 2019, p.5). Through appropriate knowledge, skills and behaviours, all AHPs should provide high quality person-centred care for people affected by cancer (HEE, 2019). These skills are highlighted within a range of competencies from Macmillan Cancer Support and include communication, building good relationships with patients, and the skills to 'provide safe, effective, high quality and accountable care for people affected by cancer' (Macmillan Cancer Support, 2017, p.8). Having established the need for person-centred caring in healthcare, and thus therapeutic radiography, it would be useful to firstly consider the context of the radiotherapy setting before then exploring how person-centred caring may be defined and delivered.

## The therapeutic radiography setting

The sister profession of therapeutic radiography holds certain similarities with that of diagnostic radiography, but subtle differences may suggest a slightly different approach to person-centredness in clinical practice.

In contrast to the often-transient patient or service user interactions within diagnostic radiography, radiotherapy patients generally receive more than one treatment from a team of therapeutic radiographers. The concept of person-centred caring in radiotherapy may therefore be slightly nuanced. All those receiving treatment have a cancer diagnosis, unlike those attending for diagnostic radiography, with a variety of conditions, or indeed be 'healthy' and attending screening. Undoubtedly, cancer is a long-term condition, and most definitely a life-changing diagnosis for both the individual and their family.

Within this chapter the concepts related to person-centred caring, and indeed what the term might mean, will be discussed. If care is truly person-centred, the differences between diagnostic and therapeutic settings may be of little significance due to the individualised concepts involved. However, this could be interesting to debate and discuss. Tensions within radiotherapy may reflect those of diagnostic settings but will be explored separately through recent research and published literature.

## Defining person-centred caring

Caring can be considered a basic human need (Maslow, 1970), a humanistic attribute that is essential in healthcare (Bolderston, Lewis and Chai, 2010). Policy situates caring at the heart of healthcare practice as previously highlighted (DHSC, 2021; Dewar, 2013) but to bring person-centred caring from visions, principles and ideals into the clinical setting may require more deconstruction of the concept. Although previous chapters will have considered such definitions in more detail, it is useful to remind ourselves of one such model, that from the Picker Institute.

In 1986 in the United States of America, Jean Picker was undergoing treatment for a long-term condition. She and her husband, Harvey, recognised that despite the provision of scientifically and technologically advanced treatment, the individual needs and preferences of patients were not routinely met. The same year they founded a not-for-profit organisation, The Picker Institute, to develop and promote patient-centred care. In 1987, the Eight Principles of Person-Centred Care were suggested by the Picker Institute and have been internationally recognised and adopted since then (Picker Institute, 2023). The Principles include access to advice, effective treatment, continuity of care, including family and carers, clear information and support, respect and involvement in decision making, empathy and emotional support, meeting physical and environmental needs (Picker Institute, 2023). Research involving patients, their families, and staff informed the Picker Principles of Person-Centred Care. Placing the person at the centre of services promotes enablement, support and care and is closely entwined with autonomy, individualisation and respect.

The Picker Principles of Person-Centred Care address every facet of care across patients' pathways and cover both practical and emotional aspects of person-centredness. In comparison to other definitions of person-centredness, this may support a more holistic approach to care, potentially including the Holistic Needs Assessment (Macmillan, 2020). For Macmillan (2020), holistic care refers to 'all the needs that you might encounter as an individual'. This is determined through dialogue with the individual and carers, through the Holistic Needs Assessment (HNA). The HNA enables patients and families to discuss what is important to them, what worries they may have, and which areas they may require more support. Although a relationship is inferred through the discussions required, it is not overt in their definition.

Similarly, The Health Foundation (2023) suggest Person Centred Care:

> supports people to develop the knowledge, skills and confidence they need to more effectively manage and make informed decisions about their own health and health care.

Their definition also focuses upon the practical aspects of care which may be suggestive of a more pragmatic, outcomes- or disease-centric model, aligned more with the biomedical than the holistic model of health. These aspects may well bring the concepts of patient-centred care verses person-centred care to the fore. Are they similar, the same, or essentially different?

## Person-centred and patient-centred caring

Comparing the concepts related to patient-centred and person-centred care, many similarities are revealed (Eklund et al., 2019). Both concepts require a holistic understanding of the individual patient and their world. This includes knowing them as a unique human being before attempting to provide a diagnosis, investigation, or treatment as part of the healthcare process. The concept of 'centredness', of knowing the person behind the illness, has an ethical perspective and sees the individual as active in their care and decision making (Hughes, Bamford and May, 2008). As such, person-centred care in radiotherapy should include inherent humanistic values, which some definitions do not overtly include. The aim of Eklund et al.'s paper was to determine differences and similarities between the concepts of patient- and person-centredness. They suggest that although the route to both concepts may differ, the meaning of person-centred, and patient-centred care are indeed similar. This may be contentious, and the meaning perhaps different. In person-centred care the goal is for a 'meaningful life' (Eklund et al., 2019, p.8) whilst in patient-centred care it is for a 'functional life' (p.8), which I propose is an extension of the biomedical model. This possible tension may best be explored through an example such as empathy.

In patient-centred care, Eklund et al. (2019, p.8) suggest empathy means to 'infer the patients' specific feelings' where their emotional state is recognised, and support is provided, delivering a functional outcome. In radiotherapy, this could relate to the experience of radiation-induced side effects. The radiographer can recognise the skin reaction, empathise and provide care to minimise that side effect. Conversely, empathy within person-centred care looks 'beyond the person's specific feelings in the present moment to the life he or she is living' (Eklund et al., 2019, p.8). This means not only understanding an individual's fears, joys and sadness at the present time, but also to see from the person's perspective their life's extension and structure. So the influence that side effect has directly upon the individual's life, enjoyment and feelings potentially affecting their hobbies, lifestyle and relationships. If humanistic values, and individualised knowledge of a person are essential to person-centred caring, then moving from a 'patient' to a 'person' focus may best enable caring in the therapeutic radiography setting.

The 'caring for patients' discourse reported by Pluut (2016, p.504) relates to the need for 'holistic caring' for the patient as a whole person. This is similar in concept to that of the holistic focus determined by Eklund et al. (2019). Pluut suggests the suffering and vulnerability of disease can be reduced and alleviated through the process of caring. Caring is constructed through patient-centeredness and the biopsychosocial model of health, so that caring is for the whole person not merely the parts. Seeing patients as individuals is part of that holistic caring process together with an understanding of the values held by that individual (Pluut, 2016). This is essential within radiotherapy, avoiding the objectification of patients by their treatment area or diagnosis. Knowing the context, personal development, and life history is part of knowing the individual and caring for them holistically (Pluut, 2016). Through this understanding, vulnerabilities can be alleviated but should not be assumed by the healthcare practitioners. Pluut (2016, p.505) further explains her findings that to 'emphasise the importance of therapeutic engagement between healthcare practitioners and patient' is indicative of caring discourse. The relational aspects of caring, alongside a humanistic approach, aim to reduce vulnerability and suffering, which may suggest patients are passive recipients of caring, being subservient to the biomedical model of health. The humanistic discourse may be appropriate but not as a reaction to distress, more as a human-to-human relational approach.

Constructing person-centred care involves a moral concept that cannot and should not be determined by a generic model (Pluut, 2016). Openness is required to explore what individuals' desire from caring, empowerment and responsiveness. This suggests individualisation for every person receiving healthcare and thus radiotherapy. Although Pluut determines patient-centeredness to be a relationally constructed concept, she provides similar ideas to those by Eklund et al. around holism, empowerment, and communication. What may be interesting to note is that empathy and compassion did not feature in Pluut's analysis, which may indicate an over-reliance, conscious or not, on the biomedical model. This may be a key difference between patient-centred and person-centred care.

Although not exclusively within therapeutic radiography, the general aspects of person-centred care provide a conceptually interesting lens through which to consider our practice. Oncology departments and hospitals frequently use person-centred and patient-centred care in vision and values statements and as such the phrases are important to all therapeutic radiographers. Empathy and compassion are also well cited and important aspects within the notion of person-centred caring in therapeutic radiography. A recent phenomenological study by Hendry (2023) explored perceptions of caring within therapeutic radiography. She suggested person-centred caring in radiotherapy included twelve intertwined concepts shown in Figure 24.1. These aspects of person-centred caring would be delivered to and experienced by people receiving radiotherapy through the radiographer having an emotional involvement with them, facilitating a therapeutic relationship. To enable this, the radiographer would need to have virtue; self-awareness, humanity and moral values which would flourish within a culture of caring. Themes emerging from Hendry's study included the therapeutic radiographer 'Being Caring' whilst 'Caring For' and 'Caring About' individuals. These conceptualisations of person-centred caring will now be considered within radiotherapy.

## Being caring in therapeutic radiography

Being a caring individual was perceived by participants in Hendry's study to be a multifaceted concept. It emerged as a way of being, intuitive to the personality of the individual and suggestive of subconscious virtue. Caring was considered as an almost innate component of the person that could be enhanced and developed. The desire to help and nurture emerged as a form

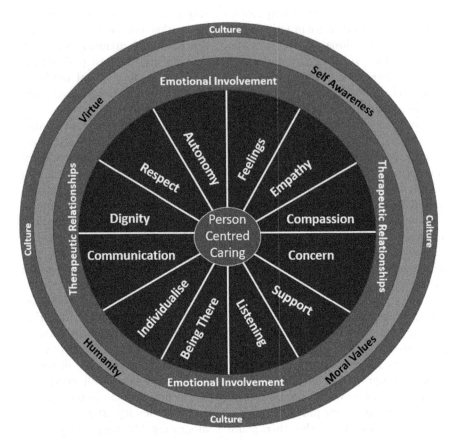

*Figure 24.1* Suggested framework of person-centred caring in radiotherapy.

Source: Hendry, 2023.

of altruism which was considered present, to various degrees, in everyone who becomes a therapeutic radiographer. All individuals could possess the capacity for caring, but for some this would be more developed than in others. Caring characteristics emerged from life experiences, both as children and adults and for some that lead them to a caring profession such as thera-peutic radiography. It can be suggested that someone with a caring person-ality would be driven to join the caring profession as a radiographer.

Considering the caring personality of the therapeutic radiographer, it may be indicative of the characteristics or virtues of that individual, acknow-ledging that concepts of personality, traits, attributes, and characteristics may be used as synonyms. Personality is understood by Gelhaus (2013, p.127) to be the balance of virtues and all the required attitudes of caring, but with emphasis upon the emotive virtues. Character is the enduring essence of an individual, which may change over time, and describes how that person is inclined to be, act, feel and to value (Gelhaus, 2013, p.126). Gelhaus further describes virtue as connected aspects of skills, habits, and ideals, integrated within an individual. In a similar way, attitudes can involve a cognitive, thinking element and an emotive, feelings element, but the important difference here is the action aspect of character. Furthermore, the emphasis of virtue is upon the agent, whilst attitude focuses upon the relationship with the patient or student.

Virtues can be considered moral dispositions linked to the intrinsic parts of an individual and related to appropriate feelings. As early as Aristotle's Nicomachean Ethics, written in 340 BCE republished in 2009, virtues have been noted to be at the core of caring, proposing them to be a desire for the wellbeing of others beyond a mere goodwill, wanting to do good as part of a cognitive process (Aristotle, 2009). Although not described as caring virtues, Aristotle's work has been subsequently developed over time. More recently, and in a current healthcare setting, Kerasidou et al. (2021) considered empathy and compassion as an expression of virtue and related professional skill. They explain how practitioners engage both cognitive and emotional dimensions alongside the intention and desire to do good. Similarly, being intuitive and 'picking up cues' from patients and students would originate from a desire to be caring as part of character and virtue (Kerasidou et al., 2021; Aristotle, 2009; Watson, 2006; Morse et al., 1990; Roach, 2002) and suggests a level of emotion. The moral imperative was revealed by Morse et al. (1990) where they describe caring as a fundamental value, not just an array of traits or behaviours, more a way of being that underpins all actions of the nurse, and I would suggest the therapeutic radiographer.

Person-centred caring in radiotherapy can be interpreted as character and virtue, being part of the radiographer as an individual. This would be shown with the 'want to care', being perceived as almost innate and altruistic, providing a natural nurturing that was beyond conscious thought. It suggests elements of intuition and an awareness of cues being the 'morally sensitive self, attuned to values' (Roach, 2002, p.56), which relates to an authentic moral consciousness, humanity and the values aspects of caring for people as a therapeutic radiographer.

The caring personality of an individual practitioner was reported as a sub-theme within the study by Bolderston, Lewis and Chai (2010), one of few studies of caring within therapeutic radiography. They also mention how the workplace setting is influential with respect to caring which is supported by Kerasidou et al. (2021); despite the importance of individuals' empathetic virtues, there is also influence from the whole healthcare environment. Although empathy and compassion can be considered a virtue rather than a skill, if the environment or culture is not supportive of expressing those virtues, both essential in therapeutic radiography, they are less likely to be developed. Habituation of the skills to facilitate empathy and compassion, in addition to the moral development of caring virtue would therefore be considered hampered (Kerasidou et al., 2021). This is important in radio-therapy where the high technical focus alongside the need for timely service delivery can sometimes work against a culture for caring. Indeed the perceived decline in caring may not be due to individual practitioner failings alone, or compassion burn out, more that institutions are creating barriers to caring through culture and workload.

Bolderston, Lewis and Chai (2010) reported that, within radiotherapy, the work environment and organisational culture reduced the ability of practitioners to care. This included time pressures when treating patients and the potential for caring to be sacrificed for technology. Murphy (2006) and Naidoo, Lawrence, and Stein (2018) revealed similar tensions around time and technology in diagnostic radiography. Whilst Williams, Kinnear, and Victor (2016) described how in high technology healthcare environments, habitual tasks may erode caring more than the technology itself. It remains essential for all radiographers to maintain their values and caring moral attributes despite practice demands. Without this commitment, person-centred caring will not be demonstrated when patients would be most in need of compassion, caring and connections. It is now timely to explore how person-centred caring may be displayed, demonstrated and experienced.

## Caring behaviours – caring for and about individuals

A therapeutic radiographer with a caring disposition and those with a greater disposition or virtue may demonstrate caring about in addition to caring for individuals, through caring behaviours.

Behaviours perceived as caring in Hendry's study included a variety of thoughts, feelings and actions believed to portray caring about the patient. Caring behaviours included making time, prioritising, and listening to patients. Knowing and connecting with patients emerged as emotional caring behaviours. It may be that in making time and listening to patients, radiographers enable 'knowing' of the individual. There is a significant importance to making time and prioritising patients and so caring about them. This was also reported in the diagnostic setting by Naidoo, Lawrence, and Stein (2018), who described caring as a humanistic way of interacting with people, making time, listening, and enabling empathy and compassion. Making time can sometimes be to the detriment of the therapeutic radiographer and their own wellbeing. Making time enabled emotional 'heartfelt' connections and a humanistic relationship with patients with radiographers going 'above and beyond' the basics of caring for patients demonstrating the importance of the 'small things'. These behaviours can be considered authentic 'caring about' individuals and an essential aspect to their role in the clinical setting. Bolderston, Lewis and Chai (2010) also suggested 'empathetic human connection' constituted aspects of caring.

Making time enables and facilitates 'empathetic connection', which may be perceived to include individualisation, and knowing patients as people.

The humanising aspects of caring emerging from Hendry's study show how humanity is borne from being authentically caring and facilitated by 'knowing' and 'individualisation' which enables a connectedness with patients. Hendry interpreted this as 'caring about' patients. Themes such as 'Empathy, Engagement, Relationships, Communication, Holistic and Individualised Focus' were demonstrated. Relationships and an individualised focus create connections that can build a partnership, resulting in mutual trust. This enables a therapeutic relationship from which both the patient and practitioner benefit; involving trust, understanding, individualisation and shared experiences that would bring a sense of fulfilment to the radiographer and wellbeing to the patient.

Connectedness and relationship aspects of caring are firmly embedded within person-centredness and human connections through 'knowing' patients holistically within physical, emotional, social, intellectual, and spiritual dimensions. This may be where therapeutic radiographers can excel in delivering person-centred care; building connections and relationships over a course of radiotherapy, a luxury our diagnostic counterparts do not routinely have with more transient patient interactions.

Caring in radiotherapy can be considered multifaceted with the functional, task-based 'caring for' and the personalised, emotionally connected, 'caring about'. In 'caring for' patients, radiographers might ensure accurate treatment delivery and management of side effects. However, when 'caring about' patients, radiographers will connect with them as individuals at a human level, involving a therapeutic relationship and heartfelt connections. In knowing the person behind the patient or indeed their diagnosis, true individualised, person-centred caring can take place. This involves an awareness of the person's family and homelife, what is important to them as individuals, beyond an empathetic response but involving true compassion, concern and feelings.

### Feelings, empathy and compassion

Empathy is an integral part of radiotherapy and an aspect of caring directed by professional codes and policy. Empathy is linked to an understanding of feelings which enables a therapeutic relationship to develop between patient and practitioner. As a suggested component of caring, empathy itself is multifaceted with affective, cognitive, and behavioural aspects that facilitate an emotional interaction. It is this emotional interaction that intrinsically links empathy and compassion.

Compassion may be the key to caring relationships (Haslam, 2015). In his debate article 'More than Kindness', Haslam links compassion, caring and empathy with person-centred and quality care. Compassion involves certain attributes or characteristics such as 'empathy, sensitivity, kindness and warmth' (Haslam, 2015, p.2), which he suggests are lacking in poor quality, impersonal 'task-based' care. He also describes how people wish for dignity, respect, and compassion, all of which are attributes that 'cost nothing'. Haslam (2015, p2) defines compassion as the 'humane quality of understanding suffering in others and wanting to do something about it'. Such aspects are essential in radiotherapy and person-centred caring, borne from the therapeutic relationship with individuals. Conversely, if there are non-existent personal relationships within radiotherapy it could be perceived as 'dehumanising' healthcare (Haslam, 2015, p.2). Arguably, from empathy comes compassion and Haslam (2015, p.2) describes how 'compassion often flows spontaneously from empathy – the ability to imagine another's experience'. The absence of empathy, compassion and caring could be addressed by finding out what matters to the patient, what is important, and what are their values. Haslam (2015, p.3) explains how 'humanising the individual will be the first step to empathy – the very opposite of the task-based care that risks getting in the way of empathy and compassion'. Thus for person-centredness within radiotherapy, every radiographer must humanise every patient as an individual. Humanity leads to dignified autonomous caring, another aspect of person-centredness.

### Dignity, autonomy and advocacy

Dignified care relates to 'individualised, patient focussed care... humanistically provided through the presence of a caring relationship' (Williams, Kinnear and Victor, 2016, p.783). The 'little things' that matter to patients are important aspects of person-centredness. Advocacy may link to the concepts of empathy and support but also involves an action which aligns more with compassion as previously discussed (Vitale et al., 2019). A recent concept analysis of patient advocacy by Abbasinia et al. in 2020 defined attributes of patient advocacy as safeguarding, apprising (information giving), valuing (enabling decision-making, individualisation and humanity), mediating and championing social justice. Although in the nursing domain, these attributes of advocacy translate well into the radiotherapy setting and may be considered as aspects of person-centred caring. Effective patient advocacy can help to preserve values and autonomy of patients (Abbasinia et al., 2020) which links to empathy, centredness and individualisation in addition to other aspects of caring.

Patient autonomy, alongside shared decision-making of treatment options, has been accepted to an extent. However, even in high technology areas of healthcare, of which radiotherapy is one, the move from passive recipient of care, to being a care partner has been a paradigm shift for patients and healthcare practitioners. Individual values and wishes being respected and enacted links with person-centred care and are arguably suggestive of 'self-rule' or autonomy. This connects with communication and listening to individual patients as part of person-centredness.

## Listening and communication

Effectively communicating encompasses actively listening, using intuition, information giving and consistency in verbal and non-verbal behaviours (Quirk et al., 2008). An important idea emerging from the study by Quirk et al. (2008, p.364) is arguably that at a behavioural level, 'caring is in the eye of the beholder', involving 'eliciting and understanding each patient's perspective' suggesting an individualisation of person-centredness. The orientation towards the patient as part of caring can be intuitive, anticipatory, and reflective to meet the other's needs. Rather than a set of behaviours or attitudes, person-centred caring involves a set of abilities that may be of most importance at an intuitive or unconscious level.

In a focus group study by Hendry (2011), patients receiving radiotherapy for breast cancer reported that a friendly face, active listening, and good communication by staff enabled relationships to be built. The patient–radiographer relationship was a form of support to patients and was perceived as more than just professionalism; the rapport and relationship were highly influential of the emotional and psychological support experienced. Similarly, dignity and individualised care were also reported as important to the women receiving radiotherapy – all key aspects of person-centred caring. More recently Probst et al. (2021) employed a participatory co-design methodology to explore women's experiences during breast radiotherapy. Feelings of disempowerment and a loss of dignity were experienced by women in the study, suggesting these aspects of person-centredness remain troublesome within radiotherapy practice.

A qualitative study by Egestad (2013) in Norway reported how patients felt less anxious when radiographers were friendly, kind and showed appropriate humour. Smiling staff and being treated as an individual were important findings. In addition, respecting patients' dignity, covering them, and 'understanding' were important to patients alongside good communication and information-giving which reduced anxiety, uncertainty, and vulnerability. Lack of conversation meant loneliness and rejection to some patients, but empathy, compassion and a 'relationship' enabled an improved experience for patients. Similarly in an earlier study, Martin and Hodgson (2006), discuss communication during the first day interviews patients have with radiographers. They promote the importance of active listening, empathy, communication, and information-giving to the extent that they suggest elements of counselling skills are needed by radiographers. These skills include reflective listening and communicated empathy, where the patient needs to know the radiographer is empathetic to build their relationship. Intertwined with this experience is compassion and 'catharsis'. Catharsis relates to the releasing of strong emotions, which logically exist with a cancer diagnosis and how patients may arrive at their first treatment with a plethora of emotions. Martin and Hodgson suggest the first day interview, if skilfully performed, can enable a trusting relationship and be empowering and cathartic for the

| Theme | Sub-Themes |
|---|---|
| Being caring in therapeutic radiography | • Personality<br>• Antecedents and experiences in life<br>• Traits and virtues<br>• Emotionality and perceptiveness<br>• Practitioner fulfilment |
| Caring for patients in therapeutic radiography | • Practical actions<br>• Signposting<br>• Information<br>• Communication<br>• Processes and tasks for patients |
| Caring about patients in therapeutic radiography | • Knowing the patient<br>• Connecting with patients<br>• Individualised practice<br>• Relationships and rapport<br>• Holism and humanity<br>• Making time and prioritising<br>• Active listening<br>• Above and beyond/small things<br>• Empathy, compassion, dignity, empowerment |

*Figure 24.2* Composite themes related to person-centred caring.

Source: Hendry, 2023.

patient. This links to previous concepts of autonomy, an aspect of centredness and caring. Yet there remains the need to better understand the elements of person-centred caring in the therapeutic radiography clinical settings to provide optimal care for every individual receiving radiotherapy. Such elements are described as themes arising from Hendry's study (2023), being summarised in Figure 24.2.

## Summary

Person-centred caring is essential in healthcare and particularly within radiotherapy where individuals will be facing their cancer diagnosis, treatment and anticipated prognosis. Delivering person-centredness involves a multitude of skills, attributes and behaviours from practitioners who are being caring. Caring characteristics alongside an almost innate altruistic desire to be caring, are borne from moral virtue. These were nurtured by experiences of caring as individuals bring therapeutic radiographers to a caring profession.

Within radiotherapy, a duality to caring is proposed which includes the practical process-oriented action of caring for, perceived as perfunctory in nature, lacking the connectedness shown by caring about. Caring about is an outcome of being caring and involves a variety of caring behaviours. These include empathy, compassion, and individualised practice that are manifested by knowing individuals, through heartfelt connections, trusting relationships and rapport.

Being able to provide individualised support during radiotherapy for people dealing with the devastating disease of cancer, involves more than practical caring to deliver treatment. Therapeutic radiographers must form humanistic connections with patients and their families despite fears and sadness that caring aspects of our profession are being eroded by overly focusing upon tasks, processes, and the ever-increasing technology. In the absence of the ideal institutional culture of caring, it becomes more challenging for practitioners to deliver levels of person-centred caring appropriate to meet their moral values and virtue.

Caring for patients within radiotherapy includes the practical, task-based aspects of caring that do not specifically include connecting with the individual, being more perfunctory in nature. Whilst caring about patients in the clinical radiotherapy department involves humanistic interactions, listening, and making time for individuals to enable empathetic human connections. From knowing and individualising patients, person-centred caring and connections would be fostered. The importance of the small things perceived as meaningful to patients also enabled caring about. These conceptualisations of caring in the radiotherapy setting are new knowledge emerging from Hendry's study.

Despite the absence of a caring culture in the clinical setting, all therapeutic radiographers must strive to avoid the over emphasis on tasks, processes, and ever-increasing technology, which possibly demonstrates a preference towards caring for. Such tensions can be somewhat balanced against the radiographer experiencing a sense of fulfilment, connection, and pride when caring about individual patients, not only because policy and codes expect it, but from our moral virtue that initially directed radiographers to their profession. It is essential that such virtues and values are nurtured within a culture of caring. As Nelson Mandela reminded us: 'There can be no greater gift than that of giving one's time and energy to help others without expecting anything in return'.

## References

Abbasinia, M., Ahmadi, F., and Kazemnejad, A. (2020) Patient advocacy in nursing: A concept analysis. *Nursing Ethics*. 27(1), 141–151.

Aristotle. (2009) *The Nicomachean Ethics*. Oxford University Press.

Bolderston A., Lewis, D., and Chai, M. (2010) The concept of caring: Perceptions of radiation therapists. *Radiography*. 16, 198–208.

Department of Health and Social Care. (2021) *NHS Constitution and Values*. Available at: www.gov.uk/government/publications/the-nhs-constitution-for-engl and/the-nhs-constitution-for-england Accessed December 2022

Dewar, B. (2013) Cultivating compassionate care. *Nursing Standard*. 27(34), 48–55.

Egestad, H. (2013) Radiographers relationship with head and neck cancer patients. *Journal of Radiotherapy in Practice*. 12, 245–254.

Eklund, J. H. et al. (2019) 'Same same or different?' A review of reviews of person-centered and patient-centered care. *Patient Education and Counseling*. 102, 3–11

Gelhaus, P. (2013) The desired moral attitude of the physician: (III) care. *Medicine Health Care and Philosophy*. 16, 125–139.

Haslam, D. (2015) 'More than kindness'. *Journal of Compassionate Health Care*. 2, 6. https://doi.org/10.1186/s40639-015-0015-2

HCPC. (2023) *Standards of Proficiency*. London: Radiographers. Health and Care Professions Council.

Health Education England (HEE). (2019) Exploring the role of allied health professionals in the care of people affected by cancer: The patient and practitioner voices project. [Online] Available at: www.hee.nhs.uk/sites/default/files/docume nts/The%20Role%20of%20AHPs%20in%20Cancer%20Care%20FINAL.pdf Accessed April 2023

Health Foundation. (2023) *Person-Centred Care*. Available at: www.health.org.uk/ topics/person-centred-care Accessed January 2023

Hendry, J. A. (2011) A qualitative focus group study to explore the information, support and communication needs of women receiving adjuvant radiotherapy for primary breast cancer. *Journal of Radiotherapy in Practice*. 10, 103–115

Hendry, J. A. (2023) '*Caring in Therapeutic Radiography – An Exploration of Academic Educator Perceptions and Experiences*'. Doctoral Thesis (unpublished).

Hughes, J., Bamford, C., and May, C. (2008) Types of centredness in health care: Themes and concepts. *Medicine Health Care and Philosophy*. 11, 455–463.

Hyde, E. and Hardy, M (2020) Patient centred care in diagnostic radiography (Part 1): Perceptions of service users and service deliverers. *Radiography*. 27(1), 8–13.

Kerasidou A., Bærøe, K, and Berger, Z, et al. (2021) The need for empathetic healthcare systems. *Journal of Medical Ethics*. 47, e27.

Macmillan Cancer Support. (2017) The macmillan allied health professions competence framework for those working with people affected by cancer. [online] Available at: www.macmillan.org.uk/_images/allied-health-professions-framewo rk_tcm9-314735.pdf Accessed April 2023

Macmillan. (2020) *Holistic Needs Assessment and Care Plan*. Available at: www. nbt.nhs.uk/sites/default/files/attachments/Holistic%20Needs%20Assessment%20 and%20Care%20Plan_NBT003095.pdf Accessed December 2022

Martin, K-L., and Hodgson, D. (2006) The role of counselling and communication skills: How can they enhance a patient's 'first day' experience? *Journal of Radiotherapy in Practice*. 5, 157–164.

Maslow, A. (1970). *Motivation and Personality*. New York: Harper and Row.

Morse, J. et al. (1990) Concepts of caring and caring as a concept. *Advances in Nursing Sciences*. 13(1), 1–14.

Murphy, F. J. (2006) The paradox of imaging technology: A review of the literature. *Radiography*. 12, 169–174.

Naidoo, K., Lawrence, H and Stein, C. (2018) The concept of caring amongst first year diagnostic radiography students: Original research. *Nurse Education Today*. 71, 163–168.

Picker Institute (2023) *Principles of Person-Centred Care*. Available at: https:// picker.org/who-we-are/the-picker-principles-of-person-centred-care/ Accessed December 2022

Pluut, B. (2016) Differences that matter: developing critical insights into discourses of patient-centeredness. *Medicine Health Care and Philosophy*. 19, 501–515

Probst, H. et al. (2021) The patient experience of radiotherapy for breast cancer: A qualitative investigation as part of the SuPPORT 4 All study. *Radiography*. 27, 352–359.

Quirk, M et al. (2008) How patients perceive a doctor's caring attitude. *Patient Education and Counseling*. 72, 359–366.

Roach, M.S. (2002) *Caring: The Human Mode of Being. A Blueprint for the Health Professions*. Ottawa, Canada: Canadian Hospital Association.

Society of Radiographers. (2013) *Code of Professional Conduct*. Available at: Source URL: www.sor.org/learning/document-library/code-professional-conduct Accessed April 2023.

Vitale, E., Germini, F., Massaro, M., and Fortunato, R. (2019) How patients and nurses defined advocacy in nursing? A review of the literature. *Journal of Health, Medicine and Nursing*. 64, 63–69

Watson, J. (2006) Watson's theory of human caring and subjective living experiences: Carative factors/caritas processes as a disciplinary guide to the professional nursing practice. Available at: http://dx.doi.org/10.1590/S0104-070720 07000100016

Williams, V. Kinnear, D. and Victor, C. (2016) 'It's the little things that count': Healthcare professionals' views on delivering dignified care: A qualitative study. *Journal of Advanced Nursing*. 72(4), 782–790. doi: 10.1111/jan.12878

Willis, P. (2015) *Raising the Bar Shape of Caring: A Review of the Future Education and Training of Registered Nurses and Care Assistants*. London: Health Education England.

# Person-centred care

The patient experience

# 25 Person-centred care

## A health professional's patient experience

*Margot McBride*

The following is an account of the author of this chapter's personal journey from diagnosis to treatment and remission.

### Diagnosis

I contracted Methicillin-Resistant Staphylococcus Aureus (*MRSA*) in 2007; the source was unconfirmed but thought to be the result of an infected skin cut. This is a serious illness causing impairment of the lymphatic system and can weaken the immune system. Shortly after recovery I was diagnosed with Barrett's Oesophagus (*2008*), which is a life-threatening inflammatory disease and then diverticulosis and irritable bowel syndrome (*IBS*) (*2009/2010*). The latter is often associated with stress which results in inflammation of the bowel. In 2009, I suffered a mild stroke which was thought to be the result of hypertension.

Before my diagnosis, I experienced psychological effects which manifested itself, in my case, with bouts of anxiety, emotionalism, depression and hypomanic attacks (*2014*). I was hospitalized for six weeks, and my Consultant Psychiatrist began to consider that Cushing syndrome (CS) was possibly the cause of my hypomania.

Cushing syndrome is a baffling portmanteau of symptoms, each often ascribed to other medical conditions, but together representing diagnostically, challenging medical conditions which can occur if the body produces too much of the hormone cortisol – sometimes, but not always, due to long-term use of steroid medicines. With an estimated incidence of around 10 to 15 million worldwide per year, and often coming, "disguised," as other stand-alone conditions such as hypertension, obesity, psychological disorders, diminished libido, infertility, osteoporosis, and even domestic abuse (*due to skin which bruises at the touch*), CS can be present but be undiagnosed for many years (McBride, 2022).

Prior to being diagnosed with CS, I was informed by my Orthopaedic Consultant that I was suffering from compression fractures of Lumber Vertebrae (*LV*) 4 and LV5, which resulted in severe sciatica. I had experienced

DOI: 10.1201/9781003310143-36

back pain for three years and my right lower leg had begun to swell, and my skin slowly developed a serious skin infection. A Radiologist's report confirmed that I was suffering from degenerative spondylolisthesis which caused my severe sciatical a swelling was also identified in my right kidney. This incidental finding on the x-ray was reported as an adenoma in my right adrenal gland and resulted in a referral to an Endocrinologist who prescribed a series of biochemical tests which included a dexamethasone blood test, low dose overnight dexamethasone test, adrenal venous sampling (AVS), urine free cortisol (UFC) test and a late-night salivary sample (LNST), ultrasound (US), computed tomography (CT) scan and a radionuclide imaging scan (RNI) (*2015*).

## Treatment

These tests confirmed that the definitive diagnosis was CS. The adenoma turned out to be benign, and was subsequently removed surgically, i.e., an adrenalectomy. My right lower leg pain persisted and, following several x-rays and eventually a magnetic resonance imaging (MRI) scan, I was diagnosed with a musculo-skeletal aggressive soft tissue sarcoma. My leg was subsequently amputated in 2016. This was not the end of my surgical journey as my follow-up biochemical tests post adrenal surgery showed I continued to have high cortisol levels. An MRI pituitary scan and bilateral inferior petrosal sinus sampling (BIPSS) procedure confirmed a diagnosis of Cushing's disease (CD). I was then referred to a Consultant Neurosurgeon (*2017*). The combination of the biochemical tests and imaging procedures therefore identified the presence of a benign pituitary adenoma which was removed by transsphenoidal surgery (TSS). I was informed one year later by my Neurosurgeon and Endocrinologist that I was in remission from CS and CD (*2018*). Slowly my medication (*Hydrocortisone*) was reduced and in January 2019, my cortisol levels returned to normal, and I was withdrawn from this medication being declared, "cured." I remain in remission (2023).

## The road to recovery

On reading the evidence presented by other Cushing's patients in numerous research papers, and my attendance at the Pituitary Foundation UK conference (*2019*), the diagnostic odysseys resembled many of my own. My Cushingoid features included truncal obesity, excess hair growth, hypertension, and skin bruising. I also experienced bone pain, insomnia, fatigue, emotional lability, depression leading to suicidal ideation and generally I was very weak. According to Law, 61% of CS patients are reported has having emotional lability, 49% of these have cognitive difficulties. Most of these patients, like me, experienced a lack of concentration and memory loss and out of 17% per cent of these patients, at least one is admitted to hospital for psychiatric care prior to their diagnosis of CS.(Law, 2017).[3] Similar to other Cushing's patients, I recognized that my proficient verbal skills were lacking, and I began to lose confidence in public-speaking and my academic activities. I was most fortunate in that I had a loving family and colleagues who supported me through the many dark days of my illness, although this sadly is not always the case. Cushing syndrome and CD can be a devastating diagnosis for anyone, but young adults are said to be particularly vulnerable to the ridicule and abuse they experience due to their personality changes and physical appearances, causing many of their friends and family to believe they are faking their illnesses. I found that when my diagnosis was finally made, I really knew little about what CS and CD were both as a patient and health professional. The brief explanation from my Endocrinologist I was given did not reassure me that I would get better nor were any of the comorbidities explained that I was likely to develop and what my Quality of Life (QoL) might be. I felt angry, confused as to, "why me"? Most of all why had it taken almost seven years to diagnose, and it wasn't until I was extremely ill that this diagnosis was made.

However, similar to many patients eventually I accepted the diagnosis and it somehow reassured me that it was all due to a hormonal imbalance, i.e., my elevated cortisol levels had driven my body to experience extreme pain, disorientation, and the development of psychosocial issues, hypomania, facial, body changes and suicidal thoughts.

Following my pituitary surgery these changes rapidly began to reverse, my self-respect and esteem returned albeit slowly. The effects of steroids and its withdrawal syndrome can affect patients in a wide variety of ways. Several published articles related to the remission of patients suggest often physicians can be inexperienced regarding the advice and importance of patient information in what to look for i.e., the expected pre illness signs and symptoms and the unexpected. Patients should not have to wait so long for a diagnosis and thus avoid the tortuous road to treatment. I continue to meet with my endocrine team yearly. Prior these appointments, I have a dexamethasone blood test, low dose dexamethasone test to assess my cortisol levels, routine blood tests include cholesterol and growth hormone deficiency, and a C-Reactive Protein (CRP) level tests; the latter test measures the level of CRP in the blood, high levels being a sign that you may have a serious health condition that causes inflammation and an MRI (*Pituitary*) scan to check that there is no indication that the pituitary gland is abnormal.

The amputation, however, relieved me of the excruciating pain that I had experienced for many years, and I was fortunate that I did not require chemotherapy or radiotherapy. However, I have found great difficulty in accepting the loss of my right lower leg and the painful, long journey of having to learn to walk with a prosthetic leg. This acceptance is slowing improving with time. Yes, there are days when I don't find it too much of a burden but other days, the phantom pains which admittedly are not as frequent as I experienced post-surgery, and not being able to participate in some of my pre-amputation sport activities, running with my grandchildren and spending most of my life in a wheelchair, is frustrating and hard sometimes to bear. My comorbidities include osteoporosis and rheumatoid arthritis. I attend the orthopaedic clinic once per year for a check-up which includes a chest x-ray and an inspection of my stump. Some amputees unfortunately can develop post-surgery infections with their stumps, so it is important to have regular check-ups, and, in the case of a limb removal caused by cancer, a chest x-ray annually is requested to assess if there is potential metastasis in the lungs. I will always have a fear of recurrence and experience bouts of anxiety prior to my appointments.

When diagnosed with CS, it is sometimes unclear as to the long-term effects and not until you fall and have a fracture you can be diagnosed with osteoporosis. It is a regular occurrence that amputees fall due to their instability normally due to muscle weakness, faulty prosthetic limbs, and general mobility challenges, taking risks and wound infections. In my case, I simply fell whilst walking on a wet surface, broke my pubic bone and my pelvic x-ray report showed that my bone mineral density was low, osteoporosis was confirmed by a dual energy X-ray absorptiometry scan (DXA). It was thought that in all probability I had developed osteoporosis due to my CS raised cortisol levels, particularly over time and causes this condition. During my painful experience of having a sarcoma, I was given large quantities of steroids, and this is another reason why I may have developed osteoporosis. If this is the case, the condition is referred to as glucocorticoid-induced osteoporosis, which can happen within the first six months of steroid treatment.

I was not offered psychological support for both my CS, CD, and my amputation. I had to learn to adopt coping skills in order to fully accept and adapt to the inevitable life changes. I often reflect on this as I wonder if it was because during my consultations, I always put my Radiographer hat on, and my clinicians perhaps concluded that I was coping well.

Post remission, I made the decision to pass on my patient experiences with the aim of turning a negative experience into a positive one.

I am generally very happy and now found an inner peace and a continuing belief in modern medicine. Reflecting on my experience, and cognizant that CS and CD are classified as rare diseases, I realised that I had built up a wealth of knowledge, which was quite unique for a patient.

For most of my career I have worked as an academic and research radiographer. The rigour and dedication required to achieve a level of knowledge for teaching and learning radiography students is high. With this in mind, I decided to undertake a Doctorate study, which I knew would allow me to study CS and CD at a high level, with the aim of passing on my knowledge in the form of publications, conference presentations and opening up opportunities to take part in patient advocacy and advisory groups. I also found myself networking and writing papers with experts in the fields of endocrinology, neurosurgery, orthopaedics, and radiology. This undoubtedly has helped me accept and understand my illnesses. One of the key findings from my study is that all health professionals must develop excellent communication skills.

Although this study was of a personal nature, driven by a desire to find answers as to why I developed CS and CD, it gave me a unique opportunity to compare my experiences with other Cushing's patients, particularly the impact factors and consequences of a CS diagnosis. My personal experience admittedly could have influenced the narrative within my Doctorate study. However, my insight into these medical conditions, provided contextualisation and support for my study and helped to broaden my medical knowledge in the form of understanding the nature of this condition within the context of the clinical evidence and outcomes, which ultimately may help inform both patients and clinicians of the need for early diagnosis, early intervention, and specialised care.

Both diagnostic and therapy radiographers during their education programmes are taught to manage their patients with compassion, understanding and care and their main aim is to diagnose and treat their patients and at the heart of their clinical services, improve their patients' QoL.

Until you are a patient yourself, you don't always acknowledge the importance of your voice which can turn the key in a clinician's decision-making during a diagnostic work-up (McBride, 2023). Unfortunately, today's challenges have increased the likelihood of spending less time on person-centred care, and many consultations are conducted by phone. Clinicians, including radiography staff are trying to find solutions to the growing pressures from rising referral lists and workforce shortages, which have led to many patients not having basic care or for some, no or delayed care which leads in many cases to death. When a patient presents for an examination, procedure, or treatment, radiographers mainly focus on giving practical information and reassurance to gain compliance. For example, according to Madi et al., "MRI-related anxiety is present in 30% of patients and may evoke motion artifacts/failed scans, which impair clinical efficiency." Madi and colleagues concluded that, "considering individual preferences for patient preparation may be more effective than a one-size-fits-all approach" (Madi et al., 2022). This philosophy should apply not only to MRI examinations but all diagnostic imaging and radiotherapy practices. Prior to defining person-centred (often referred to as patient-centred care), we must consider communication and interpersonal skills which are essential for the delivery of safe clinical practice. These skills are the backbone of person- centred care and are as equally important as understanding the anatomical, physiology and pathophysiology of the disease processes and the diagnostic and treatment methods which are practised at the heart of radiography and radiotherapy services. Radiographers apply their communication and interpersonal skills while examining and treating their patients, giving instructions for accurate patient positioning, thus improving compliance, reducing mistakes and mishaps, while giving reassurance and making them feel safe. It is well documented that clinical competence improves health care outcomes. To produce a high quality diagnostic radiographic image, radiographers require to have, like medical doctors, clinical acumen, which is a craft, but also communications skills which can be said to be the finer arts!

### Radiography education

It is important when choosing a career in Radiography that potential students understand that their chosen profession is one of caring, commitment and dedication, not only to their patients, but their colleagues and the multidisciplinary team.

Over the past 30 years radiography has evolved from diploma status to degree programmes which enables apprentices, assistant practitioners undergraduate and masters' students to study at a pace which suits their academic abilities, family and social life and financial status. This type of step on, step off education has meant that the range of radiography practitioners who emerge from these pathways, often come with many of life's experiences which adds value to peer learning and the acquisition of clinical skills.

At the commencement of a career in health, how academic staff teach these skills during our healthcare programmes defines the way in which our practitioners deliver radiology and oncology services, including how their students manage and care for their patients. The design of these programmes includes engaging with clinical staff, students, managers, employers, and patients. The latter being important steps in the understanding of patients' needs and their experiences and learning to work within a multidisciplinary healthcare team. Research is now at the heart of radiography education which is mainly scientific evidenced based and continues throughout a radiographer's career. Many programmes now incorporate complimentary placements whereby students lead patient experience clinics. This enables them to discuss their illnesses and how they have coped with not being able in many cases to do what they did prior to their illness and the different stages they experience of learning to cope. It could be suggested that as a profession we are moving at pace in embracing this form of practice, being mindful of listening to the patient's voice and acting accordingly with compassion, being non-judgemental and without prejudice.

### Compassionate leadership

When healthcare leaders are formulating strategic plans for their organisations, they rely on their ability to make decisions using their communication skills to gather relevant information, problem-solve, and apply innovative and creative thinking. Healthcare leaders require to deliver compassionate leadership, the quality of patient care being at the heart of their values which in turn empowers their staff to deliver care in an inclusive and equal manner. To enable this to happen, it is vital that the quality of healthcare is delivered by a multidisciplinary team with good leadership skills. This allows lower levels of stress within the healthcare team, whilst nurturing a culture of care, for each other and their patients. "Research in the NHS has shown that learning and innovation, in the context of psychological safety rather than a blame culture, is vital for nurturing cultures of high-quality, continually improving, and compassionate care" (West and Chowla, 2017).

It of course makes sense that if we care for one another and work as a team focused on the same aims then this in turn will reflect on the way in which we diagnose and treat our patients and deliver services within the ethical boundaries of modern medical practices.

As a patient, I found myself listening to every word that was said by my healthcare team. Sometimes drawing the wrong conclusions and hearing idle chat, which was often negative and concerning. Care and professionalism are vital as patients can misinterpret the meaning, particularly the use of abbreviations and medical jargon. In radiography education, academics teach students to adapt their radiographic technique skills for each patient; this also helps them to engage with their patients and develop effective communication skills including non-verbal. Accepting patients' vulnerabilities, ensuring that they have early feedback and their results, is all important for patient experience, satisfaction, and quality of care.

## Health-related quality of life

The introductory section in this chapter referred to QoL, stating that as health professionals this is our main aim. We could argue that choosing the correct modality to produce a high-quality image which identifies the smallest to the largest of anatomical and physiological changes combined with accurate image reporting is fundamental in diagnosing and must surely be our main aim. However, as radiographers, do we really take a moment to reflect on what we have seen on the image and if it is abnormal what impact this may have on our patient's QoL?

Before a patient leaves the department, this is a good time to take a few minutes to ask them about their general health and how their illness is impacting on their work, family, and social life. For me this was important as I felt they really cared about me as a person not just another patient. Some radiographers who took my x-rays and scans knew me as their lecturer, and this caused a little unease, probably more for me as I watched, probably like most patients, their eye contact and body language, thinking of course the worst.

These few minutes of communication may also expose further underlying medical conditions and fears which may not have been discussed during initial or follow-up consultations but may have relevance for their ongoing diagnostic and treatment pathways and of benefit to the reporting staff and their referrer. Patients are often too embarrassed, ashamed, or fearful when discussing their medical problems, particularly with Consultants and even their General Practitioners.

The concept of Health-related Quality of Life (HRQoL), since the 1980s, has evolved, and although health is one of the important domains of overall QoL, other domains have emerged, for example, cultural, social aspects of life, work status and spirituality (WHO, 1993). Measuring HRQoL is complex, particularly with the addition of these forementioned domains and requires a range of different research designs. Traditionally, biomedical testing results were the principal endpoint in research for most medical conditions. However, over the past few decades, medical and psychologist researchers have developed QoL measurement tools with the aim of improving individualised care and rehabilitation. Example of these tools are self-reported questionnaires and semi-structured interviews. These have proven to be effective in measuring HRQoL and often lead to improved treatments including the measurement of treatment outcomes.

The World Health Organisation's definition of Health is: "a state of complete physical, mental and social well-being, which is marked, not only by the absence of disease or infirmity" (WHO, 2014).[8] This definition has not significantly changed since its introduction following World War 11 and remains unchanged when writing this chapter (2023).

Although QoL and HRQoL have been used interchangeably, the latter is preferred when specific effects of a given condition on QoL are analysed. HRQoL provides an unambiguous measure of health status which merges a patient's perspective with clinical parameters, thus allowing an objective assessment of the impact of a disease or the efficacy of a medical intervention on a subject.

In everyday working practice, radiographers do not have the time, unless they are active researchers to measure their patients HRQoL, but as previously discussed it can take only a few minutes to listen to their patients' voice. This type of practice not only benefits the patients but also allows practitioners to learn. Good examples of this are cancers, diabetes, cardiovascular diseases, obesity, rare diseases, and diseases which are often diagnosed by incidental findings such as an adrenal adenoma being reported on spine or abdominal x-rays and in the case of a pituitary adenoma on a brain scan. Many patients often have undiagnosed comorbidities, for example, osteoporosis and some develop diabetes so have regular follow-ups for other medical conditions. A few minutes can save their lives by asking about their general health and thus taking a more holistic approach. Many patients are interested in the pathophysiology of disease processes and wish to know how it will manifest itself over time and if it could, for example, be a genetic disorder. It is essential that radiographers add any information gleaned from these conversations to the information sent to the reporter because sometimes little bits of information, which may not have been included in the referral can make a big difference.

### Health promotion

The use of social media platforms in most aspects of our daily lives has been increasing exponentially over the last decade. Many patients have become accustomed to accessing the internet prior to and following their diagnosis. Health promotion has therefore become embedded in our educational programmes, research activities and for patients who seek additional information and advice. The advantages of user-friendly platforms for patients who can navigate their way through the maze of platforms, allows them to take responsibility for their own wellbeing, often reduces their feelings of helplessness, and they can use this information when engaging with their clinicians giving them a chance to discuss and decide on the best course of treatments and support mechanisms.

There are, however, drawbacks to using such platforms, which includes becoming competent in computer skills, negotiating, and creating relevant information which is current and accurate for not only common illnesses but also for rare and complex medical conditions. However, some patients are not well enough or have never used or wish to use a computer.

Why it is important for health professionals to be proactive in promoting health during the delivery of person-centred care? Undoubtedly, the public have become more aware of the need for healthy living through advertising, television and the increase in dietary advice and generally easy access to sports facilities. Many however, do not have access to health promotion. During consultations there is therefore a good opportunity for health professionals to being proactive in encouraging their patients to eat healthily and stay fit. This, however, is not always possible with cost-of-living increases, expensive membership fees for gyms and recreational activities and for people with mobility challenges, including those with loss of sight and sound.

In all cases, health professionals are uniquely placed to encourage their patients where possible to seek help from physiotherapy, occupational, speech therapy and dietetics. In my case, I have been amazed at their work, dedication, and support they have given me. Helping me to walk, take showers, reduce weight, and generally give excellent advice; taking a compassionate approach when I was fearful, in pain and anxious by simply helping me to overcome my fears by listening, observing, and taking action to improve my HRQoL.

### A person-centred approach in modern radiographic and therapeutic practices

Reflecting on the importance of HRQoL and my own patient experience, all health professionals who adopt a person-centred approach within their clinical practices should therefore consider how this is measured for individual patients and what impact can the findings of such a measurement have on their decision-making. During my Doctorate study, I published a

QoL in Cushing syndrome paper which reported some of the findings from my study. I discovered that, "HRQoL is emerging as an important clinical endpoint, which complements diagnostic workup and contributes to place patients at the centre of the decision-making process through the recognition of their needs, concerns, goals and expectations." The term clinical endpoint has traditionally been applied to evaluate the impact of medical therapy and has been defined as, "characteristics that directly measure how a patient feels, functions or survives a treatment" (McBRIDE, 2022).[1] However, measuring clinical endpoints, can be challenging due to the variability of medical conditions, and are time-consuming normally due to the variety and complexities of their patients' individual needs.

It could be argued that QoL and clinical endpoints are intertwined, and the medical condition experienced in individual patients determines the psychological and physical impairment(s). Radiographers undertaking research related to the impact of illness on a patient's life, should consider the inclusion of QoL measures in their research design. It is well documented that a delay in diagnosis can be detrimental to a patient's psychological wellbeing. This means that the challenges of delayed diagnosis can further exacerbate a patient's illness, thus increasing the potential of developing additional illnesses and in some cases more complex medical conditions. For Radiographers, patients generally present with various levels of stress, and anxiety and this requires them to observe, listen and communicate effectively.

Until I was ill, I did not fully appreciate the varying and important role of a carer. I had nursed my father when he was very ill, until he died and I had often spoken to carers when I was a clinical radiographer, but now I really appreciate the responsibility and dedication they have, and how they are often the patient's voice, whether it be a paediatric a geriatric patient or a patient who has a medical condition which has caused them to be unable to communicate. Some health professionals and carers find that developing nursing, communication, and caring skills does not come naturally and takes time and experience to develop while their patients' needs are often unmet. Carers can lose their independence, have to give up work and social activities with often little financial reward. Some young adults find themselves carers and somehow have to manage.

I found my husband had to step in to care for me and this to a certain extent remains the case. Being dependent on someone is always hard, particularly if you have been used to looking after yourself. Reflecting on my own experiences, there was no induction or training on how to be a carer. My husband was thrown in to "the deep end" a few days after I was discharged with only a few days visit from a district nurse.

## Conclusion

The socio-economic aspects of healthcare services are complex. Healthcare providers are challenged with how they can provide clinical services from conception to death within the financial constraints. The integration of health and social care makes sense but is proving to be extremely challenging due to these financial constraints and divided opinions in health organisations often clouds the realisation of making it happen. Modern medicine has increased the chances of living longer but with this comes the need for specialist provision including physical and psychological support and care for those who care for them.

Education and training for all health professionals particularly over the past 30 years has changed, medicine has moved on developing new advanced technologies which have enabled an even better understanding of how the body works and the digital revolution is and will further help drive more efficient and speedier ways of working. Person-centred care is now embedded within the radiography curriculum encouraging the new generation of staff to engage more with their patients and to accept that we are all individuals from different cultural and diverse backgrounds. We need to ask the correct questions without embarrassing the patients and we must respect their rights to privacy, while ensuring we gain the correct information for their care and safety.

Multidisciplinary learning and patient, public involvement has helped to nurture a more person-centred approach, albeit this has become more of a challenge as every minute is precious when diagnosing patients and attempting to reduce waiting times. Undoubtedly my lived experience has taught me to cope with my illness and to seek advice when needed. As health professionals we have an open door to promote and deliver person-centred care, which in turn will achieve our main aim, which is to deliver the highest quality of medical care whether it is in the National Health Service or private sectors.

## References

Law, E. R. (2017). *Cushing Disease: An Often Misdiagnosed and Not So Rare Disorder*. (Edited by Law Jnr. ed.). Elservier Inc.

Madi, J. S. (2022). Preparing patients according to their individual coping style improves. patient experience of magnetic resonance imaging. *Journal of Behavioural Medicine*, 45, 841–845.

McBride, M. (2022). *Cushing Syndrome and Disease. A Study of the Diagnosis, Treatment, Clinical Consequesnces, and Health-Related Quality of Life Associated with these Medical Conditions. A Doctor of Philosophy Study*. Unpublished thesis. https://insight.cumbria.ac.uk/id/eprint/6770/

McBride, M. (2023). *The Patient's Voice is the Master Key in a Clinician's Toolbox*. United Kingdom. Imaging and Oncology Conference, 2023 Abstracts.

McBride, M. C. (2021). Quality of Life in. Cushing's syndrome. Best Practice &. Research. *Clinical Endocrinology & Metabolism*, 101, 505.

McBride, M. Crespo. I., Webb, S., Valassi. E. (2021). Quality of life in cushing's syndrome. *Best Practice &. Research. Clinical Endocrinology & Metabolism*, 101, 505.

West, M. S., & Chowla, R. (2017). Compassionate leadership for compassionate health care. In P. Gilbert (Ed.).*Compassion: Concepts, research and applications.* Routledge, 237–257.

WHO. (1993). Study protocol for the WHO project to develop a Quality-of-Life assessement instrument (WHOQOL), 2 (2), 153–159. ed. Quality of Life Research.

WHO. (2014). *Definition of Health.* World Health Organisation.

# 26 Addressing scanxiety

## Promoting person-centred care in radiology

*Shayne Chau*

### Understanding scanxiety and its impact

Scanxiety is not merely a response to the physical procedure of scanning, but rather to the anticipation and uncertainty surrounding the scan results. It is a psychological burden that can affect patients across different stages of cancer and is often inadequately addressed in clinical practice (Bauml et al. 2016). Despite the inevitability of scans in cancer care, the distress associated with them can be profound and varies among individuals. It is crucial to acknowledge the individuality of patient experiences and provide tailored support to alleviate scanxiety (Gimson et al. 2022)

#### *Promoting adaptive coping strategies*

Patients develop various adaptive coping strategies to manage scanxiety, including use of analgesia, mindfulness, breathing techniques, prayers, increased consumption of alcohol and even planning of fun activities. Recognizing these strategies and building upon them can empower patients to navigate the emotional challenges associated with scans (Bellhouse 2021). Healthcare professionals, including radiographers and the cancer multidisciplinary team, play a vital role in supporting patients by fostering open communication, empathy, and trust. By understanding patients' unique coping mechanisms and engaging them as partners in their care, radiology professionals can enhance the patient experience and mitigate scanxiety (Schoenmaekers et al. 2022).

### Targeted interventions for managing scanxiety

Addressing scanxiety requires a multidimensional approach, encompassing both psychological and operational interventions. Psychotherapeutic and educational interventions can provide patients with tools to manage anxiety and improve their overall well-being. Healthcare professionals, including psychologists, can play a crucial role in delivering supportive care and ensuring patients feel recognized as individuals rather than just cases (Bui

DOI: 10.1201/9781003310143-37

et al. 2021). Moreover, operational changes such as streamlining procedures and improving patient comfort during scans can contribute to reducing distress and anxiety.

## Patient–clinician relationship and support

The patient–clinician relationship emerges as a critical factor in managing scan-related anxiety and distress. Patients highly value a supportive relationship characterized by shared knowledge, mutual decision-making, and transparency. Radiographers, as frontline care providers, have a unique opportunity to establish a compassionate and empathetic connection with patients during the scanning process (Baun et al. 2020; Golden et al. 2017). Enhancing undergraduate and postgraduate curriculums to emphasize communication skills and empathy training can help radiographers better address patient anxieties and contribute to a more person-centred approach.

## Involving support networks

Recognizing the importance of support networks, including family and friends, is essential in managing scanxiety. Patients often develop their own self-management strategies, and a supportive network can provide emotional comfort and practical assistance. By involving and educating support networks, radiology professionals can create a collaborative care environment that extends beyond medical interventions (Tyldesley-Marshall et al. 2020).

## Closing thoughts

Addressing scanxiety is crucial for improving the overall patient experience in radiology. By recognizing the distress associated with scans and implementing targeted interventions, radiographers and healthcare professionals can alleviate anxiety and support patients throughout their cancer journey. Integrating person-centred care principles into radiology practice will ensure that patients feel valued, understood, and supported during this challenging time. By embracing a holistic approach, radiology professionals can enhance the well-being of patients and contribute to high-quality, person-centred cancer care.

As an author of this chapter and a cancer survivor myself, I have had firsthand experience with scanxiety and understand the immense impact it can have on patients undergoing medical imaging. Reflecting on my own journey, I vividly recall the overwhelming mix of emotions that would consume me leading up to each scan. The uncertainty, fear, and anticipation would often overshadow my thoughts, making it difficult to focus on anything else.

During those moments, I realized the true value of person-centred care in radiology. The radiographers and healthcare professionals who recognized my scanxiety and took the time to listen, provide reassurance, and explain the process were the ones who made a lasting impact on my overall experience.

However, not every encounter was as positive. I remember one particular incident that left a lasting impression on me. As a radiographer myself, I had a certain level of understanding about the scanning process. I approached one scan with a sense of familiarity, hoping for a compassionate and empathetic experience. But to my surprise, the radiographer I encountered seemed dismissive and detached. They assumed that my professional background meant I had all the answers and dismissed my concerns, saying, "You are a radiographer, so you know what to expect. Off you go, lay down on the scanner."

In that moment, I felt vulnerable and reduced to a mere case. The lack of empathy and understanding left me feeling isolated and unheard. It was a stark reminder that even as a healthcare professional, I was still a patient, deserving of compassionate care and support.

This personal experience reinforced my commitment to promoting person-centred care in radiology. It reminded me that every individual, regardless of their knowledge or background, deserves to be treated with dignity, respect, and empathy. It strengthened my resolve to advocate for improved education and training for radiographers, emphasizing the importance of communication skills and cultivating a compassionate approach towards patients experiencing scanxiety.

Through my journey as both a patient and a radiographer, I have come to understand that addressing scanxiety requires a comprehensive approach. It involves not only the technical aspects of imaging but also the emotional well-being of the patient. By incorporating person-centred principles into radiology practice, we integrate medical expertise and compassionate care, ensuring that patients feel supported, informed, and empowered throughout their scan experiences.

Sharing my personal experience allows me to empathize with patients and fellow healthcare professionals who may encounter similar challenges. It is a reminder that we all have a role to play in fostering a culture of empathy and understanding within radiology. By sharing our stories and advocating for change, we can create a positive and inclusive environment that prioritizes the holistic well-being of patients, transforming scanxiety into a more manageable and supportive experience.

## References

Bauml JM, Troxel A, Epperson CN, et al. Scan-associated distress in lung cancer: quantifying the impact of "scanxiety". *Lung Cancer*. 2016;100:110–113.

Baun C, Vogsen M, Nielsen MK, Høilund-Carlsen PF, Hildebrandt MG. Perspective of patients with metastatic breast cancer on electronic access to scan results: Mixed methods study. *Journal of Medical Internet Research*. 2020;22(2):e15723.

Bellhouse S, Brown S, Dubec M, et al. Introducing magnetic resonance imaging into the lung cancer radiotherapy workflow–An assessment of patient experience. *Radiography*. 2021;27(1):14–23.

Bui KT, Blinman P, Kiely BE, Brown C, Dhillon HM. Experiences with scans and scanxiety in people with advanced cancer: a qualitative study. *Supportive Care in Cancer*. 2021;29(12):7441–7449.

Gimson E, Dottori MG, Clunie G, et al. Not as simple as " fear of the unknown": A qualitative study exploring anxiety in the radiotherapy department. *European Journal of Cancer Care*. 2022;31(2):e13564.

Golden S, Thomas C, Slatore C. It wasn't as bad as I thought it would be: A prospective, qualitative longitudinal study of early stage non–small cell lung cancer patients after treatment. *International Journal of Radiation Oncology, Biology, Physics*. 2017;98(1):241–242.

Schoenmaekers JJ, Bruinsma J, Wolfs C, et al. Screening for brain metastases in patients with NSCLC: A qualitative study on the psychologic impact of being diagnosed with asymptomatic brain metastases. *JTO Clinical and Research Reports*. 2022;3(10):100401.

Tyldesley-Marshall N, Greenfield S, Neilson S, English M, Adamski J, Peet A. Qualitative study: Patients' and parents' views on brain tumour MRIs. *Archives of Disease in Childhood*. 2020;105(2):166–172.

# 27 The journey to person-centred care in radiology

*Shayne Chau, Christopher Hayre,*
*Karen Knapp and Emma Hyde*

### Person-centred care: A paradigm shift in healthcare

As we conclude this comprehensive journey, it is crucial to reflect on the key themes and insights presented in the preceding chapters and consider the future directions for embracing person-centred care in radiology.

First and foremost, it is evident that person-centred care represents a paradigm shift in the healthcare landscape in radiology. We have previously observed this in other specialist fields, such as nursing (McCormack and McCance, 2010), with fewer evidence or texts emerging in radiology, in general. Here, this approach looks to challenge health professionals to transcend the traditional focus on technical aspects and pathology, encouraging them to recognize the unique needs, preferences, and aspirations of individuals for them to become more "personalised", vis-à-vis "personalised radiology". By fostering collaborative partnerships, active engagement, and open communication, radiographers and other healthcare providers can create an environment that empowers patients, making them equal partners in their own care journey. This approach not only enhances the quality of care but also fosters a sense of dignity, respect, and trust between the healthcare team and the individuals they serve. Throughout this textbook, we have emphasized the importance of cultural competence in person-centred care. Healthcare systems are becoming increasingly diverse, with individuals from various cultural backgrounds seeking radiology services. Cultural competence entails a deep understanding and appreciation of cultural norms, beliefs, and values, enabling healthcare professionals to provide tailored and sensitive care. By embracing cultural diversity and promoting inclusivity, radiographers can establish a foundation of trust and promote positive health outcomes for all individuals, irrespective of their cultural backgrounds (Ahmed et al. 2018).

### Balancing technology and the human touch in care provision

The integration of technology and artificial intelligence (AI) in radiology presents both opportunities and challenges for person-centred care. While advancements in imaging technologies and AI algorithms have revolutionized

DOI: 10.1201/9781003310143-38

diagnosis and treatment planning, they must be implemented with careful consideration for maintaining the human touch in care provision. Radiographers should remain mindful of the potential emotional and psychological impact of technological advancements on individuals undergoing imaging or radiotherapy. Balancing technical expertise with empathy, compassion, and emotional support will ensure that person-centred care remains at the forefront of radiology practice, even in the era of rapid technological progress (Champendal et al. 2023).

## Education and continuous professional development for person-centred care

Education and continuous professional development play a pivotal role in fostering person-centred care in radiology. As the research conducted by the authors of each chapter demonstrates, evidence-based practice serves as a foundation for improving care quality and meeting the expectations of individuals and their carers. It is essential for radiographers to stay updated with the latest research, guidelines, and best practices related to person-centred care. By investing in their professional development, radiographers can enhance their knowledge and skills, ultimately translating into better care experiences and outcomes for their patients.

## Prioritizing person-centred care: Organizational considerations

It is imperative that radiology departments and healthcare organizations prioritize person-centred care as a core principle. Leadership support, policy frameworks, and resource allocation are essential for fostering a culture that values and prioritizes the individual's well-being throughout the imaging and radiotherapy journey. Collaboration between radiographers, radiologists, nurses, administrators, and other healthcare professionals is crucial to ensure the integration of person-centred care principles in the organizational fabric. By creating an enabling environment that supports and encourages person-centred care, healthcare institutions can truly embody the concept and elevate the overall quality of care provided.

## Closing remarks: Synthesizing insights and looking to the future

Through 26 chapters, this textbook has explored the multifaceted dimensions of person-centred care in radiology from international perspectives. We have delved into various areas, including cultural competence, patient care in different radiographic modalities, and the experiences of healthcare professionals themselves. By synthesizing the knowledge and evidence presented in this book, radiographers and other healthcare professionals can adopt a comprehensive approach to person-centred care, enriching the lives of individuals and their carers. As we look to the future, it is crucial that radiology continues to evolve and adapt to the changing healthcare landscape while keeping person-centred care as a guiding principle. By embracing innovation, promoting cultural competence, investing in education, and fostering collaboration, we can ensure that person-centred care remains at the forefront of radiology practice. Let us strive together to create a future where the delivery of high-quality care in radiology is not only technologically advanced but also deeply compassionate, empathetic, and person-centred.

With each interaction, radiographers have the power to make a positive impact on the lives of those they serve. By placing individuals at the heart of their care, we can truly transform the radiology experience, promoting healing, and enhancing the overall well-being of each person. Let us embark on this journey together, embracing person-centred care as the cornerstone of radiology practice and fulfilling our commitment to providing high-quality care to all.

## References

Ahmed, S., Siad, F.M., Manalili, K., Lorenzetti, D.L., Barbosa, T., Lantion, V., Lu, M., Quan, H. and Santana, M.J., 2018. How to measure cultural competence when evaluating patient-centred care: A scoping review. *BMJ Open*, 8(7), p.e021525.

Champendal, M., Marmy, L., Malamateniou, C. and Dos Reis, C.S., 2023. Artificial intelligence to support person-centred care in breast imaging-A scoping review. *Journal of Medical Imaging and Radiation Sciences*, 54(3), 511–544. doi: 10.1016/j.jmir.2023.04.001. Epub 2023 May 12. PMID: 37183076.

McCormack, B. and McCance, T., 2010. *Person-Centred Nursing: Theory and Practice*. Wiley-Blackwell: London.

# Index

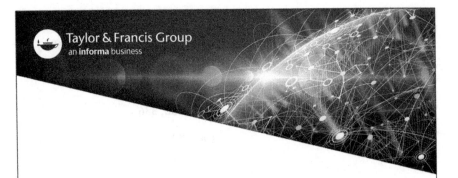

Printed in the United States
by Baker & Taylor Publisher Services